XXV Fluid Mechanics Conference (FMC 2022)

Journal of Physics: Conference Series Volume 2367

Rzeszow, Poland
7 – 9 September 2022

Editors:

Anna Kucaba-Piętal
Adrian Kordos
Ewa Rejwer-Kosińska

ISBN: 978-1-7138-6475-2
ISSN: 1742-6588

Printed from e-media with permission by:

Curran Associates, Inc.
57 Morehouse Lane
Red Hook, NY 12571

Some format issues inherent in the e-media version may also appear in this print version.

Institute of Physics
Dirac House, Temple Back
Bristol BS1 6BE UK

Phone: 44 1 17 929 7481
Fax: 44 1 17 920 0979

techtracking@iop.org

Additional copies of this publication are available from:

Curran Associates, Inc.
57 Morehouse Lane
Red Hook, NY 12571 USA
Phone: 845-758-0400
Fax: 845-758-2633
Email: curran@proceedings.com
Web: www.proceedings.com

TABLE OF CONTENTS

Preface

Peer Review Statement

Bottom Drag Measurements in Experimental Wind Tunnel Tests 1
A Krzysiak

Rotorcraft Thickness Noise Control by Tip Ventilation 7
O Szulc, T Suresh, P Flaszyński

Influence of Tip Speed Ratio on the Efficiency of Savonius Wind Turbine with Deformable Blades 15
Emil Marchewka, Krzysztof Sobczak, Piotr Reorowicz, Damian Obidowski, Krzysztof Jóźwik

Modeling of Free-Surface Shape in Unbaffled Tanks 23
T. Płusa, P. Duda

CFD Modelling of a Fan with a Cycloidal Rotor 30
T Staśko, M Majkut, S Dykas, K Smołka

Statistical Analysis of Flow Velocity Field Inside Random Packed Beds 38
Maciej Marek

Application of High Accuracy Penalized Vortex in Cell Method for the High Reynolds Number Turbomachinery Flows. 46
Dominik Błoński, Henryk Kudela, Katarzyna Strzelecka

Conservative Weakly Compressible Smoothed Particle Hydrodynamics Applied to Flows with High Vorticity 54
Adam Kajzer

A Parametric Analysis of Concentration Losses in an Anode of a Solid Oxide Fuel Cell 62
T Prokop, G Brus, J Szmyd

CFD Analysis of a Multi-Rotor Flying Robot for Air Pollution Inspection 69
G Suchanek, R Filipek

Numerical Analysis of the Influence of Wall Roughness on the Turbulent Boundary Layer Separation 77
P Kaminski, A Tyliszczak

Numerical Analysis of Heat and Mass Transfer Through Beds of Spherical and Non-Spherical Elements 85
Ewa Szymanek, Artur Tyliszczak, Maciej Marek

Liquid Phase Identification in Wet Steam Transonic Flows by Means of Measurement Methods 93
S Shabani, M Majkut, S Dykas, K Smołka, X Cai

Development of the LES-ADM Approach for Combustion Modelling Using High-Order Filters 101
Lena Caban, Artur Tyliszczak

CO_2 Methanation: On the Modeling of Reacting Laminar Flows in Structured Ni/$MgAl_2O_4$ Catalysts 109
J A Medina Méndez, M González-Castaño, F M Baena-Moreno, H Arellano-García

Numerical Analysis of Influence of Various Bluff-Body Shapes on Diffusion Flame Dynamics 117
Robert Kantoch, Agnieszka Wawrzak, Artur Tyliszczak

Strouhal Number and Vortices in Far Wake of Cylinder 125
D Duda, V Yanovych, V Horáček, V Uruba

Experimental Study of Laminar-To-Turbulent Transition in an Adverse Pressure Gradient Flow 133
K Gumowski, S Kubacki

Wind Tunnel Investigations of Aircraft Airfoil in Cruise Conditions 141
P Kaczyński, R Szwaba, M Piotrowicz, P Flaszyński, P Doerffer

Microphysical Measurements of Hydrometeors Using Shadowgraph Imaging Technique 149
Moein Mohammadi, Jakub L Nowak, Szymon P Malinowski

Experimental Analysis of the Impact of an Acoustic Excitation on a Separated Boundary Layer Behaviour 157
V Sokolenko, W Elsner, A Dróżdż, R Gnatowska, Z Rarata, S Kubacki

Effect of the Upstream Cylinder Shape on the Flow Around the Downstream Rectangular Object in Tandem Configuration 164
Karolina Gajewska, Paweł Niegodajew, Renata Gnatowska, Witold Elsner

On Dynamics of Flow Past a Stage Axial Air Turbine 171
V Uruba, D Duda, V Yanovych

Least-Order Representation of Control-Oriented Flow Estimation Exemplified for the Fluidic Pinball 179
Songqi Li, Wenpeng Li, Bernd R Noack

Large Eddy Simulations of Excited Flames Issuing from Rectangular Nozzles 187
Jakub Stempka, Artur Tyliszczak

On the Generalization of Calogero-Ahmed Summation Formulas 195
K Urbanowicz, M Stosiak, A Bergant

Density Diffusion in Low Mach Number Flows 203
J Pozorski, A Kajzer

Simulations of Johnson-Segalman Viscoelastic Fluids Flows in Shear-Thinning Regime 211
Tomáš Bodnár, Adélia Sequeira

Kinematic Simulations as a Subgrid-Scale Model for Particle Motion: A Priori LES of Homogeneous Isotropic Turbulence 219
Michał Rajek, Jacek Pozorski

Simulation of Sloshing in the Moon Pool with Smoothed Particle Hydrodynamics 227
E Spricigo, J Pozorski

Analysis of the Moist Air Transonic Flow in the Symmetric and Asymmetric Nozzle of the Low Expansion 235
P Wiśniewski, S Dykas, M Majkut, K Smołka

Non-Equilibrium Dissipation Scaling in Atmospheric Turbulence 243
M Wacławczyk, J L Nowak, S P Malinowski

Author Index

Journal of Physics: Conference Series **2367** (2022) 011001 doi:10.1088/1742-6596/2367/1/011001

2022 FMC

RZESZOW UNIVERSITY OF TECHNOLOGY

XXV FLUID MECHANICS CONFERENCE

Rzeszów, Poland
7-9 September 2022

PROCEEDINGS

Editors: Anna Kucaba-Pietal
Adrian Kordos
Ewa Rejwer- Kosinska

Journal of Physics: Conference Series **2367** (2022) 011001 doi:10.1088/1742-6596/2367/1/011001

XXV Fluid Mechanics Conference

Rzeszow University of Technology, Poland

7 - 9 September 2022

PROCEEDINGS

Editors:

Anna Kucaba-Piętal
Adrian Kordos
Ewa Rejwer-Kosińska

Journal of Physics: Conference Series **2367** (2022) 011001 doi:10.1088/1742-6596/2367/1/011001

XXV FLUID MECHANICS CONFERENCE,
Rzeszow University of Technology, 7 - 9 September 2022, Poland

PREFACE

This volume of the Journal of Physics: Conference Series contains proceedings of the jubilee XXV Fluid Mechanics Conference (FMC2022) https://www.fmc2022.prz.edu.pl/, which was organized by the Rzeszow University of Technology (Poland) under the auspices of the Polish Academy of Sciences, Committee of Mechanics. All submitted papers in this volume have been reviewed and approved by anonymous reviewers, eminent specialists in fluid mechanics, whose names are listed later in the Preface.

This Conference, hold in Rzeszow, was the twenty-fifth in the series started forty-eight years ago (1974) in Jaszowiec, Poland. Meetings, the idea of which is to create a forum for discussing the latest results and experiences in the field of fluid mechanics, are held in Poland and are organized every two years by a Polish university under the auspices of the Polish Academy of Sciences, Committee of Mechanics.

The scientific program of the Conference consisted of plenary lectures, sectional lectures an poster presentations. Conference topics were devoted to: aerodynamics, combustion, computational fluid dynamics, experimental fluid mechanics and measurement techniques, flow machinery, general fluid dynamics, interdisciplinary areas in heat and fluid flow, micro-, nano- and bio-flows, multi-phase flows, turbulence.
Prior to the Conference, all contributed extended abstracts of papers were peer reviewed. Recommendations had been received from FMC2022 International Scientific Committee. Accordingly, of the 131 eligible submissions 95 were invited for their presentations. The abstracts have been printed in the FMC2022 Book of Abstracts (ISBN 978-83-7934-590-8).

Over 130 scientists from Poland and abroad (China, USA, Sweden, Czech Republic, Denmark, Germany, Australia), representing 30 research centers, Universities, scientific institutions and companies from around the world, took part in the three-day Conference sessions (7-9 September 2022). The results of the latest researches were delivered during 94 presentations: 70 sectional lectures in frames of 15 fluid mechanics thematic sessions and 24 presented during the poster session. Furthermore five plenary lectures were given by invited outstanding scientists: prof. Ali Beskok (USA), prof. Beverly McKeon (USA), prof. Nicholas Lawson (Australia), prof. Andrzej Herczynski (USA) and prof. Bernd Noack (China). Also, an industrial session was organized, where the possibilities of the latest hardware and software for conducting research in the field of fluid mechanics were presented by the following companies: Amecam - LaVision, Eurotek International, Symkom, UniSens.

Within the general subject of the Conference, the J. W. Elsner Competition for the best fluid mechanics paper presented during the Conference was organized under the auspices of ERCOFTAC and PTMTiS. The winners of the Competition has been: Michał Rajek (1-st place), Tomasz Prokop (2-nd place) and Lena Caban (3-rd place).

Prize of the Rector of Rzeszow University of Technology, for the best poster presented during the poster session of the Conference, received Emil Marchewka *et al*.

We would like to congratulate the winners !

The conference was held in a stationary manner. Additionally the first day sessions and the FMC2022 opening ceremony were broadcasted on the official YouTube channel of the Rzeszow University of Technology to disseminate knowledge among scientists and students.

Contemporary fluid mechanics posses both , the fundamental problems from the area of pure science, and its strong links with modern technology.

We hope that the FMC2022 conference stimulated the endeavors to find the solutions to open problems if fluid mechanics, and possessed new questions, to be answered in future. The number of submitted and accepted papers provided evidence of vitality of contemporary fluid mechanics.

We believe that the FMC2022 Proceedings will be an inspiration for your further research!

Anna Kucaba-Piętal
Adrian Kordos
Ewa Rejwer-Kosińska

Journal of Physics: Conference Series **2367** (2022) 011001 doi:10.1088/1742-6596/2367/1/011001

ACKNOWLEDGMENTS

We would like to express grateful appreciation to the members of the FMC2022 International Scientific Committee, especially for their hard work in doing revisions. Their advices and efforts have helped us to overcome the problems normally associated with organizing international meetings.

Special thanks goes to the Reviewers of Proceedings papers for their work. Their contribution cannot be overestimated.

We appreciate the honorary patronage of the XXV Fluid Mechanics Conference given by: Minister of Education and Science of the Republic of Poland, Rector of the Rzeszow University of Technology, the Mayor of the City of Rzeszow, Marshal of the Podkarpackie Region, Voivode of the Podkarpackie Voivodeship and scientific societies: European Research Community On Flow, Turbulence and Combustion (ERCOFTAC) and the Polish Society of Theoretical and Applied Mechanics (PTMTiS).

The Conference was co-financed by the Ministry of Education and Science of the Republic of Poland on the basis of the agreement no. DNK/SP/513188/2021 of 17.12.2021, entitled "Organization of the international conference XXV Fluid Mechanics Conference carried out as part of the Excellent Science program". Funding of MEiN: 85 000.00 PLN, total project value: 300 700.00 PLN. The support of the Ministry of Education and Science made it possible i.a. to publish this Volume, we appreciate it.

We thank the Sponsors of the FMC2022 Conference: Amecam – LaVision, Eurotek International, Symkom, UniSens, for presenting their current fluid mechanics technology and financial support.

We would like to cordially thank all the contributors of this Conference. Your contribution of up to minute research made the Conference a success!

Thank you all so much !

Anna Kucaba-Piętal
Adrian Kordos
Ewa Rejwer-Kosińska

FMC 2022 INTERNATIONAL SCIENTIFIC COMMITTEE

Chairman: Prof. Jacek Pozorski, Szewalski Institute of Fluid-Flow Machinery, Polish Academy of Sciences
Prof. Janusz Badur, Szewalski Institute of Fluid-Flow Machinery, Polish Academy of Sciences
Prof. Ali Beskok, Southern Methodist University, USA
Prof. Tomas Bodnar, Institute of Mathematics, Czech Academy of Sciences
Prof. Andrzej Bogusławski, Czestochowa University of Technology, Poland
Prof. Tadeusz Chmielniak, Silesian University of Technology, Poland
Prof. Piotr Doerffer, Szewalski Institute of Fluid-Flow Machinery, Polish Academy of Sciences
Prof. Stanisław Drobniak, Czestochowa University of Technology, Poland
Prof. Sławomir Dykas, Silesian University of Technology, Poland
Prof. Maria Ekiel-Jeżewska, Institute of Fundamental Technological Research, Polish Academy of Sciences
Prof. Witold Elsner, Czestochowa University of Technology, Poland
Prof. Mohamed Gad-el-Hak, Virginia Commonwealth University, USA
Prof. Bernard Geurts, University of Twente, the Netherlands
Prof. Andrzej Herczyński, Boston College, USA
Prof. Krzysztof Kempa, Boston College, USA
Prof. Tomasz Kowalewski, Institute of Fundamental Technological Research, Polish Academy of Sciences
Prof. Anna Kucaba-Piętal, Rzeszow University of Technology, Poland
Prof. Henryk Kudela, Wroclaw University of Technology, Poland
Prof. Nicholas Lawson, The University of Sidney, Australia
Prof. Duncan Lockerby, Warwick University, UK
Prof. Marek Morzyński, Poznan University of Technology, Poland
Prof. Bernd Noack, Harbin Institute of Technology, Shenzhen, China
Prof. Alfredo Soldati, TU Wien, Austria
Prof. Janusz Szmyd, AGH University of Science and Technology, Poland
Prof. Jacek Szumbarski, Warsaw University of Technology, Poland
Prof. Ewa Tuliszka-Sznitko, Poznan University of Technology, Poland
Prof. Artur Tyliszczak, Czestochowa University of Technology, Poland
Prof. Markus Uhlmann, KIT Karlsruhe, Germany
Prof. Luc Vervisch, INSA Rouen, France
Prof. Włodzimierz Wróblewski, Silesian University of Technology, Poland
Prof. Weiwei Zhang, Northwestern Polytechnical University, Xi'an, P.R.China

Journal of Physics: Conference Series **2367** (2022) 011001 doi:10.1088/1742-6596/2367/1/011001

ORGANISER

RZESZOW UNIVERSITY OF TECHNOLOGY **Rzeszow University of Technology**

ORGANIZING COMMITTEE

Prof. Anna Kucaba-Piętal - Conference Chair, Rzeszow University of Technology, Poland

DSc Liliana Rybarska-Rusinek - Secretary, Rzeszow University of Technology, Poland
PhD Małgorzata Kmiotek - Secretary, Rzeszow University of Technology, Poland
PhD Adrian Kordos - Technical Secretary, Rzeszow University of Technology, Poland
PhD Ewa Rejwer-Kosińska - Technical Secretary, Rzeszow University of Technology, Poland

DSc Andrzej Majka, Rzeszow University of Technology, Poland
PhD Irena Nowotyńska, Rzeszow University of Technology, Poland
PhD Krzysztof Marzec, Rzeszow University of Technology, Poland
PhD Michał Piętal, Rzeszow University of Technology, Poland
PhD Łukasz Święch, Rzeszow University of Technology, Poland

VOLUNTEERS HELPING DURING FMC2022 SESSIONS

PhD Robert Jakubowski - Volunteers Leader, Rzeszow University of Technology, Poland
PhD Arkadiusz Bednarz, Rzeszow University of Technology, Poland
MSc Marcin Borkowski, Rzeszow University of Technology, Poland
MSc Bartłomiej Filip, Rzeszow University of Technology, Poland
MSc Maciej Kalwara, Rzeszow University of Technology, Poland
MSc Natalia Marszałek, Rzeszow University of Technology, Poland

Journal of Physics: Conference Series **2367** (2022) 011001 doi:10.1088/1742-6596/2367/1/011001

PAPERS REVIEWERS

Prof. Janusz Badur - Szewalski Institute of Fluid-Flow Machinery, Polish Academy of Sciences
Prof. Ali Beskok - Southern Methodist University, USA
Prof. Ewa Błazik-Borowa - Lublin University of Technology, Poland
Prof. Tomas Bodnar - Institute of Mathematics, Czech Academy of Sciences
Prof. Andrzej Bogusławski - Czestochowa University of Technology, Poland
Prof. Tadeusz Chmielniak - Silesian University of Technology, Poland
Prof. Piotr Doerffer - Szewalski Institute of Fluid-Flow Machinery, Polish Academy of Sciences
Prof. Stanisław Drobniak - Czestochowa University of Technology, Poland
Prof. Sławomir Dykas - Silesian University of Technology, Poland
Prof. Maria Ekiel-Jeżewska - Institute of Fundamental Technological Research, Polish Academy of Sciences
Prof. Witold Elsner - Czestochowa University of Technology, Poland
Prof. Paweł Flaszyński - Szewalski Institute of Fluid-Flow Machinery, Polish Academy of Sciences
Prof. Mohamed Gad-el-Hak - Virginia Commonwealth University, USA
Prof. Bernard Geurts - University of Twente, the Netherlands
Prof. Andrzej Herczyński - Boston College, USA
Prof. Marek Jaszczur - AGH University of Science and Technology, Krakow, Poland
Prof. Krzysztof Kempa - Boston College, USA
Prof. Tomasz Kowalewski - Institute of Fundamental Technological Research, Polish Academy of Sciences
Prof. Sławomir Kubacki - Warsaw University of Technology, Poland
Prof. Anna Kucaba-Piętal - Rzeszow University of Technology, Poland
Prof. Henryk Kudela - Wroclaw University of Technology, Poland
Prof. Nicholas Lawson - The University of Sidney, Australia
Prof. Duncan Lockerby - Warwick University, UK
Prof. Maciej Matyka - University of Wroclaw, Poland
Prof. Marek Morzyński - Poznan University of Technology, Poland
Prof. Bernd Noack - Harbin Institute of Technology, Shenzhen, China
Prof. Jacek Pozorski - Szewalski Institute of Fluid-Flow Machinery, Polish Academy of Sciences
Prof. Krzysztof Sobczak - Lodz University of Technology, Poland
Prof. Wojciech Sobieski - University of Warmia and Mazury in Olsztyn, Poland
Prof. Alfredo Soldati - TU Wien, Austria
Prof. Ryszard Staroszczyk - Institute of Hydro-Engineering, Polish Academy of Sciences
Prof. Janusz Szmyd - AGH University of Science and Technology, Poland
Prof. Jacek Szumbarski - Warsaw University of Technology, Poland
Prof. Krzysztof Tesch - Gdansk University of Technology, Poland
Prof. Ewa Tuliszka-Sznitko - Poznan University of Technology, Poland
Prof. Artur Tyliszczak - Czestochowa University of Technology, Poland
Prof. Markus Uhlmann - KIT Karlsruhe, Germany
Prof. Kamil Urbanowicz - West Pomeranian University of Technology in Szczecin, Poland
Prof. Luc Vervisch - INSA Rouen, France
PhD Marta Wacławczyk - Institute of Geophysics University of Warsaw, Poland
PhD Tomasz Wacławczyk - Warsaw University of Technology, Poland
Prof. Włodzimierz Wróblewski - Silesian University of Technology, Poland
Prof. Weiwei Zhang - Northwestern Polytechnical University, Xi'an, P.R.China

Journal of Physics: Conference Series **2367** (2022) 011001 doi:10.1088/1742-6596/2367/1/011001

PARTICIPANTS OF THE FMC2022 CONFERENCE

MSc Ahmad Ababaei - Institute of Meteorology and Water Management, National Research Institute, Poland
PhD Arkadiusz Antonowicz - Eurotek International Sp. z o.o., Poland
PhD Paweł Baj - Warsaw University of Technology, Poland
Prof. Ali Beskok - Southern Methodist University, USA
MSc Dominik Błoński - Wroclaw University of Technology, Poland
PhD Tomasz Bobiński - Warsaw University of Technology, Poland
Prof. Tomas Bodnar - Institute of Mathematics, Czech Academy of Sciences
Prof. Andrzej Bogusławski - Czestochowa University of Technology, Poland
PhD Karol Bukowski - Warsaw University of Technology, Poland
MSc Lena Caban - Czestochowa University of Technology, Poland
PhD Guy Cornejo Maceda - Harbin Institute of Technology, China
PhD Nan Deng - Harbin Institute of Technology, China
Prof. Stanisław Drobniak - Czestochowa University of Technology, Poland
PhD Artur Dróżdż - Czestochowa University of Technology, Poland
DSc Daniel Duda - University of West Bohemia in Pilsen, Czech Republic
Prof. Sławomir Dykas - Silesian Univeristy of Technology, Poland
Prof. Maria Ekiel-Jeżewska - Institute of Fundamental Technological Research, Polish Academy of Sciences
Prof. Witold Elsner - Czestochowa University of Technology, Poland
PhD Jakub Fabisiak - Warsaw University of Technology, Poland
PhD Wojciech Faliński - Amecam – LaVision, Poland
MSc Bartłomiej Filip - Rzeszow University of Technology, Poland
Prof. Paweł Flaszyński - Szewalski Institute of Fluid-Flow Machinery, Polish Academy of Sciences
PhD Stanisław Gajek - Warsaw University of Technology, Poland
PhD Karolina Gajewska - Czestochowa University of Technology, Poland
MSc Jakub Gałecki - Warsaw University of Technology, Poland
PhD Stanisław Gepner - Warsaw University of Technology, Poland
PhD Maciej Ginalski - Symkom sp. z o.o., Poland
DSc Renata Gnatowska - Czestochowa University of Technology, Poland
PhD Bartosz Górecki - QuickerSim Sp. z o.o., Poland
MSc Konrad Gumowski - Warsaw University of Technology, Poland
MSc Ahmed Hanfy - Szewalski Institute of Fluid-Flow Machinery, Polish Academy of Sciences
MSc Emad Hasani Malekshah - Silesian University of Technology, Poland
Prof. Andrzej Herczyński - Boston College, USA
PhD Marzena Iwaniszyn - Institute of Chemical Engineering, Polish Academy of Sciences
MSc Idalia Jagodzińska - Warsaw University of Technology, Poland
PhD Łukasz Jeziorek - Lukasiewicz Research Network Institute of Aviation, Poland
MSc Arun Joseph - Szewalski Institute of Fluid-Flow Machinery, Polish Academy of Sciences
MSc Piotr Kaczyński - Szewalski Institute of Fluid-Flow Machinery, Polish Academy of Sciences
PhD Adam Kajzer - Szewalski Institute of Fluid-Flow Machinery, Polish Academy of Sciences
MSc Piotr Kaminski - Czestochowa University of Technology, Poland
MSc Robert Kańtoch - Czestochowa University of Technology, Poland
PhD Karsten Dörner - Dantec Dynamics A/S, Denmark
Prof. Natalia Kizilova - Warsaw University of Technology, Poland
MSc Michał Klamka - Warsaw University of Technology, Poland

PhD Łukasz Klotz - Warsaw University of Technology, Poland
PhD Małgorzata Kmiotek - Rzeszow University of Technology, Poland
DSc Piotr Korczyk – Institute of Fundamental Technological Research, Polish Academy of Sciences
PhD Adrian Kordos - Rzeszow University of Technology, Poland
PhD Mateusz Korpyś - Institute of Chemical Engineering, Polish Academy of Sciences
Prof. Piotr Koszelnik - Rzeszow University of Technology, Poland
PhD Marek Kraskowski - Maritime Advanced Research Centre CTO S.A., Poland
PhD Andrzej Krzysiak - Lukasiewicz Research Network Institute of Aviation, Poland
DSc Sławomir Kubacki - Warsaw University of Technology, Poland
Prof. Anna Kucaba-Piętal - Rzeszow University of Technology, Poland
Prof. Henryk Kudela - Wroclaw University of Technology, Poland
Eng. Anna Lancmanova - Institute of Mathematics, Czech Academy of Sciences
Prof. Nicholas Lawson - University of Sydney, Australia
PhD Songqi Li - Harbin Institute of Technology, China
PhD Yiqing Li - Harbin Institute of Technology, China
Prof. Aleksandr Linkov - Rzeszow University of Technology, Poland
DSc Andrzej Majka - Rzeszow University of Technology, Poland
DSc Mirosław Majkut – Silesian University of Technology, Poland
Prof. Szymon Malinowski - University of Warsaw, Poland
MSc Emil Marchewka - Lodz University of Technology, Poland
DSc Maciej Marek - Czestochowa University of Technology, Poland
PhD Krzysztof Marzec - Rzeszow University of Technology, Poland
Prof. Beverley McKeon - California Institute of Technology, USA
PhD Juan Medina - BTU Cottbus-Senftenberg, Germany
MSc Abdelhakim Merdjani - Warsaw University of Technology, Poland
PhD Antoine Michel - Institute of Meteorology and Water Management,
National Research Institute, Poland
MSc Moein Mohammadi - University of Warsaw, Poland
Prof. Marek Morzyński - Poznan University of Technology, Poland
MSc Jan Muchowski - Rzeszow University of Technology, Poland
PhD Paweł Niegodajew - Czestochowa University oof Technology, Poland
PhD Yadav Nikesh - Warsaw University of Technology, Poland
Prof. Bernd Noack - Harbin Institute of Technology, China
PhD Irena Nowotyńska - Rzeszow University of Technology, Poland
Prof. Grzegorz Ostasz - Rzeszow University of Technology, Poland
PhD Mohammadsadegh Pahlavanzadeh - Silesian University of Technology, Poland
MSc Michele Pellegrino - Kungliga Tekniska Hogskolan, Sweden
PhD Adam Piechna - Symkom sp. z o.o., Poland
PhD Michał Piętal - Rzeszow University of Technology, Poland
MSc Michał Piotrowicz - Szewalski Institute of Fluid-Flow Machinery, Polish Academy of Sciences
MSc Marek Ples - Silesian University of Technology, Poland
MSc Łukasz Pleskacz - AGH University of Science and Technology, Poland
PhD Tomasz Płusa - Cracow University of Technology, Poland
PhD Marta Poćwierz - Warsaw University of Technology, Poland
Prof. Jacek Pozorski - Szewalski Institute of Fluid-Flow Machinery, Polish Academy of Sciences
PhD Tomasz Prokop - AGH University of Science and Technology, Poland
MSc Michał Rajek - Szewalski Institute of Fluid-Flow Machinery, Polish Academy of Sciences
PhD Zbigniew Rarata - Warsaw University of Technology, Poland
PhD Ewa Rejwer-Kosińska - Rzeszow University of Technology, Poland

PhD Michał Remer - Warsaw University of Technology, Poland
DSc Bogdan Rosa - Institute of Meteorology and Water Management - National Research Institute, Poland
PhD Krzysztof Rusin - Silesian University of Technology, Poland
MSc Mariusz Rutkowski - Warsaw University of Technology, Poland
DSc Liliana Rybarska-Rusinek - Rzeszow University of Technology, Poland
Prof. Jarosław Sęp - Rzeszow University of Technology, Poland
MSc Sima Shabani - Silesian University of Technology, Poland
PhD Krystian Smołka - Silesian University of Technology, Poland
MSc Marta Sobkowiak - Silesian University of Technology, Poland
MSc Vasyl Sokolenko - Czestochowa University of Technology, Poland
MSc Eleonora Spricigo - Szewalski Institute of Fluid-Flow Machinery, Polish Academy of Sciences
PhD Hubert Stańczyk - UniSens - Anna Stańczyk, Poland
PhD Jakub Stańczyk - UniSens - Anna Stańczyk, Poland
MSc Tomasz Staśko - Silesian University of Technology, Poland
MSc Jakub Stempka - Czestochowa University of Technology, Poland
MSc Kamila Stryczniewicz - Warsaw University of Technology, Poland
PhD Wit Stryczniewicz - Lukasiewicz Research Network Institute of Aviation, Poland
MSc Grzegorz Suchanek - AGH University of Science and Technology, Poland
MSc Thanushree Suresh - Szewalski Institute of Fluid-Flow Machinery, Polish Academy of Sciences
Prof. Janusz Szmyd - AGH University of Science and Technology, Poland
PhD Maciej Szudarek - Symkom sp. z o.o., Poland
PhD Oskar Szulc - Szewalski Institute of Fluid-Flow Machinery, Polish Academy of Sciences
DSc Jacek Szumbarski - Warsaw University of Technology, Poland
MSc Ewa Szymanek - Czestochowa University of Technology, Poland
Prof. Artur Tyliszczak - Czestochowa University of Technology, Poland
DSc Kamil Urbanowicz - West Pomeranian University of Technology, Poland
Prof. Vaclav Uruba - Institute of Thermomechanics, Czech Academy of Sciences
PhD Marta Wacławczyk - University of Warsaw, Poland
PhD Tomasz Wacławczyk - Warsaw University of Technology, Poland
DSc Zbigniew Walenta - Institute of Fundamental Technological Research, Polish Academy of Sciences
PhD Filip Wasilczuk - Szewalski Institute of Fluid-Flow Machinery, Polish Academy of Sciences
PhD Agnieszka Wawrzak - Czestochowa Univeristy of Technology, Poland
MSc Karol Wawrzak - Czestochowa University of Technology, Poland
DSc Iwona Włoch - Rzeszow University of Technology, Poland
DSc Wojciech Wolański - Silesian University of Technology, Poland
Prof. Włodzimierz Wróblewski - Silesian University of Technology, Poland
DSc Jerzy Żółtak - Lukasiewicz Research Network Institute of Aviation, Poland
MSc Daria Żyła - Warsaw University of Technology, Poland

Journal of Physics: Conference Series **2367** (2022) 011001 doi:10.1088/1742-6596/2367/1/011001

HONORARY PATRONAGE AND SUPPORT

Minister of Education
and Science

Co-financed by the Ministry of Education and Science of the Republic of Poland on the basis of the agreement no. DNK/SP/513188/2021 of 17.12.2021, entitled "Organization of the international conference XXV Fluid Mechanics Conference carried out as part of the Excellent Science program", Funding of MEiN 85 000.00 PLN, Project value 300 700.00 PLN.

Journal of Physics: Conference Series **2367** (2022) 011001 doi:10.1088/1742-6596/2367/1/011001

HONORARY PATRONAGE

Minister of Education and Science
DSc Przemysław CZARNEK

Minister of Education and Science

Rector of Rzeszow University of Technology
Prof. Piotr KOSZELNIK

RZESZOW UNIVERSITY OF TECHNOLOGY

PAN
POLISH ACADEMY OF SCIENCES

ptmts

ERCOFTAC
European Research Community On Flow, Turbulence And Combustion

XXV Fluid Mechanics Conference (FMC 2022) IOP Publishing
Journal of Physics: Conference Series **2367** (2022) 011001 doi:10.1088/1742-6596/2367/1/011001

Mayor of the City of Rzeszów
Konrad FIJOŁEK

Marshal of the Podkarpackie Region
Władysław ORTYL

MARSZAŁEK
WOJEWÓDZTWA
PODKARPACKIEGO

Voivode of the Podkarpackie Voivodeship
Ewa LENIART

WOJEWODA PODKARPACKI

Journal of Physics: Conference Series 2367 (2022) 011001 doi:10.1088/1742-6596/2367/1/011001

SPONSORS AND EXHIBITORS

uniSENS
Mierz wysoko

www.unisens.pl

Symkom Ansys
CHANNEL PARTNER

www.symkom.pl

DANTEC DYNAMICS
Turn Measurements into Knowledge
Eurotek International

www.dantecdynamics.com, www.lasery.net.pl

LaVision
FOCUS ON IMAGING
AMECAM
Technologie laserowe dla nauki i przemysłu

www.amecam.pl

Journal of Physics: Conference Series **2367** (2022) 011002 doi:10.1088/1742-6596/2367/1/011002

Peer Review Statement

All papers published in this volume have been reviewed through processes administered by the Editors. Reviews were conducted by expert referees to the professional and scientific standards expected of a proceedings journal published by IOP Publishing Publishing.

- **Type of peer review:** Double Anonymous
- **Conference submission management system:** Morressier
- **Number of submissions received:** 52
- **Number of submissions sent for review:** 48
- **Number of submissions accepted:** 32
- **Acceptance Rate (Submissions Accepted / Submissions Received × 100):** 61.5
- **Average number of reviews per paper:** 2.03
- **Total number of reviewers involved:** 28
- **Contact person for queries:**
  Name: Prof. Anna Kucaba-Pietal
  Email: anpietal@prz.edu.pl
  Affiliation: Rzeszow University of Technology, Poland

Journal of Physics: Conference Series **2367** (2022) 012001 doi:10.1088/1742-6596/2367/1/012001

Bottom drag measurements in experimental wind tunnel tests

A Krzysiak[1]

[1] Łukasiewicz Research Network – Institute of Aviation, al. Krakowska 110/114, 02-256 Warsaw, Poland

E-mail: andrzej.krzysiak@ilot.lukasiewicz.gov.pl

Abstract. The paper presents the results of wind tunnel tests aimed at determining the model bottom drag in the case of rocket model tests. The balance measurement technique of the rocket model fixed in the wind tunnel test section by the rear sting was discussed. The model was equipped with the two parallel boosters. Based on the wind tunnel test of the rocket twin model the values of the bottom pressure was determined for tested Mach numbers. An algorithm of wind tunnel corrections was shown, which allowed the total drag determination in a case of the rocket active or passive rocket flight. The test results showed the necessity of the bottom drag measurements in wind tunnel tests.

1. Introduction

In wind tunnel tests, a very important issue is the accuracy of the test results. One of the basic problems that must be solved is the correct modeling of the tested object [1 - 4]. Experimental study of all kinds of objects, both flying and moving on the ground, requires taking into account corrections related to the impact of fixing these objects in the testing area. These corrections result from the fact that in real objects such fixing devices do not exist, and their presence in model tests may significantly affect the aerodynamic characteristics of the tested object.

In the case of experimental studies of the aerodynamic characteristics of whole aircraft models there are usually used two basic methods of mounting the model in the wind tunnel test section, i.e. with the rear sting (**Figure 1**) or with the bottom or dorsal support (**Figure.2**).

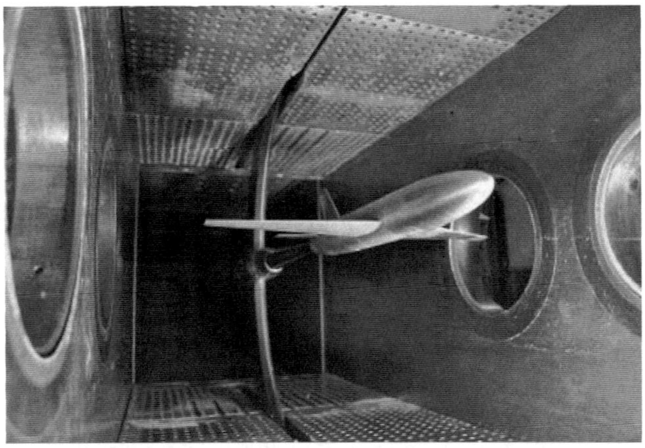

Figure 1 The ONERA calibration model with rear sting in the N-3 trisonic wind tunnel.

Figure 2 The F-16 model with the bottom support in the T-3 low-speed wind tunnel.

Past research [5, 6] has shown that the effects on the aerodynamic characteristics of airplane models supported by a rear or bottom sting were small in magnitude compared with the values for the uncorected test results.

In the case of rocket or other objects with rocket propulsion, additionally to the sting-interference effects, the various phases of their flight, i.e. with the rocket engines operating (active flight) or inoperative (passive flight) should be considered [6]. In each of these flight phases, different corrections for the bottom (base) drag are applied. These corrections significantly affect the total drag of the tested object.

The available literature describes the numerical and experimental research of the rockets and missiles bottom drag [7 – 11] as isolated bodies-of-revolution . However, these works do not contain any information on the bottom drag of the complex objects such as rockets equipped with boosters. This paper fills that gap.

This paper presents both the technique for testing rocket objects in wind tunnels and the method for determining corrections,. The corrections use the measured value of the rocket model bottom pressure. The presented research technique was used to determine the aerodynamic characteristics of the rocket model. tested in the N-3 wind tunnel (Institute of Aviation). The experimental tests were carried out in the range of Mach numbers $M = 0.4 \div 1.15$.

2. Research methodology

The experimental tests of the rocket model were conducted in the N-3 trisonic wind tunnel, i.e. can operate in subsonic, transonic and supersonic flow regimes, $M = 0.3 - 1.15$ and at $M = 2.3$. The N-3 wind tunnel is a blow-down type with partial recirculation of the flow and equipped with the closed test section (cross section 0.6m x 0.6m, length 1.58 m).

The rocket model consisted of the core rocket part (of the 599 mm length and 29 mm diameter) and two removable rear boosters (of the 361 mm length and 21 mm diameter) [12].

The tested rocket model was mounted in the wind tunnel testing chamber on the rear sting, **Figure 3**. Measurements of the aerodynamic characteristics of the rocket model were performed using a six-component internal strain gauge, which also enables static pressure measurement the model rear part. It can be assumed that the pressure measured on the aerodynamic balance is also the model bottom pressure (P_B) and its value is usually lower than the static pressure of the undisturbed flow (P_∞).

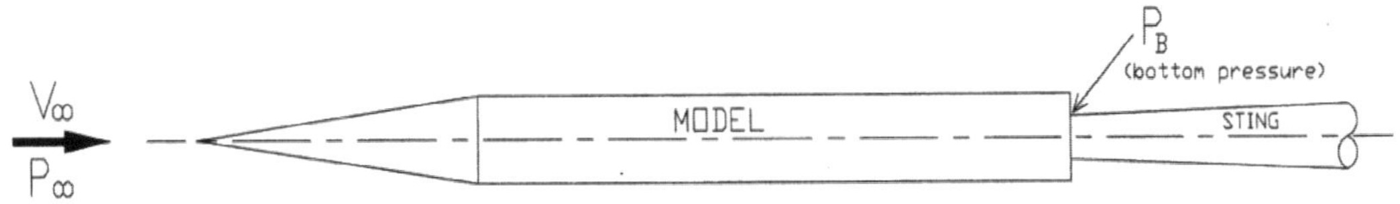

Figure 3 The rocket model attached to the rear sting.

In the case of rocket flight with operating engine, we deal with the outflow of exhaust gas flow from the rocket bottom. Therefore, the measured aerodynamic characteristics should not include the flow around the rocket bottom. For that to happen, the diameter of the sting on which the rocket model is mounted should be equal to the diameter of the rocket bottom (which method is rarely used), or the pressure at the rocket bottom should be equal to the static pressure of undisturbed flow. As this is not the case in presented wind tunnel tests, there is a need to introduce an appropriate correction in relation to the measured value of the rocket drag (in rocket configuration without boosters), as follows;

$$Fx = Fx_m - (P_\infty - P_B) * (S_B - S_S) \tag{1}$$

where;

Fx – rocket drag,
Fx_m – rocket drag measured by the internal strain gauge balance,
P_∞ - static pressure of the undisturbed flow,
P_B - model bottom pressure measured by the internal strain gauge balance,
S_B - model bottom area,
S_S – sting cross-sectional area in the model bottom plane.

In the case of rocket flight with the bottom of the rocket engine off, at the rocket there is a pressure (total bottom pressure - P_{BT}) which value differs (due to the presence of the sting) from the pressure value measured by the internal strain gauge balance (P_B). To determine the exact value of the total bottom pressure, a twin rocket model was constructed. The twin model had the same dimensions as the main model and was fixed to test section walls using cables (at a zero angle of attack), without the sting that normally holds the model in the wind tunnel testing chamber, as shown in **Figure 4**. The twin rocket model was tested in the N-3 wind tunnel in the same range of Mach numbers as the real rocket model ($M = 0.4 – 1.15$). During these investigations the rocket core unit bottom pressure (P_{BC}) was measured. Due to the fact that the rocket model tests in the N-3 wind tunnel were carried out in a small range of the angle of attack, i.e. $\alpha = -4^0 \div 4^0$, the influence of the bottom pressure of the angle of attack on the rocket core unit was neglected.

Figure 4 The twin rocket model in the N-3 trisonic wind tunnel.

In the case of rocket flight with the rocket engine off (in rocket configuration without boosters), it is necessary to introduce an appropriate correction in relation to the measured value of the rocket drag, as follows.

$$Fx = Fx_m - (P_{BC} - P_\infty) * S_S - (P_{BC} - P_B)*(S_B - S_S) \tag{2}$$

where;

P_{BC} – rocket core unit bottom pressure measured at the rocket twin model bottom.

In case where rocket core engine is supported by the side engines (boosters), as it was in the presented research, then the twin model should also be equipped with such boosters (**Figure 4**) and in each of these boosters the bottom pressure should be measured (P_{BB}). The wind tunnel tests of the rocket model equipped with boosters refer to the case when boosters are not active, so the appropriate corrections, presented below, should be introduced only when the rocket flight with operating boosters is under consideration.

$$Fx = Fx_m - (P_\infty - P_B) * (S_B - S_S) - N*(P_\infty - P_{BB}) * S_{BB} \tag{3}$$

where;

N – number of boosters,

P_{BB} – booster bottom pressure measured at the rocket twin model booster bottom,

S_{BB} - booster bottom area.

The rocket drag coefficient (C_d)was determined by dividing the measured values of the rocket drag force (Fx) by the reference area (cross-sectional area of the model core unit - 0.00062 m^2) and by the dynamic pressure value.

The uncertainty of the measurements presented in this paper was established using data obtained during aerodynamic balance calibration and datum measurements taken throughout the campaign. The resulting uncertainties for drag coefficient was 0.049 (the 95% confidence interval).

3. Examples of test results

Wind tunnel tests showed that the value of the bottom pressure was significantly dependent on the undisturbed flow velocity (Mach number). In **Figure 5** the value of the static pressure of the undisturbed flow, bottom pressure measured by the strain gauge balance and bottom pressures measured at the rocket twin model (rocket core unit and booster) versus Mach number are presented.

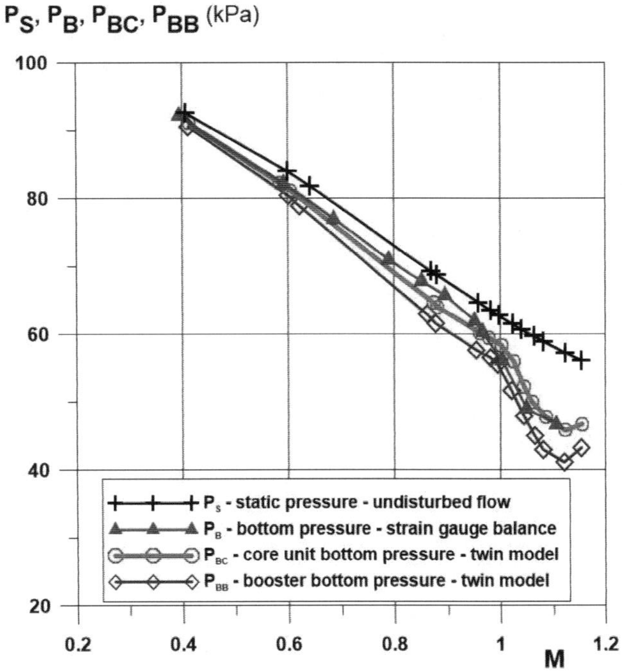

Figure 5 Bottom pressures measured at the rocket model and the twin model versus Mach number.

Based on the measured values of the bottom pressures, the appropriate corrections of the rocket drag for its active and passive flight (chapter 2) were calculated. The corrected values of the drag force were used to calculate the aerodynamic coefficients of the rocket model. In **Figure 6** the difference in the drag coefficient value between the active and passive flight (that is, the influence of the bottom drag) of the core rocket unit (without boosters) and of two parallel boosters (active rocket core unit active) versus Mach number for $\alpha = 0°$ is presented.

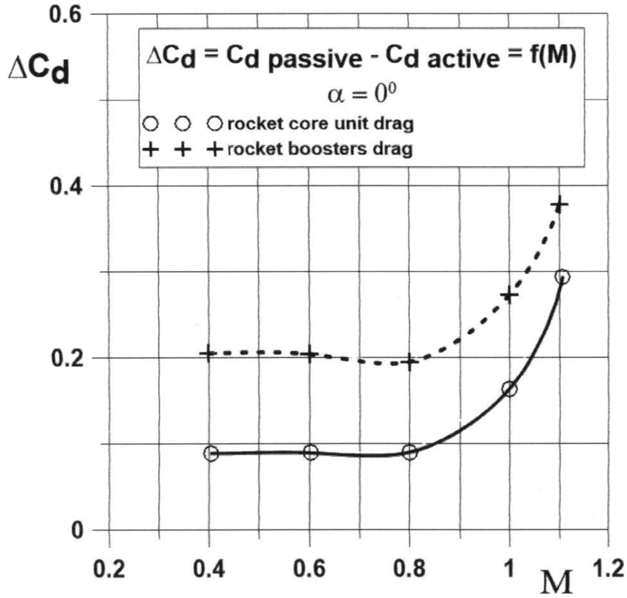

Figure 6 The influence of Mach number on the rocket core unit and the boosters bottom drag.

Journal of Physics: Conference Series **2367** (2022) 012001 doi:10.1088/1742-6596/2367/1/012001

4. Conclusions

Experimental measurements of bottom pressure of the rocket model equipped with two parallel boosters in the trisonic wind tunnel N-3 were performed. Wind tunnel tests were carried out in the range of Mach numbers M = 0.4 ÷ 1.15. Based on the test results, the appropriate corrections of the rocket drag for its active and passive flight were calculated. The corrected values of the drag force were used to determine the drag coefficient of the rocket model. Finally, the influence of the bottom drag of the core rocket unit (without boosters) and of two parallel boosters (active rocket core unit) versus Mach number for $\alpha = 0°$ was defined. The results showed that:

The bottom pressure of the rocket core unit as well as the boosters is less than the static pressure of the undisturbed flow in the range of tested Mach numbers. This difference in pressure values increases with flow velocity rise, especially in the range of supersonic Mach numbers (M > 1). This results from the appearance of supersonic flow in the entire rocket model surroundings accompanying with the shock waves system.

The bottom pressure of the rocket boosters is less than the rocket core unit one throughout the range of tested Mach numbers.

In the range of Mach numbers up to M ≈ 1, the presence of the sting behind the model causes a slight increase in the bottom pressure value, while in the supersonic range its decrease.

In the range of Mach numbers up to M ≈ 0.8 both the rocket core unit and the boosters bottom drag have approximately the constant value, i.e. $\Delta Cd \approx 0.09$ in the case of rocket core unit and $\Delta Cd \approx 0.2$ in the case of two boosters.

The test results showed the necessity of the bottom drag measurements in wind tunnel tests as bottom drag is an important component of a rocket's total drag.

References

[1] Tropea C, Yarin L and Foss J 1998 *Springer Handbook of Experimental Fluid Mechanics*

[2] Pope A 1966 *Low-Speed Wind Tunnel Testing*, John Wiley & Sons, New York

[3] Gebbink R, Wang G and Zhong W 2018 High-Speed Wind Tunnel Tests of the CAE aerodynamic validation Model, *Chinese Journal of Aeronautics* vol 31 pp 439-447

[4] Ristic S, Isakovic J, Ilic B and Ocokoljic G 2004 Review of methods for flow velocity measurement in wind tunnels *Scientific-Technical Review* vol. LIV No 3

[5] Loving D and Lnomu A 1967 Sting-Support Interference on Longitudinal Aerodynamic Characteristics of Cargo-Type Airplane Models at Mach 0.70 to 0.84, *Report NASA TN D-4021*

[6] Cyran F 1981 Dynamic, and Base Pressure Measurements of the Standard Dynamics Model Aircraft at Mach Numbers 0.3 through 1.3 *Report AEDC-TR-81-3*

[7] Suttles J 1964 Aerodynamic characteristics from Mach 0.22 to 4.65 of a two-stage rocket vehicle having an unusual nose shape, *Report NASA TN D-2163*

[8] Brazzel C, Henderson J 1966 A correlation of the base drag of bodies-of-revolution with a jet exhausting through the base, *Report Army Missile Command, Redstone Arsenal*

[9] Moore F 2001 An improved semi-empirical method for power-on base drag prediction, *AIAA Atmospheric Flight Mechanics Conference and Exhibit*, AIAA-2001-328

[10] Krasnov N 1970 *Aerodynamics of Bodies of Revolution,* Deane N. Morris, Rand Corporation

[11] Ruchała P, Placek R, Stryczniewicz W, Matyszewski J, Cieśliński D and Bartkowiak B 2017 Wind tunnel tests of influence of boosters and fins on aerodynamic characteristics of the experimental rocket platform, *Transactions of the Institute of Aviation*, No. 4(249), pp. 82-102.

[12] Krzysiak A, Cieśliński D, Placek R and Kekus P Experimental Study of the Boosters Impact on the Rocket Aerodynamic Characteristics, *Aircraft Engineering and Aerospace Technology* (accepted for publication)

Rotorcraft thickness noise control by tip ventilation

O Szulc[1], T Suresh[1] and P Flaszyński[1]

[1] Institute of Fluid-Flow Machinery, Fiszera 14, 80-231 Gdańsk, Poland

E-mail: oskar.szulc@imp.gda.pl

Abstract. The paper explains further developments of a new concept called TNC (Thickness Noise Control) of the application of surface ventilation to the reduction of helicopter rotor low-frequency in-plane harmonic (LF-IPH) noise. The TNC method is based on introduction of four cavities covered by perforated plates (connected to low and high pressure reservoirs) and positioned symmetrically at the front and rear extremities of the blade tip. Two operational modes are analyzed: constant (steady) and periodically varying (unsteady) transpiration mass-fluxes. For exemplary two-bladed model helicopter rotor of Boxwell et al. (Bell UH-1H Iroquois helicopter) in hover conditions, the results of numerical simulations, based on the CFD code SPARC (Spalart-Allmaras turbulence and Bohning-Doerffer transpiration models), suggest that the acoustic pressure fluctuations are significantly reduced in the near-field of the blade tip. Moreover, the unsteady approach (designed for forward located observers) is almost equally efficient to the basic (steady) operation mode, while substantially lowering penalties in terms of the aerodynamic performance and required transpiration flow intensity. Finally, it is proven that when TNC is activated, the twisted blade (operating in low-thrust conditions) exhibits lower deterioration of the aerodynamic performance compared to its straight (untwisted) counterpart.

1. Introduction

In helicopter rotor tip path plane the acoustic signal has a distinct shape, with large negative pressure pulse that radiates mostly in the direction of forward flight, often dominating all other sources of rotorcraft noise [1]. For helicopters operating with the advancing tip Mach numbers in the $0.75-0.80$ range (many contemporary designs) the shape of the impulse can be explained by considering only the monopole (thickness) term of the Ffowcs-Williams and Hawkings far-field solution of the inhomogeneous linear wave equation (FW-H analogy) [2]. At a fixed observer position $(x,\, y,\, z)$ in space the acoustic pressure value $p' = p - p_0$ (p – static pressure, p_0 – ambient pressure) at any given instant of time t is computed as a time derivative of the sum of all acoustic sources distributed over the entire surface of the rotor blade S:

$$p'\,(x,\, y,\, z,\, t) \;=\; \frac{1}{4\pi}\,\frac{\partial}{\partial t}\,\iint_S \left[\frac{\rho_0\, U_{\mathrm{n}}}{r\,|1 - Ma_r|}\right]_\tau dS\;. \tag{1}$$

The acoustic potential (integrand) of each source panel dS is evaluated at the retarded (emission) time: $\tau = t - r/a$ (a – speed of sound), and is dependent on four parameters: the density of fluid ρ_0, the normal velocity due to blade motion U_{n}, the distance to the observer r, and finally the Mach number in the observer direction Ma_{r}. As can be seen in equation (1) the thickness acoustic pressure signal is determined by the blade shape and kinematics only, being independent of the aerodynamics. In effect a very characteristic acoustic signal is predicted, as shown in figure 4. It is obvious from equation (1) that to reduce the thickness noise intensity it is necessary to consider all mentioned components of the monopole term. For modern helicopter main rotors it is primarily achieved passively by changing the blade tip shape (S), introducing thinning and sweep [3, 4], accompanied by a lower hover tip Mach number Ma_{T} [5]. If the rotational speed of the rotor and the blade shape are considered as fixed, the available option is to affect the normal velocity component U_{n} of the source panel dS. It may be achieved by high mass-flow rate suction and blowing through perforated surfaces placed directly at the blade tip [6, 7]. Pneumatic methods of rotor noise attenuation may be effective in damping the main source of thickness noise, but introduce an important coupling. Surface mass transfer influences not only the acoustic radiation, but also the surrounding flow-field, modifying the magnitude of the aerodynamic forces acting on the blade.

During approach the aerodynamic noise originating from the main rotor blades dominates the acoustic signature of a helicopter. Due to its high amplitude, low-frequency spectral contents, and focused forward (in-plane) directivity pattern, the main rotor noise is weakly attenuated and propagates for long distances in the atmosphere [8]. It consists primarily of deterministic (discrete frequency) contributions from thickness (monopole) and non-impulsive (steady) loading (dipole), and is called low-frequency in-plane harmonic (LF-IPH) noise. The rotational noise is present in all flight conditions, even in hover. In some specific circumstances tonal impulsive sources may also appear, such as blade-vortex interaction (BVI) noise (unsteady loading) or high-speed impulsive (HSI) noise (quadrupole) due to development of transonic flow and shock waves [9]. In contrast to thickness, loading noise spectra may exhibit not only harmonic, but also broadband character. The stochastic (non-deterministic) nature is attributed to turbulence ingestion, blade-wake interaction, or blade self-noise phenomena.

Passive helicopter rotor noise attenuation methods, such as limitation of rotational speed or blade shape optimization, while being cost-effective, are strongly depending on flight conditions and sound radiation directivity. On the other hand, active rotor technologies reveal a great potential for in-plane, low-frequency noise and vibration control [1]. On-blade trailing edge flaps (TEF) [10–12], individual blade control (IBC) [13], higher harmonic control (HHC) [14], and active twist (AT) [15] provide a viable option to reduce the acoustic emissions. However, the advanced active rotor technologies (i.e. TEF, IBC, HHC, and AT) require constant research and funds, being still not reliable enough to be directly integrated in the current rotorcrafts. A similar in-plane rotor harmonic noise cancelling effect may be achieved by applying an on-blade control source of mass and momentum (unsteady spanwise tip blowing) that constitutes a promising option for thickness noise reduction [16].

The primary objective of the paper is to numerically assess the effect of application of the proposed Thickness Noise Control (TNC) method by tip surface ventilation for a twisted rotor blade. The current results are compared with previous straight blade analysis (published in [6, 7]) in terms of low-frequency in-plane harmonic (LF-IPH) noise attenuation and associated aerodynamic performance loss. As before, steady and cyclic (1/rev) modes of operation of TNC are considered for near-field observer location, where thickness acoustic pressure impulse is resolved by CFD. The efficiency of noise reduction (overall sound pressure level, peak amplitude, and sound spectra), the required mass-flow rates, and the penalty in rotor shaft torque are quantified for straight and twisted blades operating in full-scale hover conditions. It is presented that thickness acoustic signature (the main contributor to annoyance and long-range detection of helicopters) may be significantly attenuated by TNC. Moreover, for the same operating conditions, basic blade planform, and TNC method implementation, the twisted blade exhibits two times lower aerodynamic penalty (shaft torque increase) compared to its untwisted counterpart [6, 7].

2. Thickness Noise Control (TNC) by blade tip ventilation

The proposed Thickness Noise Control (TNC) method is based on inclusion of four cavities, linked to either compressed air or vacuum supplies and covered by perforated plates, placed directly at the tip of the rotor blade (figure 1). Two front cavities (closed by high porosity panels, green color in figure 1) allow intensive suction of air from the main stream, while two rear cavities (closed by low porosity panels, red color in figure 1) permit mild blowing of fluid into the main stream. The corresponding extraction and injection surfaces are positioned in a symmetrical manner (according to the chord), at the outer portion of the blade span. If the pressure difference between both sides of the perforation holes is sufficiently large, the flow in the orifices becomes sonic (choked). According to the Bohning and Doerffer (BD) transpiration model (valid for porosity values up-to 30%), the effective Mach number of 0.57 constitutes a limiter for maximum achievable ventilation mass-flow rate [17]. For such conditions, the transpiration (normal) velocity U_t for each perforated plate is uniform and dependent on the aerodynamic porosity of the panel only (irrespectively of the current external flow state).

To suppress thickness noise radiation at its origin, the number, dimensions, and location of each suction or blowing surface is adapted to areas of maximum normal velocity U_n (equation (1)). Moreover, the direction of U_t at each surface is set opposite to U_n. The magnitude of U_t conforms to the maximum achievable mass-flow rate through the perforated surface of selected porosity. For given arrangement of TNC, the final noise canceling capability of the system is highly dependent on the balance between suction and blowing streams which are directly affected by the porosity value of each panel. Steady and unsteady (periodic) activation of surface transpiration is considered. Permanent ventilation (hover conditions) delivers perfect noise canceling directivity pattern, while having a very detrimental effect on the rotor aerodynamic performance. On the other hand, the cyclic (1/rev) approach (designed for forward flight) delivers similar efficiency in terms of the peak amplitude reduction, while allowing for multiple times lower cost regarding the required mass-flow rates and the augmentation of the rotor shaft torque. For rear surfaces subjected to blowing, the spanwise centrifugal pumping may be readily applied, as described in [18]. However, the implementation of a suction system on a helicopter rotor poses a significant engineering challenge, mainly because the flow direction within the blade extraction lines is opposite to the direction of the centrifugal force.

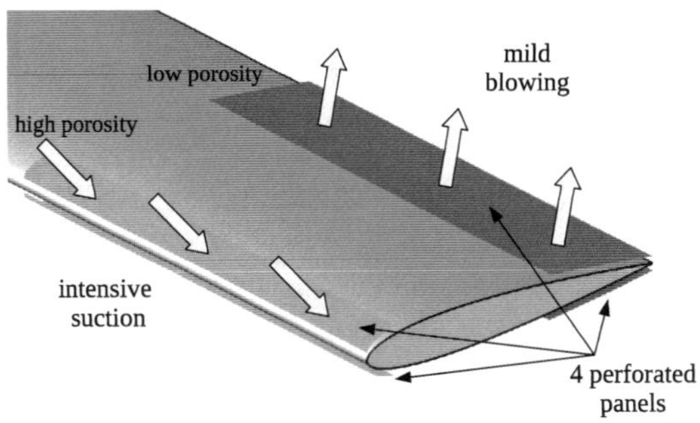

Figure 1. Helicopter rotor blade tip with TNC system.

3. Physical and numerical modelling (SPARC solver)

The research has been carried out with SPARC – a cell-centered, block-structured CFD solver developed at the University of Karlsruhe [19]. The code solves numerically the Favre-averaged mass, momentum, and energy conservation equations derived for compressible turbulent flow simulations. The set of transport equations is supplemented by a classical eddy-viscosity turbulence closure (1-equation) of Spalart and Allmaras (SA), applied in reference conditions and for configurations involving surface mass transfer [20]. The 2nd order discretization of space and time is based on the finite volume algorithm (central scheme with SLIP artificial dissipation model of Jameson) and implicit dual time-stepping approach (explicit Runge-Kutta method of integration in pseudo-time). To increase the convergence rate to steady state of the explicit iterations, the implicit residual smoothing, local time-stepping and full multigrid techniques are implemented. The equations are formulated in the absolute reference frame, therefore the rotation is imposed as motion of the entire computational domain. Moreover, the TNC method in cyclic mode of operation is also time-dependent and requires unsteady solution.

A perforated wall boundary condition was implemented in the SPARC solver for stationary and rotating surfaces [21], based on the transpiration model of Bohning and Doerffer [17]. The prediction capabilities of the code were thoroughly validated against the experimental data for basic internal (flat and convex wall channels) and external (airfoil) flows with surface mass transfer [22]. It has also been applied for investigations of new methods of rotorcraft thickness [6, 7] and high-speed impulsive [9] noise reduction. Finally, the SPARC code has already proven its accuracy in numerous investigations of isolated helicopter rotors in hover and forward flight conditions (e.g. [9, 23]). It is worth emphasizing that the CFD output (flow-field) data may be used for direct acoustic post-processing, but for near-field observers and for low-frequency contents only. The dispersion (pulse shape deformation) and dissipation (pulse amplitude damping) errors prevent any meaningful analysis of signals collected in the far-field. Such limitations are a known consequence of application of CFD algorithms and computational grids designed for fast stabilization of rotor forces.

4. Caradonna-Tung model helicopter rotor in hover (experimental validation)

The validation of the CFD model in terms of the aerodynamic performance is presented against the well known experimental database of Caradonna and Tung (NASA Ames) [24]. The model C-T rotor (radius R of $1.143\,\mathrm{m}$) consisted of two blades and operated in hover with $650\,\mathrm{RPM} - 2540\,\mathrm{RPM}$ (tip Mach number Ma_{T} between 0.23 and 0.89) and collective pitch θ settings of $0° - 12°$. Both blades were straight (untwisted) and rectangular in planform, with the classical NACA 0012 section of constant chord length c of $0.1905\,\mathrm{m}$ (aspect ratio of 6.0). Pressure coefficient c_p distributions at 5 cross-sections of the blade ($r/R = 0.5$, 0.68, 0.80, 0.89, and 0.96, r – radial position) were measured and based on the acquired data rotor thrust coefficient c_{t} was estimated. Full-scale subsonic hover conditions were selected for validation ($Ma_{\mathrm{T}} = 0.61$ and $\theta = 5°, 8°$, and $12°$ for tip Reynolds number $Re_{\mathrm{T}} = 2.6 \cdot 10^6$).

4.1. Computational model

The computational model of the C-T rotor consists of a single blade and an artificial hub surface. The blunt trailing edge of the blade is interchanged with a sharp contour. The details of the tip cap geometry are unknown, therefore a flat surface approximation is adopted. The outer boundaries of the computational domain are positioned at a distance of $3\,R$ from the rotation axis and above and below the rotor plane. The C-H-H type grid topology consists of 80 blocks and is depicted in figure 2. For mesh dependency studies, three grid densities are analyzed: coarse ($0.7 \cdot 10^6$ cells), medium ($5.7 \cdot 10^6$ cells) and fine ($45.5 \cdot 10^6$ cells) for each collective $\theta = 5°$, $8°$, and $12°$ (created by multiplying consecutively the number of cells in each direction by a factor of 2). For example configuration designed for $\theta = 8°$, 87% of volumes are characterized by the most desired orthogonality level of $75° - 90°$, with 9% in the range $60° - 75°$. The remaining 4% of the lowest quality cells ($< 60°$) cannot be improved due to sharp trailing edge corner location. The maximum aspect ratio does not exceed the value of 4800. The quality metrics of the computational grids designed for $\theta = 5°$ and $\theta = 12°$ are similar. The near-wall grid normal spacing of $3.9 \cdot 10^{-5}\,c$ (medium mesh) is chosen to ensure proper resolution of the boundary layer viscous zones (i.e. $y^+ = \mathcal{O}(1)$). The time step of the unsteady numerical scheme is fixed at $0.25°$ of azimuth for all considered cases (convergence criterion of 4 orders of magnitude density residual reduction for internal iterations in pseudo time). At the outer edges of the computational domain, the ambient pressure/temperature boundary condition is utilized with automatic back-flow (inlet/outlet) control. The rotational periodicity boundary condition removes the necessity to simulate the second blade of the hovering rotor, resulting in significant savings in terms of the required computational resources. The blade surface is modelled with the standard no-slip (impermeable) wall boundary condition, locally interchanged with the perforated (permeable) wall formulation for configurations involving transpiration. The artificial hub surface is treated as an impermeable wall with free-slip.

4.2. Discussion of numerical results

The spatial accuracy of the CFD solution is quantified in terms of total thrust $c_{\mathrm{t}} = F_{\mathrm{t}}/(\rho_0 A\, U_{\mathrm{T}}^2)$ and torque $c_{\mathrm{q}} = M_{\mathrm{q}}/(\rho_0 A\, U_{\mathrm{T}}^2 R)$ coefficients of the rotor ($F_{\mathrm{t}}\,[\mathrm{N}]$ – thrust, $M_{\mathrm{q}}\,[\mathrm{N\,m}]$ – shaft torque, $A\,[\mathrm{m}^2] = \pi R^2$ – rotor disc area, and $U_{\mathrm{T}}\,[\mathrm{m\,s^{-1}}]$ – tip speed). The relative difference in aerodynamic coefficients between coarse ($0.7 \cdot 10^6$ cells) and fine ($45.5 \cdot 10^6$ cells) mesh solutions is large for all considered collective angles $\theta = 5°$, $8°$, and $12°$ ($3.8\% - 13.6\%$ for c_{t} and $6.3\% - 7.2\%$ for c_{q}). However, the relative differences between medium ($5.7 \cdot 10^6$ cells) and fine grid computations are more acceptable ($0.2\% - 1.9\%$ for c_{t} and $0.4\% - 2.1\%$ for c_{q}). The medium mesh proves to deliver sufficient

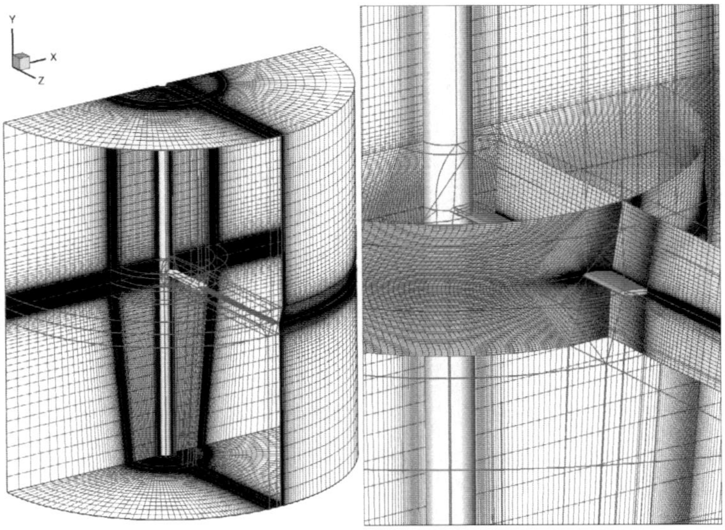

Figure 2. Caradonna-Tung rotor computational grid ($\theta = 8°$, $5.7 \cdot 10^6$ cells).

accuracy (approx. 2%) with 8-times less computational resources required compared to the refined model of $45.5 \cdot 10^6$ of volumes, therefore similar grid densities are used for all further studies presented in the paper. Moreover, it has been verified in the past that for corresponding computational set-ups and flow conditions, reducing the time step size of the numerical scheme from $\Delta t = 0.5°$, through $0.25°$, to $0.125°$ of azimuth has negligible impact on thrust c_t and torque c_q coefficients of the rotor (below 0.1%). Nevertheless, the medium time step of $0.25°$ is adopted for all subsequent investigations due to improved solution accuracy for cases involving the unsteady variant of the TNC method.

For comparison purposes, the medium grid ($5.7 \cdot 10^6$) and medium time step ($0.25°$) solutions are analyzed in figure 3. For blade collective angles $\theta = 5°$, $8°$, and $12°$, the experimental rotor thrust coefficients $c_t = 2.18 \cdot 10^{-3}$, $4.55 \cdot 10^{-3}$, and $8.07 \cdot 10^{-3}$ are predicted by CFD with discrepancies equal to: 10.8% ($c_t = 2.42 \cdot 10^{-3}$), 9.1% ($c_t = 4.96 \cdot 10^{-3}$), and 9.2% ($c_t = 8.82 \cdot 10^{-3}$), respectively. Pressure coefficient c_p distributions at 5 sections of the blade ($r/R = 0.5$, 0.68, 0.80, 0.89, and 0.96) are compared between measurements and numerical modelling in figure 3 for collective angles $\theta = 5°$ and $\theta = 8°$ ($\theta = 12°$ is omitted here due to too high level of c_t, beyond the operating conditions of the Boxwell et al. rotor investigated in Section 5). Generally, a satisfactory correlation is depicted. However, on the upper surface of the blade a slight underprediction of the leading edge suction peak is identified for $r/R = 0.5$ and $r/R = 0.8$ at $\theta = 8°$ setting. It is worth to emphasize that the quoted experimental values of c_t were not directly measured with an aerodynamic balance, but were estimated (with unknown accuracy) based on the available sectional pressure coefficient c_p data.

5. Boxwell et al. model helicopter rotor in hover (application of TNC method)

The numerical (CFD) investigation of the TNC method is conducted with the 1/7-scale model of the main rotor of the Bell UH-1H Iroquois general utility helicopter, tested experimentally at NASA in low-thrust hover by Boxwell et al. [25] and later by Purcell [26]. The model rotor (radius $R = 1.045\,\mathrm{m}$) consisted of two linearly twisted blades (root-to-tip wash-out of $-10.9°$) of rectangular planform, equipped with the classical NACA 0012 section of constant chord length $c = 0.0762\,\mathrm{m}$ (aspect ratio of 13.71). Operating conditions were chosen according to an exploratory experimental acoustic study by Sargent and Schmitz of an equivalent rotor system in the University of Maryland Acoustic Chamber (full-scale hover tip Mach number of $M_T = 0.661$ and tip Reynolds number of $Re_T = 1.6 \cdot 10^6$) [16]. A minimum amount of collective pitch ($\theta = 1.5°$ at the tip) was employed to exhaust the shed wake. Exactly the same operating conditions have been previously investigated for the untwisted variant of the UH-1H model rotor [7] and the results of this analysis are included here for reference.

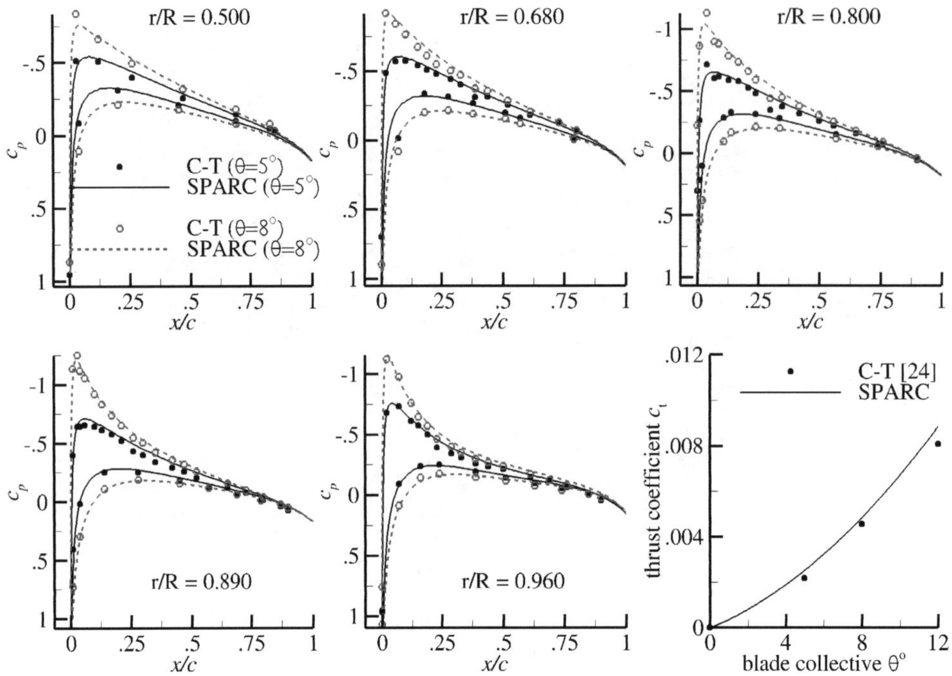

Figure 3. Sectional pressure coefficient c_p distributions and rotor thrust coefficient c_{t}.

5.1. Computational model

The computational model of the Boxwell et al. rotor is equivalent to the Caradonna-Tung rotor set-up described in Section 4.1 in details. However, the geometry is up-scaled to aspect ratio of 13.71 and -10.9° root-to-tip twist is implemented. The size of the computational domain is fixed to $3\,R$ in each direction. As for the C-T rotor, the C-H-H type grid consists of 80 blocks and $5.7 \cdot 10^6$ of volumes per blade. The collective pitch is set to $\theta = 1.5^\circ$ at the tip (low-thrust conditions). The near-wall grid normal spacing is increased to $6.3 \cdot 10^{-5}\,c$ in accordance with the reduced tip Reynolds number ($Re_{\mathrm{T}} = 1.6 \cdot 10^6$). The time step of the unsteady numerical scheme is fixed at 0.25° of azimuth (convergence criterion of 4 orders of magnitude density residual reduction). Identical number and type of boundary conditions are applied.

The computational model of the reference rotor blade is modified by incorporation of two suction and two blowing systems (according to figure 1), located at the front (between the leading edge and $0.1\,\mathrm{c}$) and at the rear (between $0.65\,\mathrm{c}$ and the trailing edge) parts of the tip (last 20% of the span). Two perforated panels of equal, relatively high porosity of 30% cover both forward cavities. In contrast, two perforated panels of equal, low porosity of 5% close the remaining rearward cavities. Sufficiently low (or high) plenum pressure assures reaching the sonic velocity in the holes. Resultant uniform suction ($60\,\mathrm{m\,s^{-1}}$) and blowing ($10\,\mathrm{m\,s^{-1}}$) rates are independent of the local flow conditions present at the ventilated surfaces. Although in the current arrangement of TNC the surface area subjected to injection of fluid is 3.1 times larger than the surface area intended for bleeding of air, the final mass-flux of suction surpasses the blowing intensity by a factor of 2.4. It is known that surface suction is advantageous for aerodynamic performance, while blowing is detrimental to boundary layer state. Hence, while choosing ventilation intensities for the TNC system a primary objective is to introduce significant noise attenuation with minimum injection of fluid involved. Moreover, depending on the distance to the observer, the optimal balance of the suction/blowing streams may differ. In contrast to the presented near-field acoustic signals, the far-field pulse shape is more symmetric in nature and might require a more equalized distribution of mass flow rates (i.e. 1:1 ratio). Two variants of the TNC method are explored here: steady (constant) transpiration velocity V_{t} and unsteady (cyclic,

1/rev) activation within a limited range of the azimuth angle (75°, equivalent to $1/5$ of the rotation period T). For the unsteady schedule, the activation/deactivation times are selected to keep the main pulse amplitude reduction similar to steady TNC, but with minimum azimuthal range (lower flow loses and required mass flow rates) and introduced acoustic disturbances. Even though the implementation of the TNC system is similar for untwisted [7] and twisted blade designs, its impact on the acoustic attenuation and aerodynamic performance may not be the same.

5.2. *Discussion of numerical results*

The efficiency in LF-IPH noise attenuation by the TNC method is quantified in figure 4 for an in-plane, near-field observer location of $r/R = 1.11$. The reference acoustic pressure p'/p_0 fluctuation in time t/T has a typical thickness pulse shape, almost symmetrical and of large negative peak amplitude (black color). In steady mode of operation (fixed suction and blowing) the resultant signal (red color) is positively affected by TNC. Not only the peak amplitude is decreased by 41%, but also the overall sound pressure level (*OASPL*) is reduced by almost 3 dB. Additionally, the resultant spectra, presented for the lowest 20 rotor harmonics (n), reveal significant sound pressure level (*SPL*) damping (by 0.9 dB – 8.0 dB). However, the reference rotor thrust coefficient $c_t = 1.56 \cdot 10^{-3}$ is depreciated by 21% and the torque coefficient $c_q = 1.08 \cdot 10^{-4}$ is increased by 20%. Moreover, the required transpiration mass-flow rates are very demanding ($2.7\,\mathrm{kg\,s^{-1}}$ for suction and $1.1\,\mathrm{kg\,s^{-1}}$ for blowing). Therefore, steady sources and sinks of mass are not considered as a viable option for rotorcraft TNC.

The TNC method in unsteady (cyclic) mode of operation (designed for forward flight conditions) delivers a comparable acoustic pressure signal (blue color in figure 4) to steady approach (peak amplitude reduction by 42%) with 3.8 times lower torque penalty (5%) and 4.8 times lower required transpiration mass-flow rates ($0.56\,\mathrm{kg\,s^{-1}}$ for suction and $0.23\,\mathrm{kg\,s^{-1}}$ for blowing). It is important to emphasize that for steady operation of TNC the suction and blowing mass flow rates are constant in time and very high (difficult to implement). On the other hand, in the unsteady (cyclic) mode, the suction/blowing streams are activated only in a certain range of azimuth (75°). For the remaining time the transpiration is blocked. Consequently, the resulting ventilation mass flow rates (averaged over full period of rotation) exhibit much lower (feasible) values. Even though the main LF-IPH noise pulse shape is the same as for steady TNC, two new acoustic disturbances are introduced due to periodic (1/rev) activation and deactivation of transpiration. Hence, the *OASPL* is decreased slightly less (i.e. by 2.6 dB). The resultant *SPL* spectrum also exhibits uneven and wavy character (damping by 0.0 dB – 10.4 dB).

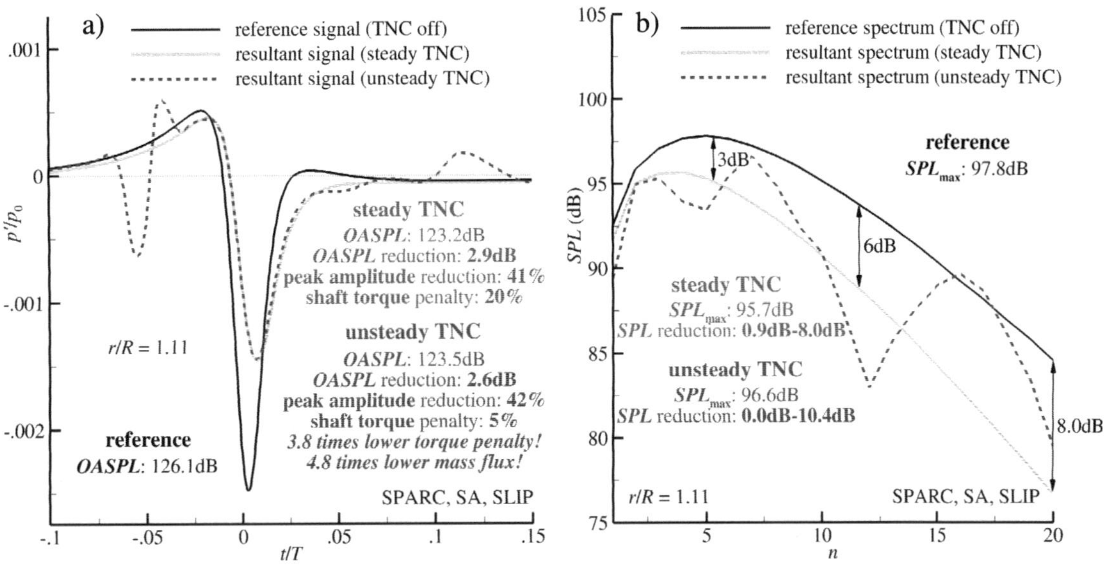

Figure 4. In-plane acoustic pressure p' signal analysis in time (a) and frequency (b) domains.

The efficiency of LF-IPH noise attenuation by TNC (cyclic mode) is similar for twisted (currently investigated) and untwisted (previously analyzed in [7]) rotor blade designs, both operating in hover with the same pitch angle of $\theta = 1.5^{\circ}$ at the tip (fixed planform and flow conditions). Nevertheless, due to higher thrust coefficient c_{t} value, the twisted blade exhibits more that 2 times lower penalty of implementation in terms of the aerodynamic performance (5% for shaft torque) compared to its straight (untwisted) counterpart (11%).

6. Conclusions

The paper describes numerical (CFD/SPARC code) study of the Thickness Noise Control (TNC) system designed for helicopter rotor low-frequency in-plane harmonic (LF-IPH) noise reduction. It relies on incorporation of multiple cavities into the blade tip that are covered by perforated panels and connected to low and high pressure supplies. Intensive surface ventilation (suction and blowing of air) is introduced to counteract the main source of rotorcraft thickness noise radiation. The computational model is validated against the experimental data of Caradonna and Tung for a model helicopter rotor in hover. Similar set-up is used consequently for the investigation of the TNC method implemented for Boxwell et al. twisted blades of the UH-1H helicopter rotor. The unsteady (cyclic) mode of operation of TNC proves to be effective in rotational noise attenuation (in-plane, near-field observer location), with reasonable price in terms of the required transpiration mass-flow rates and lower related aerodynamic performance penalty compared to the basic untwisted blade design.

A detailed analysis of the TNC method efficiency and directivity pattern in the far-field is envisaged with designated acoustic analogy tools, such as FW-H (Farassat's Formulation 1A) solvers. Still, the air extraction/feeding system poses a significant engineering challenge for implementation.

Acknowledgments

This research was supported by CI TASK and in part by PL-Grid Infrastructure.

References

[1] Schmitz F H 2016 *J. Am. Helicopter Soc.* **61** 4 doi:10.4050/JAHS.61.041001
[2] Ffowcs Williams J E and Hawkings D L 1969 *Phil. Trans. Roy. Soc. Lond.* **264** 1151
[3] Mosher M 1984 *NASA Ames Technical Memorandum* NASA-TM-85878-VOL-1
[4] Mosher M 1984 *NASA Ames Technical Memorandum* NASA-TM-85878-VOL-2
[5] Splettstoesser W R et al. 1999 *Proc. 25th European Rotorcraft Forum* (Rome) B7
[6] Szulc O and Doerffer P 2018 *J. Phys.: Conf. Series* **1101** 012042 doi:10.1088/1742-6596/1101/1/012042
[7] Szulc O 2021 *Arch. Mech.* **73** 4 10.24423/aom.3756
[8] Schmitz F H 1991 *Aeroacoustics of Flight Vehicles: Theory and Practice* (NASA Langley)
[9] Szulc O et al. 2016 *J. Phys.: Conf. Series* **760** 012031 doi:10.1088/1742-6596/760/1/012031
[10] Hagerty B and Kottapalli S 2012 *NASA Ames Technical Memorandum* NASA-TM-2012-216048
[11] Sim B W et al. 2014 *J. Am. Helicopter Soc.* **59** 2 doi:10.4050/JAHS.59.022002
[12] Shi Y et al. 2019 Appl. Sci. **9** 7 1351 doi:10.3390/app9071351
[13] Norman T R et al. 2009 *Proc. American Helicopter Society* 65th *Annual Forum* (Grapevine)
[14] Kessler Ch 2011 *CEAS Aeronaut. J.* **1** doi:10.1007/s13272-011-0005-9
[15] Booth E R and Wilbur M L 2004 *J. Am. Helicopter Soc.* **49** 1 doi:10.4050/JAHS.49.3
[16] Sargent C D and Schmitz F H 2012 *Proc. AIAA* 33rd *Aeroacoustics Conference* (Colorado Springs) 2012-2140 doi:10.2514/6.2012-2140
[17] Doerffer P P and Bohning R 2000 *Aerosp. Sci. Technol.* **4** doi:10.1016/S1270-9638(00)01063-4
[18] Platzer S et al. 2016 *Proc. 42nd European Rotorcraft Forum* (Lille)
[19] Magagnato F 1998 *TASK Quarterly* **2** 2
[20] Spalart P R and Allmaras S R 1992 *Proc. AIAA 30th Aerospace Sciences Meeting and Exhibit* (Reno) 92-G439 doi:10.2514/6.1992-439
[21] Doerffer P and Szulc O 2011 *Int. J. Eng. Syst. Model. Simul.* **3** 1-2 doi:10.1504/IJESMS.2011.038751
[22] Szulc O et al. 2020 *Comput. Fluids* **200** 104435 doi:10.1016/j.compfluid.2020.104435
[23] Doerffer P and Szulc O 2008 *TASK Quarterly* **12** 3
[24] Caradonna F X and Tung C 1981 *NASA Ames Technical Memorandum* 81232
[25] Boxwell D A et al. 1979 *Vertica* **3** 1
[26] Purcell T W 1988 *Proc. 14th European Rotorcraft Forum* (Milano) 2

Journal of Physics: Conference Series **2367** (2022) 012003 doi:10.1088/1742-6596/2367/1/012003

Influence of Tip Speed Ratio on the efficiency of Savonius wind turbine with deformable blades

Emil Marchewka[1], Krzysztof Sobczak[1], Piotr Reorowicz[1], Damian Obidowski[1] and Krzysztof Jóźwik[1]

[1]Lodz University of Technology, Institute of Turbomachinery, ul. Wólczańska 217/221, 93 - 005 Łódź, Poland

corresponding author: emil.marchewka@dokt.p.lodz.pl

Abstract. Improving machines efficiency and searching for their new applications are the main topics in the development of the renewable energy industry. In the case of Savonius type wind turbines, the works aim at the improvement of aerodynamic performance. The CFD simulations of a turbine equipped with deformable blades showed a significant positive impact of this enhancement on the machine aerodynamic efficiency. Previously, the investigation was carried out for a *TSR* (Tip Speed Ratio) equal to 0.8, typically recognized as the point of maximal efficiency for conventional Savonius wind turbines with rigid blades. However, the continuously altering shape of blades during their rotation can influence the optimal *TSR*. Therefore, the efficiency of the deformable blade turbine was investigated in a wide range of *TSR*. In this paper, the previously developed quasi-2D model with a two-way Fluid-Structure Interaction method was employed to obtain turbine efficiency characteristics as a function of TSR. The maximum power coefficient *Cp* was achieved at *TSR* = 0.9. Obtained characteristic was compared with data for a conventional rigid blades turbine, gathered with a comparable sliding mesh model.

1. Introduction

The rising awareness of climatic changes and environmental pollution stimulates the transition of energetics into renewable energy sources like wind, water and sunlight. Already existing devices producing clean energy undergo research aimed at raising their efficiency and broadening application possibilities. One of them is a Vertical Axis Wind Turbine (VAWT) of Savonius type, known for starting at lower wind speeds than conventional Horizontal Axis Wind Turbines (HAWTs). A variety of development efforts have a mutual objective of increasing device energy production capabilities. They focus on modifications of blade geometry and the application of shielding elements or ducts [1].

In this work, the application of deformable blades [2] is investigated. A dedicated mechanism constantly alters the frontal areas of both turbine blades. For the advancing blade, its area is increased, while for the returning one a reduction takes place. It results in an enhanced turbine torque. The previous research focused on the development of the CFD (Computational Fluid Dynamics) model to evaluate the aerodynamic efficiency of the novel machine. As it turned out, the use of a conventional sliding mesh method [3] to simulate an operation of such a rotor was difficult and required a high computational cost. In this method, the flow domain is divided into two non-overlapping regions with an interface where solution data are exchanged between them. The internal, circular region contains the rotor geometry with the adjacent surrounding environment and rotates with specified angular velocity. The external, stationary one, is the further turbine surroundings. To simulate the deformable blades rotor, the method requires complicated deforming and remeshing of the internal mesh in the subsequent time steps [2].

Hence, a simpler, more stable and robust computational model was developed with the overset mesh method [4]. In this case, the background computational domain and the adjacent surroundings of the solid bodies placed in the flow are discretized with separate independent overlapping meshes (overset meshes). The solver automatically generates a data interface between the background and boundaries of overset meshes and chooses donor and receptor cells. The solution data is interpolated and transferred between them to obtain the overset meshes locally merged with the background. In this method, the quality of each overset mesh, their movement, deflection and remeshing can be controlled independently before the merging. Additionally, they can be of more complex contours than circular as in the sliding mesh method.

So far, the new model was used to investigate the novel turbine only at *TSR* optimal for a conventional Savonius turbine with rigid blades, equal to 0.8 [4]. However, the blade deforming mechanism, present in the improved machine, alters the velocities of blade tips in various revolution phases. Hence, the turbine might operate with its peak efficiency at a different *TSR* value.

The paper presents a CFD aerodynamic performance assessment of Savonius wind turbine with deformable, overlapped blades. Its efficiency is investigated in a range of *TSRs* typical for Savonius wind turbines. The created computational model couples fluid dynamics and structural solvers to obtain a two-way FSI (Fluid-Structure Interaction) simulation. The former solver utilizes an overset mesh method to model turbine rotor in the flow. The latter replicates a mechanism defining blade deformation. As a reference, a standard turbine is also investigated with a conventional CFD sliding mesh model in alike conditions. Obtained *TSR* characteristics are presented for both machines to show the efficiency advantage provided by the application of deformable blades.

Table 1. Nomenclature applied in the paper

Cp	Power coefficient	***U***	Wind velocity, m/s
D	Turbine diameter, m	***y+***	Non-dimensional wall distance
e	Blade overlap, m	ρ	Air density, kg/m^3
FGCI	Fine-Grid Convergence Index	ω	Turbine angular velocity, rad/s
H	Turbine section height, m	***CFD***	Computational Fluid Dynamics
m	Rotation axis displacement, m	***FSI***	Fluid-Structure Interaction
T	Torque, Nm	***HAWT***	Horizontal Axis Wind Turbine
TSR	Tip Speed Ratio	***SST***	Shear Stress Transport
t	Turbine blade thickness, mm	***VAWT***	Vertical Axis Wind Turbine

2. Methodology

The studied reference turbine rotor was of a Savonius type with semi-circular blades, shown in Figure 1a. Its diameter D was equal to 1 m, the blades thickness t was 1 mm and the overlap between the blades e was 0.1 m. The prototype with deformable blades (Figure 1b) had the same diameter and overlap. The blades rotation axis was offset by a distance $m = 0.1$ m ($m/D = 10\%$ of eccentricity) from the turbine center. The deforming mechanism employed the eccentricity to continuously alter frontal areas of both blades and thus, rise the turbine torque. Its operation and design were thoroughly explained in [2,5].

Two task types were solved. The two-way FSI model with the overset mesh method was used to calculat the performance of the deformable blades turbine. The resultant blade deformation was an effect of combined loads from fluid flow and structural models. It was impossible to create a full 3D turbine simulation due to excessive computational power demands of the fluid flow and structural solvers. Thus, a quasi-2D transient model was created in Ansys Workbench 2020 R2, by coupling the Ansys Fluent fluid flow and the Ansys Transient Structural solvers. Application of the latter one excluded the fully 2D model creation. Blades deformation determined in the structural solver was transferred to the flow solver and the aerodynamic loads at blades were returned to the structural module via a two-way FSI coupling. To reduce the computational time, in simulations of the reference turbine rotor with rigid blades a regular sliding mesh approach was [3,6]. The previous work [4] confirmed that the overset approach generates similar results as this method.

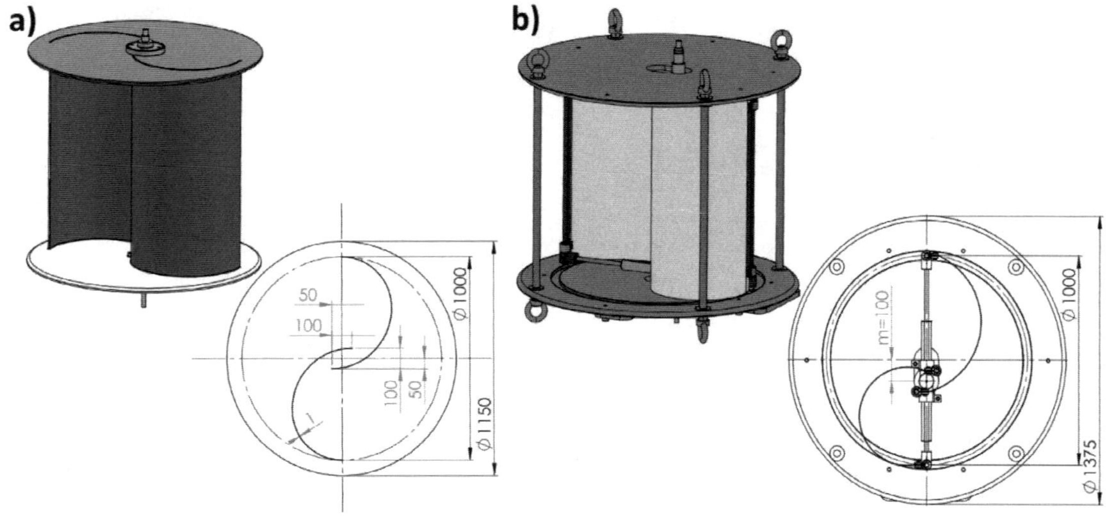

Figure 1. a) A reference Savonius turbine with indicated basic dimensions, b) Savonius turbine of the same diameter and overlap, equipped with deformable blades and the rotation axis eccentricity of 10%.

2.1. Definition of fluid flow models

The performance of turbines was investigated for *TSRs* (Tip Speed Ratios) between 0.5 and 1.2 with the step of 0.1. The specified *TSRs* (1) corresponded to the rotor angular velocity ω from 3.2 to 9.6 rad/s (38.20 to 91.67 rpm).

$$TSR = \frac{\omega D}{2U} \tag{1}$$

Both turbine configurations were investigated inside the 25 m by 20 m flow domains to limit the blockage effect. Domain thickness was 5 mm, implied by the usage of the quasi-2D approach. The deforming mechanism was not considered, as it affects the airflow only at the top and bottom of the turbine rotor. The inner domain of the sliding mesh model was a 1.5 m cylinder containing the rotor. In the overset mesh model, the rotor domain was created from two separate overlapping meshes, with boundaries offset 60 mm from the rotor blades. Figure 2 shows the clockwise rotating rotor in its initial position.

The unstructured hexahedral meshes with one layer of elements along the domain thickness were used. The region of blade rotation was discretized with 4 mm elements for both background and overset meshes (Figure 3). The same size was applied to the inner domain of the sliding mesh model. Away from the rotor zone the mesh elements gradually increased their size up to 100 mm at domain borders. Around the blade walls 4 mm elements were used with an additional refinement at blade tips. At the blade walls 31 (the sliding mesh model) or 37 (the overset mesh approach) inflation layers were applied with a 0.05 mm first layer height, which allowed for fulfilling the $y+ < 1$ condition. All meshes met the Fluent quality criterion (orthogonal quality > 0.2 and aspect ratio < 100).

Discretization uncertainty for both presented turbine models was evaluated with Richardson extrapolation [7] during the model development for TSR = 0.8, and comprehensively presented in [4]. The fine-grid convergence index (FGCI) in the sliding mesh model with 316617 elements was 0.24%, whereas in the overset mesh approach with 467684 elements was negligibly low (< 0.00%).

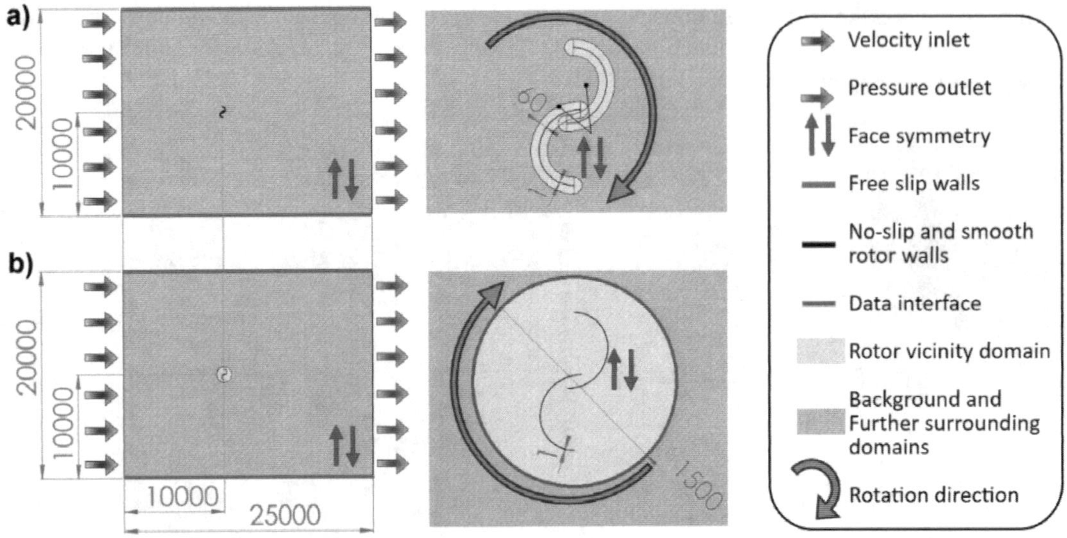

Figure 2. Boundary conditions at the domains with close-ups of rotor regions. a) Overset mesh approach; b) Sliding mesh approach.

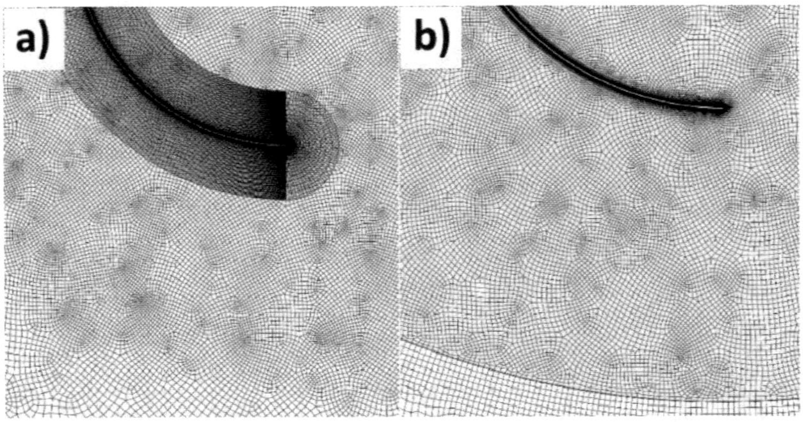

Figure 3. The close-ups on the rotor surrounding and background meshes. a) Overset mesh approach; b) Sliding mesh approach.

All simulations were transient with 1080 time steps per rotor revolution. The Reynolds Averaged Navier Stokes (RANS) pressure solver of Ansys Fluent was utilized. The k-ω SST (Shear Stress Transport) turbulence model of Menter [8] was employed, as it is commonly used in Savonius wind turbine simulations [9–12]. The air properties were defined for standard atmospheric conditions (15°C and 1 atm). Air velocity at the domain inlet (Figure 2) was 4 m/s, which corresponded to the Reynolds number of $2.7x10^5$, based on the turbine diameter. The specified inlet turbulence intensity was 5% with a viscosity ratio of 10. At the domain outlet, the atmospheric pressure was assigned. Both top and bottom of the domains were defined as free slip walls (without shear), while turbine blades were no-slip, smooth walls. The quasi-2D approach implied the symmetry condition at both sides of the mesh slice. Appropriate interfaces passing solution data between stationary and rotating meshes were applied in both methods. Deformation of the rotor overset meshes (green in Figure 2a) was enabled. The first-order implicit scheme and the Coupled pressure-velocity coupling method were used with the second-order spatial discretization schemes for mass, momentum and energy, as well as for the turbulence model transport equations. The residual target 1e-3 for all equations was reached in almost all timesteps. Other details of the solver settings can be found in [4].

2.2. *Structural model definition*

The blade deforming mechanism shown in Figure 4 was modelled in Ansys Transient Structural solver. The 5 mm thick geometry contained the guiding ring (a), being the endplate track, the blades (b) attached to the rotor hub (c) and two sliders (d). The red dot (e) indicated the turbine (guiding ring) centre, while the green one (f) was the blades (rotor) rotation axis. The displacement between (e) and (f) indicated as *m* defines the maximal magnitude of the blade deformation. The mechanism was meshed with one layer of unstructured hexahedral elements, varying in size from 10 to 3 mm. Two layers of 0.5 mm edge size cells covered the blades. To reduce model complexity, the hub was defined as rigid and excluded from meshing and structural analysis. The standard Ansys library provided the material properties. A 316 steel was applied to the blades and structural steel with Young's modulus of 200 GPa and 0.30 Poisson's ratio to the remaining elements. The ring eccentricity $m = 0.1$ m (10% eccentricity) and thus, pre-tension of blades, was achieved during the first rotor revolution. More details of the configuration and operation of the structural model were provided in [4].

2.3. *Coupling of both solvers*

The two-way FSI coupling method, providing data flow shown in Figure 5, led to the final blades deformation. Aerodynamic loads calculated at walls in the fluid flow solver were passed to the structural solver. It joined them with blade loads imposed by the deforming mechanism and sent back the determined blade deflections. The combination of both load types also provided data for a future fatigue strength evaluation of turbine blades. The overset meshes in the flow solver were deformed in an implicit way every five sub-iterations and at the end of each time step. A spring mesh smoothing method [13] was enabled in the CFD solver. It prevented negative cell volumes and kept a good quality of the thinnest inflation layer elements in the deformed meshes.

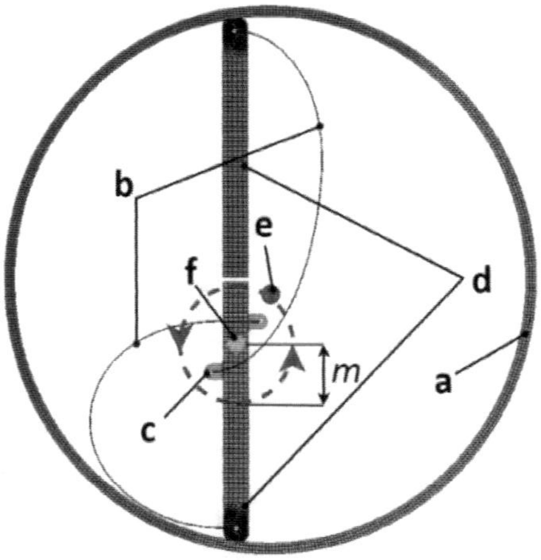

Figure 4. The blade deformation mechanism with the stationary turbine hub and the ring center traveling along the red dashed circle.

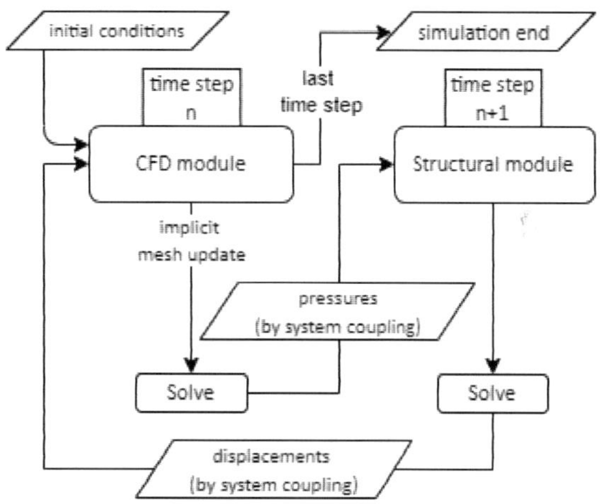

Figure 5. A flow of data between the coupled solvers.

3. Results and discussion

To obtain a fully developed flow, at least 15 rotor revolutions were solved in each task. The torque *T* on the turbine shaft due to aerodynamic loads was recorded at each time step. Data from the last five full revolutions were averaged and the coefficient of power *Cp* (2) was calculated.

$$Cp = \frac{2T\omega}{\rho U^3 DH} \tag{2}$$

The performance characteristics of turbines with deformable and rigid blades are presented in Figure 6. The reference turbine with rigid blades achieved the maximum power coefficient $Cp = 0.272$ at $TSR = 1.1$. The power coefficients were no lower than 90% of the maximal value for *TSRs* between 0.8 and

1.2, being the most efficient operation range. The maximal value of *Cp* was about 20% higher than usually reported for the classic Savonius rotors and was reached at *TSR* higher than in experimental data [14,15]. However, such results are often encountered for data obtained with 2D computational models, where the endplates influence, bearing friction losses and inertial effects are omitted [6,9]. Application of the blade deforming mechanism with 10% eccentricity resulted in peak performance at *TSR* = 0.9 with the power coefficient *Cp* = 0.412. The optimal operation range extended for *TSRs* between 0.7 and 1.15. In the entire studied range, the deformable blade turbine performed superior to the rigid blade variant. The benefit approached 56% for the peak performance point of the enhanced turbine. For the remaining *TSRs* the gain was not lower than about 40%. The results are similarly overestimated as for the conventional turbine, due to the same physical limitations in both numerical models and a lack of friction caused by the deforming mechanism. A prototype in development will be used for test measurements to emphasise the possible differences between a purely aerodynamic numerical performance and the one of a real machine, where mechanical losses play a significant role.

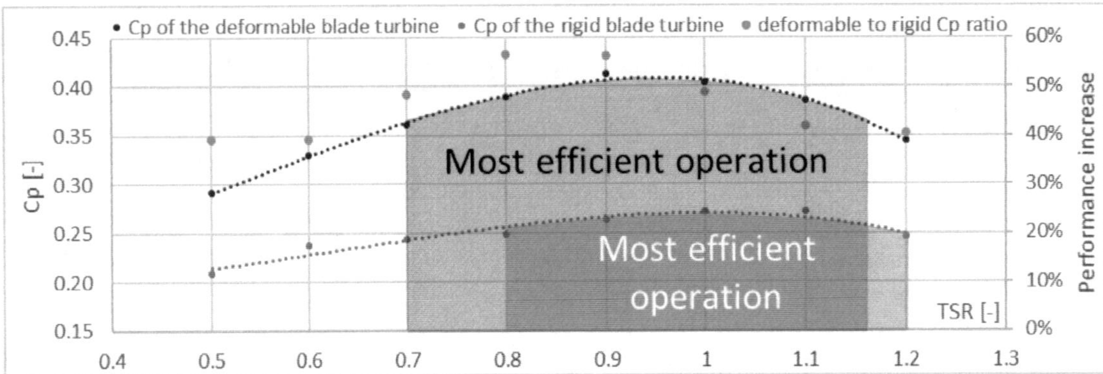

Figure 6. Efficiency characteristics – *Cp* as a function of *TSR* for both turbines and performance increase comparison.

The power coefficient changes during one revolution of the deformable and rigid blade rotors are presented in Figure 7 for *TSR* = 0.9 and additionally for the rigid blade rotor for *TSR* = 1.1, where it reached its maximal efficiency. Deformation of the mesh as well as interpolation errors, as the overset method merged the cells of background and deformed blade surroundings meshes in each time step to pass the solution data between them, caused local random changes of the blade load. Therefore, the pressure coefficient averaged for the corresponding rotor positions over the last 5 revolutions is presented. Curves for *TSR* = 0.9 show the advantage of the deformable blade variant in each revolution phase. Its peak efficiency exceeded 0.7 at revolution angles equal to about 125° and 305°, while the minimal value equal to about 0.1 corresponded to about 25° and 205°. The rigid rotor achieved the minimal, near 0, efficiency at similar angles, whereas the peaks approaching 0.6 were at revolution angles about 105° and 285°. The rotor operating at its optimal *TSR* = 1.1 outperformed the deformable blade one for phases from 60° to 120° and from 240° to 300°. However, the negative coefficient values between 350° and 35° and also from 170° to 215° deteriorated its global efficiency.

Figure 8 shows the static pressure distributions for analyzed cases at several revolution angles. The returning blade compression and expansion of the advancing blade observed for the position of 0°, 45° and 135° caused a substantial increase in rotor performance mainly due to the change of the blade frontal areas. The change of the blade shapes also influences the pressure distribution. A smaller stagnation region at the convex side of the returning blade can be noticed at the 0° position. Furthermore, an increased pressure difference between both sides of the stretched advancing blade at 45° and especially 135° implied higher positive torque. The same size of the deformable blades at the 90° position resulted in a lack of performance increase. Some changes can be also observed in the size and intensity of the advancing blade outer tip vortex and the overlap flow between both blades which was partially jammed due to their deformation.

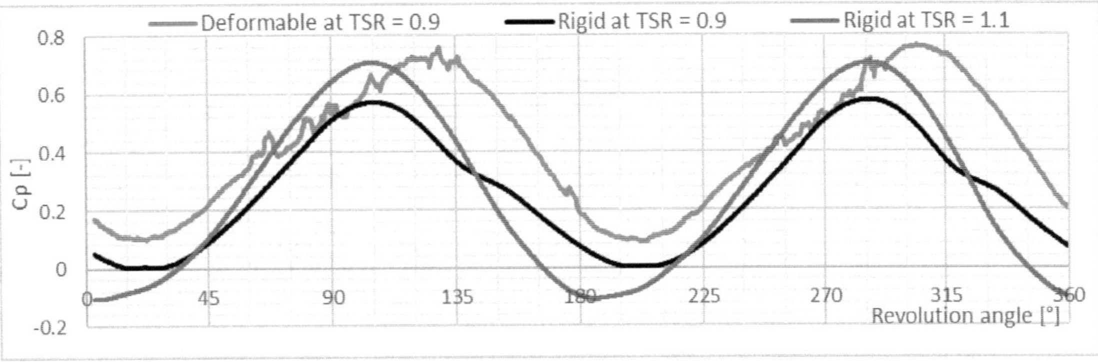

Figure 7. Power coefficient as a function of rotor revolution angle.

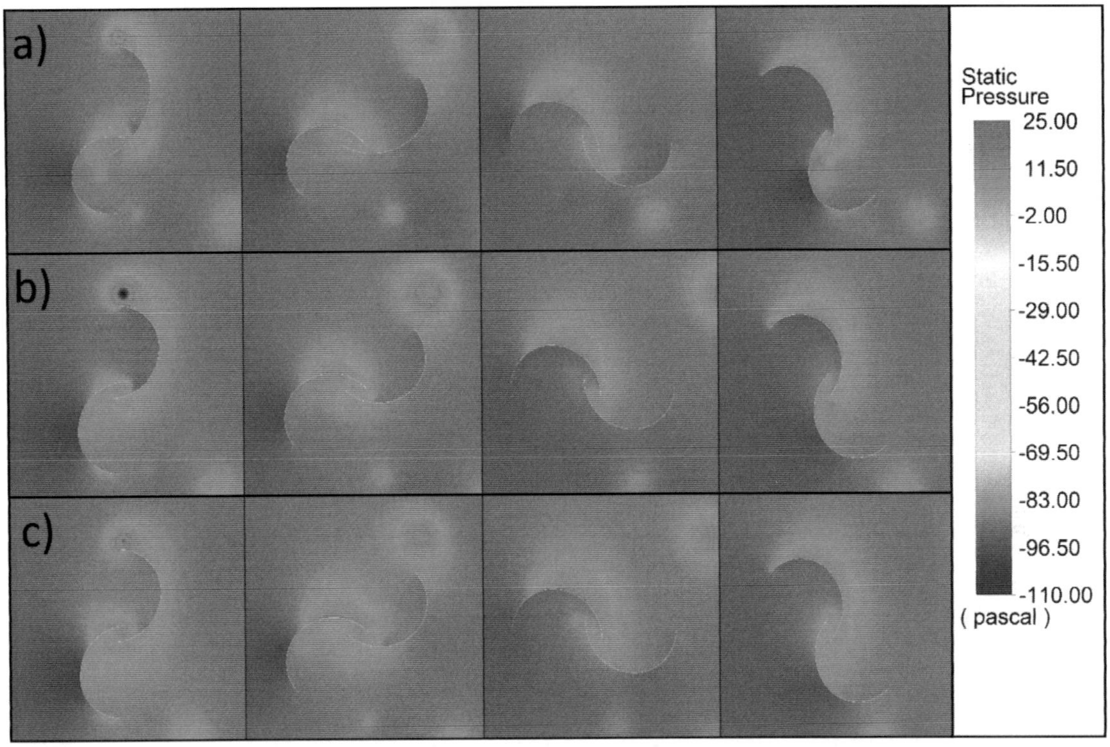

Figure 8. Static pressure distributions at rotor angles of 0°, 45°, 90° and 135°, consecutively. a) the deformable blade turbine at $TSR = 0.9$; b) the rigid blade turbine at $TSR = 0.9$; c) the rigid blade turbine at $TSR = 1.1$.

4. Conclusions

The paper presents the numerical study of the Savonius turbine with deformable blades. The quasi-2D numerical model using the overset mesh approach and a two-way FSI coupling between CFD and structural solvers was used to simulate the rotor with blade deformation. The performance of this turbine was determined in the wide range of *TSR* and compared with the performance of the Savonius wind turbine with rigid blades where the sliding mesh model was used.

The following conclusions summarize the outcomes of works presented in this paper:

- The conventional turbine with rigid blades achieved the maximal $Cp = 0.272$ at $TSR = 1.1$. The turbine with deformable blades reached its maximal $Cp = 0.412$ for $TSR = 0.9$ which corresponds to a 51% increase in the turbine aerodynamic performance.

- In the entire investigated *TSR* range, the aerodynamic performance of the deformable blade turbine was significantly higher than of the conventional rotor. The maximal efficiency gain of the enhanced turbine was 56% for *TSR* equal to 0.8 and 0.9.
- The changes of the power coefficients during turbine revolution show the deformable blade rotor generated energy in each revolution phase, contrary to the rigid one. Deformation of blades increases the advancing blade positive torque and reduces the returning blade negative torque.
- A rise in turbine efficiency can be achieved by an increase in the eccentricity and modifications of blade shape. It requires further flow and also structural studies as the blades subjected to numerous periodic deformations require a fatigue-resistant material. The deformation mechanism needs to be further developed. It lowers the overall turbine performance by using a part of the produced energy.

Acknowledgements
This paper was created as a part of a project "WDROŻENIOWA SZKOŁA DOKTORANCKA" (POWR.03.02.00-00-I042/16-00) co-financed by the EU under the European Social Fund.

References

[1] Dewan A, Gautam A and Goyal R 2021 Savonius wind turbines: A review of recent advances in design and performance enhancements Materials Today: Proceedings **47** 2976–83
[2] Obidowski D, Sobczak K, Jozwik K and Reorowicz P Vertical Axis Wind Turbine with a variable Geometry of Blades. European Patent EP3702610B1, 22 December 2021
[3] Alom N and Saha U K 2019 Influence of blade profiles on Savonius rotor performance: Numerical simulation and experimental validation Energy Conversion and Management **186** 267–77
[4] Marchewka E, Sobczak K, Reorowicz P, Obidowski D and Jozwik K 2021 Application of Overset Mesh approach in the investigation of the Savonius wind turbines with rigid and deformable blades Archives of Thermodynamics **42** 1–16
[5] Sobczak K, Obidowski D, Reorowicz P and Marchewka E 2020 Numerical investigations of the savonius turbine with deformable blades Energies (Basel) **13** 3717
[6] D'Alessandro V, Montelpare S, Ricci R and Secchiaroli A 2010 Unsteady Aerodynamics of a Savonius wind rotor: A new computational approach for the simulation of energy performance Energy **35** 3349–63
[7] Celik I B, Ghia U, Roache P J, Freitas C J, Coleman H and Raad P E 2008 Procedure for estimation and reporting of uncertainty due to discretization in CFD applications Journal of Fluids Engineering, Transactions of the ASME 130 0780011–4
[8] Menter F R, Langtry R, Völker S and Huang P G 2005 Engineering Turbulence Modelling and Experiments 6 ed W Rodi and M Mulas (Elsevier Science)
[9] Ferrari G, Federici D, Schito P, Inzoli F and Mereu R 2017 CFD study of Savonius wind turbine: 3D model validation and parametric analysis Renewable Energy **105** 722–34
[10] Kacprzak K and Sobczak K 2014 Numerical analysis of the flow around the Bach-type Savonius wind turbine Journal of Physics: Conference Series **530** 012063
[11] Sobczak K 2018 Numerical investigations of an influence of the aspect ratio on the Savonius rotor performance Journal of Physics: Conference Series **1101** 012034
[12] Marzec Ł, Buliński Z and Krysiński T 2021 Fluid structure interaction analysis of the operating Savonius wind turbine Renewable Energy **164** 272–84
[13] ANSYS Inc. 2020 ANSYS Fluent 20.2 User's Guide (Canonsburg, PA, USA: ANSYS Inc.)
[14] Akwa J V, Vielmo H A and Petry A P 2012 A review on the performance of Savonius wind turbines Renewable and Sustainable Energy Reviews **16** 3054–64
[15] Kamoji M A, Kedare S B and Prabhu S v. 2009 Experimental investigations on single stage modified Savonius rotor Applied Energy **86** 1064–73

Journal of Physics: Conference Series **2367** (2022) 012004 doi:10.1088/1742-6596/2367/1/012004

Modeling of free-surface shape in unbaffled tanks

T. Płusa[1] **, P.Duda**[1]

[1] Cracow University of Technology, Institute of Thermal and Process Engineering, al. Jana Pawła II 37, 31-864 Kraków

E-mail: tomasplusa@gmail.com

Abstract. This paper aims to present a numerical investigation of the two-phase flow in unbaffled uncovered stirred tanks equipped with a Rushton turbine. The shaft is positioned eccentrically. As simulation software, the Simcenter Star CCM+ 2021.2 is used to carry out multiphase unsteady simulations. Next, the numerical results are compared with results obtained on a test bench.

1. Introduction

In the industry, there are some restrictions of using baffled vessels, e.g., in some catalytic processes, crystallization, and pharmaceutical applications where cleanliness of the equipment is important [1]. Baffles can cause the appearance of dead zones, which can result in damage of cells in some biotechnological processes of mixing liquids. Some researchers [2] present how the standard *k-ε*, shear-stress transport, and the differential Reynolds-stress model are combined with the volume-of-fluid (VOF) method to predict free-surface flows in an unbaffled dish-bottom stirred tank reactor equipped with a centrally positioned Rushton turbine. The predicted mean flow field is in good agreement with Laser Doppler velocimetry measurements. The experimental testing of parameters of the flow of an agitated liquid in a stirred tank with an eccentrically positioned shaft and with a Rushton turbine are shown in [3]. The results of LDA measurements in uncovered baffled and unbaffled tanks with a centrally positioned Rushton turbine are presented in [4]. Three turbulence models are verified to simulate mixing dynamics in these tanks. Some researchers focus also on measuring and analyzing vortex shape and depth in centrically positioned shaft. They use digital camera to collect a set of pictures. After, these pictures are averaged and analyzed. [5][6]. This paper presents the results of simulations performed for the uncovered unbaffled stirred tank with an eccentrically positioned Rushton turbine.

2. Geometry and measurements

A cylindrical tank with an eccentrically mounted Rushton turbine was assembled on a test bench. The inner diameter of the tank is D=0.286m. The vessel was made of Duran glass. The turbine was mounted directly to the shaft, which was connected to the electric motor. The impeller consists of 6 equally spaced blades mounted to the disc. The external diameter of the rotor is $d=D/3$, and disc diameter and thickness are $d_d=d/3$ and d_t=0.004m, respectively. The blade width, height and thickness are: w=d/4, b=d/5, t=0.003 m. The off-bottom clearance (distance between disc center and bottom of the vessel) is

$h=d$. The rotational speed of the Rushton turbine was equal to 225 RPM and positioned 0.25D from the vessel center. The tank was filled with dimethyl sulfoxide to the level of $H=D$. The following fluid properties were assumed: density – ρ=1100 kg/m^3, dynamic viscosity – μ=0.0023 Pa·s. Fig. 1 presents the scheme of the stirred tank.

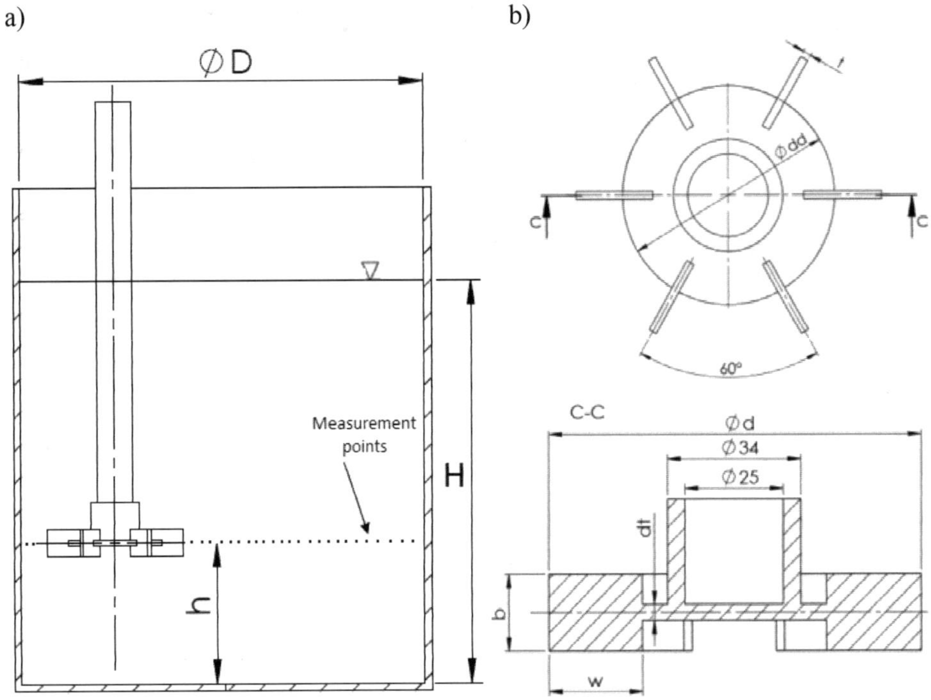

Figure 1. Test stand: a) stirred tank assembly; b) Rushton turbine.

The Laser Doppler Anemometry (LDA) was applied as a measurement method. The Dantec laboratory equipment was used to get average values of axial and radial velocities [4]. Two types of results were obtained: quantitative and qualitative. A flow field is presented in Figure 2:

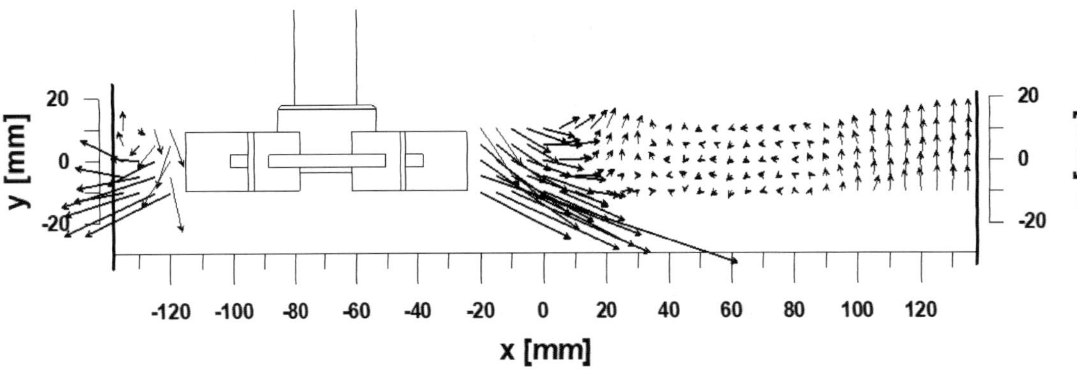

Figure 2. Flow field obtained by LDA.

Positive axial velocities occur close to the right wall. A recirculation zone can be seen between the right wall and the impeller. In the vicinity of the turbine (right side), a robust inclined stream towards the bottom of the vessel is observed. A strong vortex can also be detected on the left side of the turbine. The rotation of this vortex has an opposite direction to recirculation on the right side. A split of the flow can be seen in this area – one part of the stream goes upward and the second downward.

Additionally, a photo of the vortex, which is formed on the air-liquid interface, was taken using the digital camera Canon EOS M50 II. Pictures were captured on two planes to show the shape and position of the vortex. They are shown in Figure 3.

a) b)

Figure 3. Vortex shape: a) side view; b) front view.

3. Numerical approach

The two-phase flow in an unbaffled uncovered stirred tank is simulated by the commercial CFD software Simcenter Star CCM+ 2021.2. All calculations are performed on a high-performance computer – Prometheus (PL-Grid infrastructure) using 380 cores.

Modeling is performed using a transient approach in multiphase mode. As a multiphase model, the VOF option is applied. A sliding mesh interface is used to connect the rotational and stationary domains. The following equations are applied:

$$\frac{\partial \rho}{\partial t} + \nabla \cdot (\rho \vec{v}) = 0 \tag{1}$$

$$\frac{\partial}{\partial t}(\rho \vec{v}) + \nabla \cdot (\rho \vec{v}\vec{v}) = -\nabla p + \nabla \cdot \left[\mu \left(\nabla \vec{v} + \vec{v}^T \right) \right] + \rho \vec{g} + \vec{F} \tag{2}$$

$$\rho = \beta_L \rho_L + (1 - \beta_L) \rho_G \tag{3}$$

where: v is velocity, ρ is density, p is pressure, μ is dynamic viscosity, β_L is the liquid fraction, t is time, g is gravity acceleration, and F stands for the force.

Simulations are carried out using the Shear Stress Transport k-ω model. It is a viscous model because fluid applied on the test bench is viscous and fluid is a Newtonian type (linear dependency between stress and strain). Two types of the grid are generated: a polyhedral mesh in a rotating domain and a trimmed mesh in a stationary region. The most significant advantage of using a trimmed mesh is that cells are easy to divide. The positions of the whirl's axis for the side view in the experiment and the numerical model (Fig. 3a, Fig. 5a) are 53 mm and 58 mm from the vessel center, respectively. For the front view (Fig. 3b, Fig. 5b) the locations 21 mm and 23 mm are recorded. An adaptive mesh is applied to resolve the interface between air and liquid properly. In the simulation software, there are some features that allow to detect an interface between two fluids and densify the mesh on it. If the free surface moves to the other position in space, the adaptive meshes coarse the mesh cells to the initial size. On the air-liquid interface, cells are divided two times, meaning that each cell can be transformed after a mesh adaption process to sixteen cells. The initial mesh has 5.5 MLN cells, while after the adaptation process between 9 – 11 MLN of cells. The adaptation process is performed within seven steps. Fig. 4 shows the initial mesh and the final mesh for calculations.

a) b)

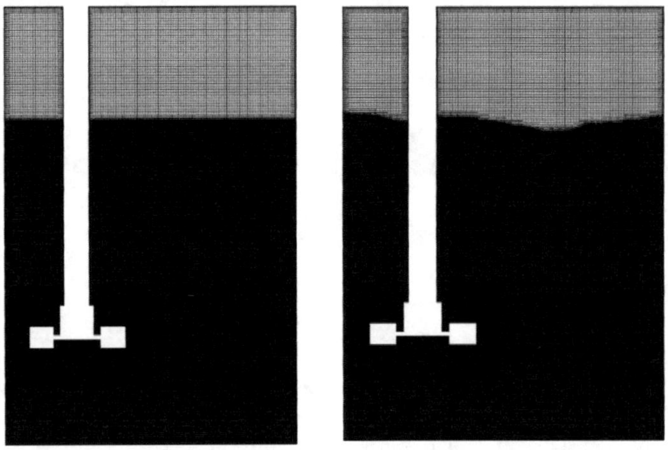

Figure 4. Mesh: a) initial; b) after adaptation process.

A time step size is found iteratively using the Courant–Friedrichs–Lewy (CFL) condition. It is calculated to limit the CFL value at the air-liquid interface by the value of 0.5. After some tries, the value of 4E-04 s is found. It ensures a good convergence of mass, momentum, and phase equations. Simulations are done using second-order discretization schemes and a segregated solver type.

4. Results

Two-phase numerical calculations make it possible to determine the vortex shape. Fig. 5 shows that the calculated vortex shape is similar to the shape observed on the test bench. Unfortunately, the aeration process at the bottom of the vortex is not visible in the numerical results due to limitations in mesh compaction. The mean polyhedral cell size was set as 2 mm and 1 mm on the rotor surface. The trimmed mesh was generated using 4 mm and 2 mm on the walls. Regarding the prism layers, fifteen cells across were generated to properly resolve a boundary layer.

Fig. 6 presents the calculated velocity distribution in a cylindrical tank. A map of vectors projected tangentially is applied to visualize the flow field. The calculated flow field is similar to the observed on the test bench, as it is shown in Fig. 2. A strong inclined stream on the right side of the impeller, a recirculation zone between a turbine and the right wall, and a vortex on the left side of the impeller are simulated.

Additionally, the calculated axial and radial velocity components are compared to the values measured by the LDA. All velocities are averaged over 10 sec frame after a quasi-state flow in the stirred tank is achieved.

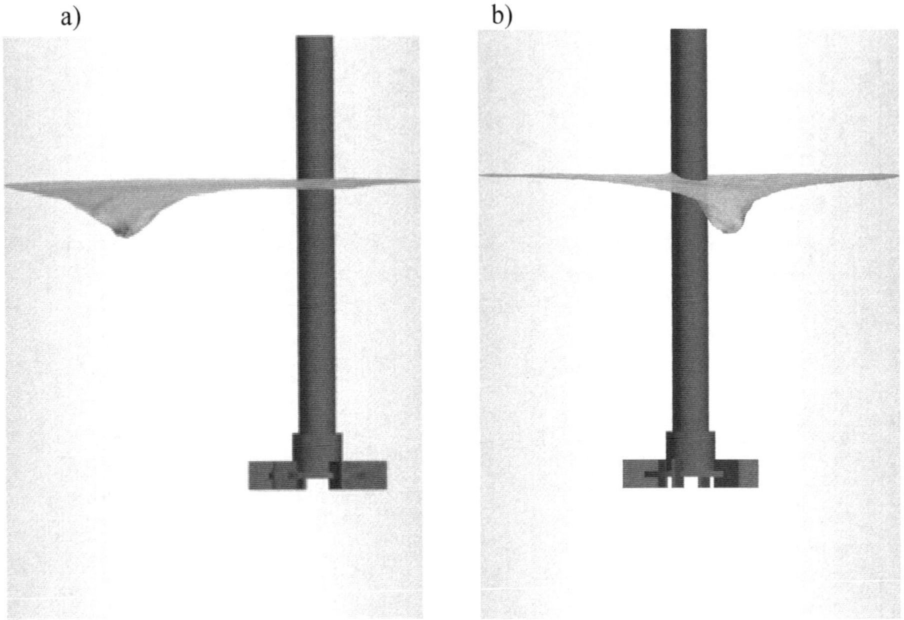

Figure 5. Numerical results of vortex: a) side view; b) front view.

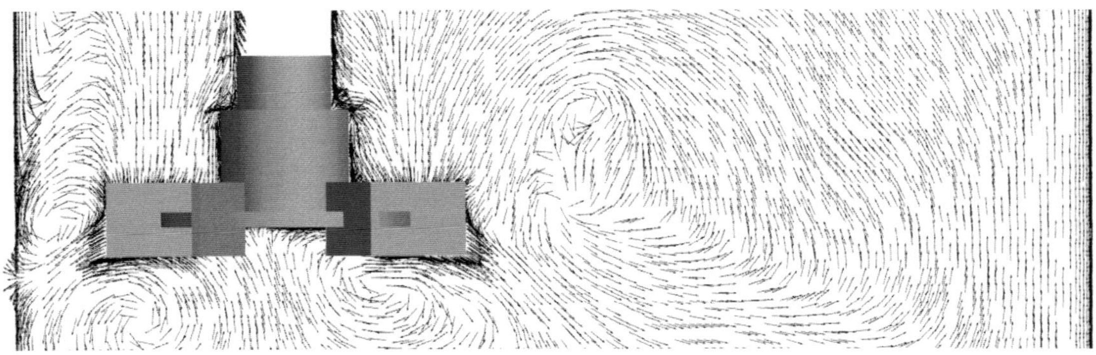

Figure 6. Tangential projection of the velocity field presented as a map of vectors.

Fig. 7 demonstrates a flow field on the right side of the tank. The axial velocity distribution presents a good agreement between measurements and simulations. Only in the area of a recirculation zone between the impeller and right wall can some discrepancies be observed. A similar situation occurs for the radial velocity. Additionally, in the vicinity of the impeller, a radial velocity is underestimated.

Figure 8 shows velocity distribution on the left side of the turbine. Velocity components are measured only in 4 points because of a small gap between the impeller and the wall.

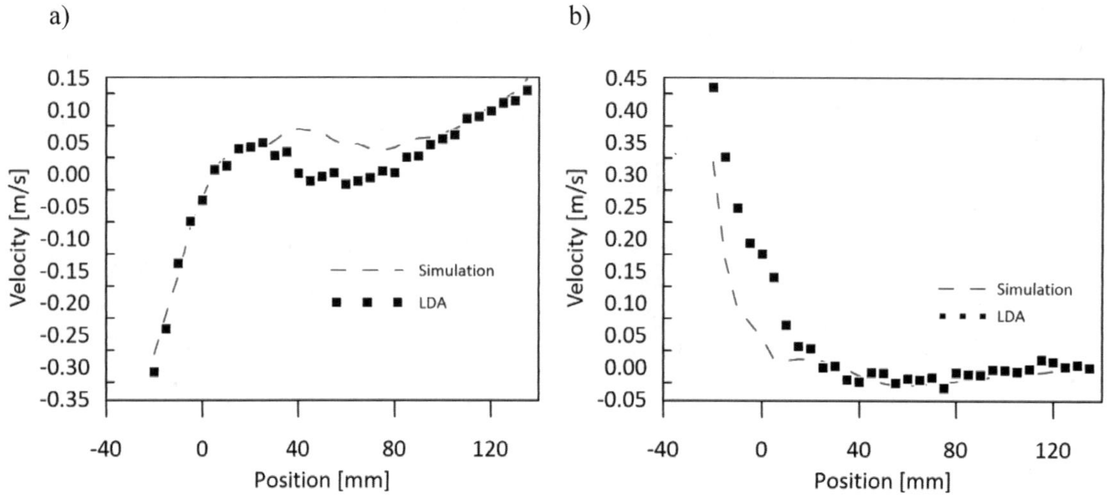

Figure 7. Velocity components distribution on the right side of the turbine: a) axial; b) radial.

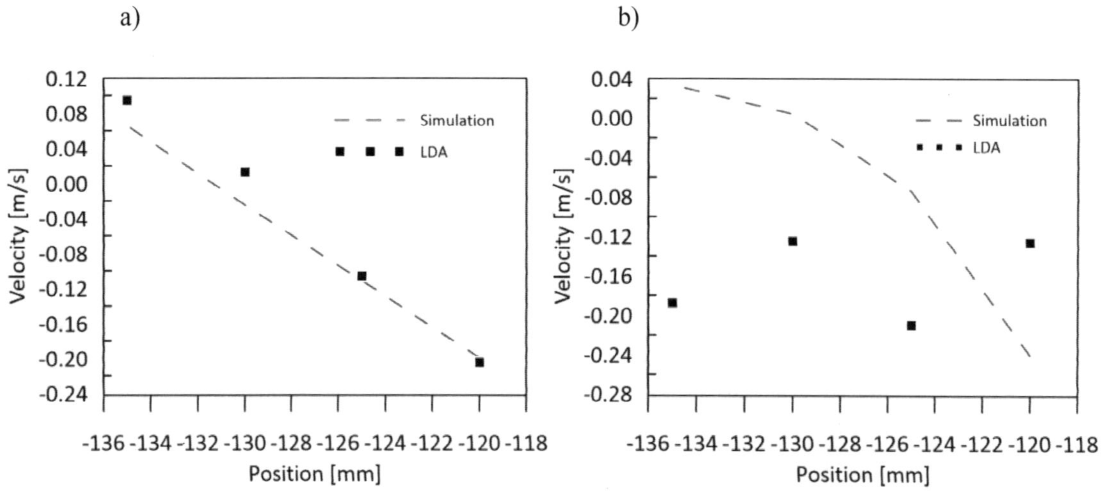

Figure 8. Velocity components distribution on the left side of the turbine : a) axial; b) radial.

Fig. 8 presents a good convergence of the calculated axial velocity to the LDA measurements. The simulated radial velocity does not agree with measured values. Not all phenomena are registered in the simulation. The aeration and the vortex at the cylinder's lowest part are not visible. Also, the mesh influences the critical zones like the gap between the impeller and left wall. A more advanced turbulence model should also be applied like the LES instead of the RANS approach.

5. Conclusions

A correct free surface shape can be obtained even using the RANS turbulence modelling and appropriate mesh size on the liquid-air interface. The accuracy of the numerical results is proved by the data obtained on a test bench. A comparison of the free surface shape and the distribution of radial and axial velocities shows that the proper settings of the numerical models can provide accurate results.

Acknowledgment
This research was partly supported by the PL-Grid Infrastructure.

References

[1] Kazemzadeh A, Elias C, Tamer M, Lohi A, Ein-Mozaffari F. 2020 *Chem Eng Sci.* **219** 115606
[2] Haque JN, Mahmud T, Roberts KJ, et al. 2011 *Can J Chem Eng.* **89** (4) 745-753
[3] Talaga J, Duda P. 2020 *Arch Thermodyn.* **41** 3–14.
[4] Płusa T, Talaga J, Duda A, Duda P. 2021 *AIChE J.* **67** 9 e17322
[5] Torre J, Fletcher D, Lasuye T, Xuereb C, 2007 *Chem Eng Sci.* **219** 1915-1926
[6] Busciglo A, Caputo G, Scargiali F, 2013 *Chem Eng Sci.* **104** 868-880

CFD modelling of a fan with a cycloidal rotor

T Staśko[1], M Majkut[1], S Dykas[1] and K Smołka[1]

[1]Department of Power Engineering and Turbomachinery, Silesian University of Technology, Gliwice, Akademicka 2a, Gliwice, Poland

Corresponding author's email: tomasz.stasko@polsl.pl

Abstract. Despite the wide possibilities of using a cycloidal rotor in the form of propulsion systems for unmanned aerial vehicles (UAV), cycloidal propellers for sea-going ships, rotors of wind turbines or sea or river cycloidal energy converters, there is practically no research on the use of this solution in the form of a cycloidal rotor fan (CRF) for HVAC (heat, ventilation & air conditioning) applications. The main features of such a machine, compared to conventional solutions, is a possibility of changing the flow direction only by changing pitch angles of the rotor blades. This study analysed two variants of the fan, first equipped with an asymmetrical CLARK Y profile, and other with a symmetrical NACA0012. Numerical simulation of cycloidal rotor fan developed in Ansys CFX was presented, that enables the simulation of fan operation. The results obtained from CFD for both variants were compared with those obtained during experimental measurements made with the use of Laser Doppler Anemometry (LDA). The comparison showed good agreement between the numerical analysis and the performed experiment. Despite the operation based on the same cycloidal regulation settings and rotation speed, the fan equipped with symmetrical blades slightly more curved the flow angle than the asymmetrical one.

1. Introduction

Even though the cycloidal rotor concept is more than a century old [1], it is still not researched widely enough despite its advantages. It is commonly used in aviation as a propulsion for MAVs (micro air vehicles) [2], and in marine industry as a cycloidal propeller for sea – going ships [3]. There are also an attempts to use a cycloidal rotor in river and offshore energy converters [4]. Some scientists also try to implement this idea in vertical axis wind turbines such as Darrieus wind turbines [5]. The first complete and well documented researches was carried out by Kirsten and Boeing in late 1920s. They developed a Kirsten - Boeing rotor that was capable of directing thrust [6]. The aim of this research was to develop and manufacture a propulsion system for airships, as well as to build a cycloplane (an airplane powered by two Kirsten-Boeing propellers). Due to construction difficulties and insufficient technology at this time, the idea was abandoned. The Kirsten – Boeing rotor is used in marine industry as cycloidal propeller. Another important step was made by Wheatley [7,8], when he built an experimental stand on which he tested a 4-blade cycloidal rotor equipped with blades with the NACA 0012 profile. Author described the distribution and the way in which forces are generated on the rotor with determination of the relationships between them Important work was also performed in relation to the European CROP [9-12] (Cycloidal Rotor Optimized for Propulsion) project of 2015, which aimed to create a "radically different propulsion system for aerial vehicles". In a series of papers, the authors examined the influence of the geometrical properties of the cycloidal rotor itself on its aerodynamic efficiency and determined the optimal angles for MAVs equipped with a cycloidal rotor while hovering. Experimental tests were also carried out and the behaviour of the machine during flight and hovering was examined. Adams [13] developed a fully flying cyclocopter (UAV) that uses 2 cycloidal rotors for flight. To make this possible, it was necessary to significantly reduce the device mass, develop a mechanism for changing the angle of the blades and design a control system.

As it can be seen, there is no systematic research on the possibility and use of such a solution in stationary applications, e.g. in HVAC (heat, ventilation and air conditioning) systems. The presented article will show the CFD analysis of a fan with a cycloidal rotor. Two variants of the CRF was compared, the first equipped with asymmetrical CLARK Y blades and the other equipped with a symmetrical NACA0012. The results obtained from the numerical analysis were compared with those obtained with the LDA technique.

2. Subject of investigation

The subject of the research was a cycloidal rotor fan (CRF) operating in a tunnel (Fig. 1). Based on the available scientific studies [10, 11], it was decided to test a relatively small machine with a diameter of *R*=7cm and a span s=25 cm. A 4-blade variant of the fan was selected due to the popularity of this application (Fig. 2)[11]. Two variants of CRF were tested, for which the symmetrical NACA 0012 and CLARK Y were selected as blade profiles, both with the same chord of *c*=5cm. The proposed symmetrical NACA 0012 profile is very often used as a comparative profile for other blade profiles, while CLARK Y is a typical aviation profile used in small aircraft. The fan was placed in the middle of a 4-meter tunnel with internal dimensions of 25x25cm made of polycarbonate. The channel used was slightly higher than the built-in fan, and the losses it generated were neglible.

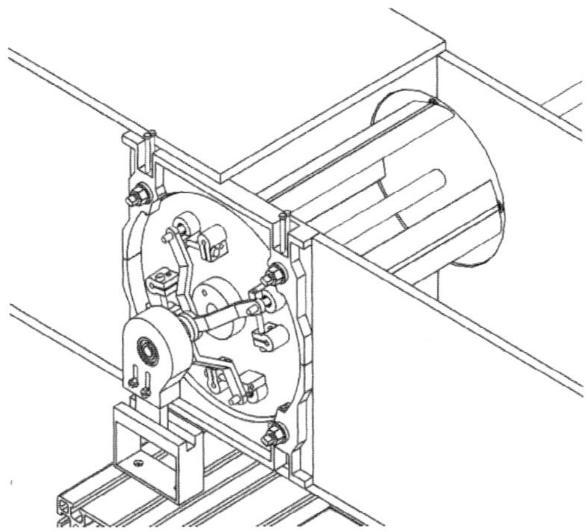

Figure 1. CRF test bench

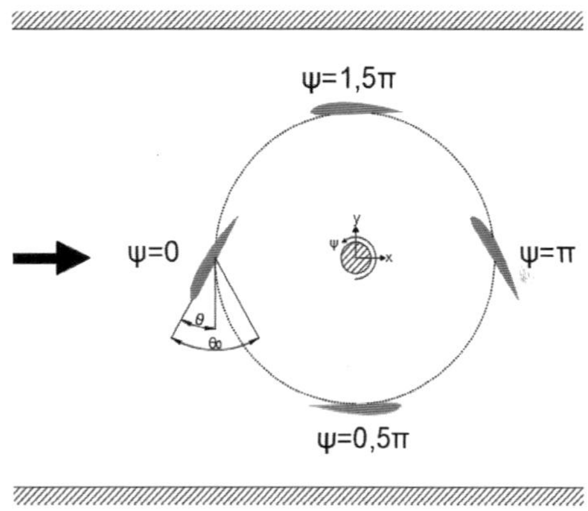

Figure 2. Scheme of the CRF

Main feature of the CRF is the possibility of changing the blade inclination angle θ depending on its position ψ during the rotation of the rotor. This function can be described as follows [10]:

$$\frac{\pi}{2} - \sin^{-1}\left[\frac{e}{a}\cos(\psi + \varepsilon)\right] - \cos^{-1}\left[\frac{a^2 + d^2 - L^2}{2ad}\right] \tag{1}$$

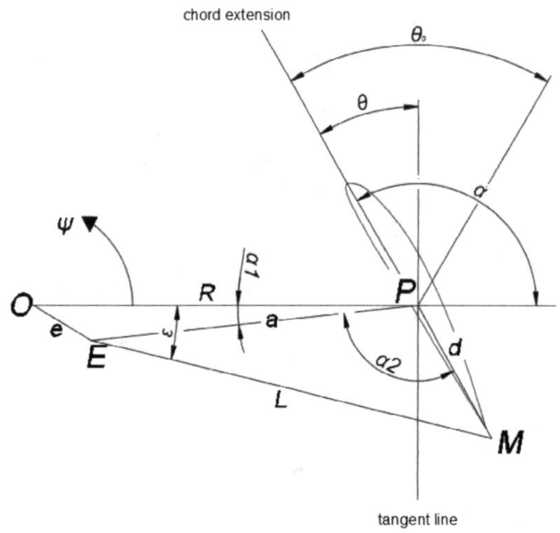

Figure 3. CRF with elements of cycloidal regulation [14]

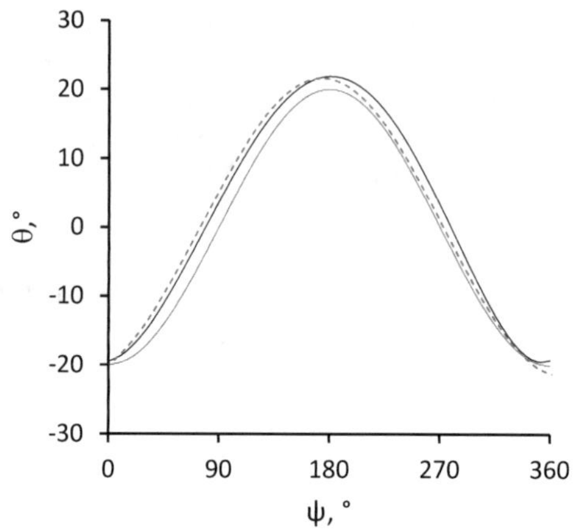

Figure 4.Comparison of the course of functions for θ_0=40°

In figure 3 a scheme of a CRF cylcloidal regulation was shown, where; *O*-geometric center of the rotor; *E*-geometric center of cycloidal regulation; *M*-cycloidal regulation connection point to the blade; *P*-blade pivot point; *L*-length of control link; *d*-distance between the blade pivot point and the cycloidal control link; *e*-moving the cycloidal control away from the rotor center; ε-cycloidal control shift angle

It can be noticed that the change of the pitch angle θ is cyclical, and each blade should perform the same function shifted by the value of π/2. Due to the characteristic of the rotor operation, it is possible to determine three cycloidal functions. The theoretical function is a cosine function shifted by an angle. If the fan is operating in accordance with this, the change of blade pitch angle θ in each quarter of revolution would be the same, allowing almost vibration-free operation. Using equation (1), the software in Labview was developed, which made it possible to determine a certain model function for which the cycloidal fan control was set, and thus its amplitude θ_0 (understood as the difference between the highest and the lowest pitch angle of the blade). The model function has been optimized in terms of rotor operation, i.e. the azimuthal angles of the blades were selected so that the difference between their absolute value θ was as low as possible. After setting up and calibrating the stand, another measurement was carried out and the last, actual cycloidal function was measured and adjusted with the Matlab software. Figure 4 shows the course of cycloidal functions for the amplitude θ_0=40°. When analyzing the course of the function, it can be noticed that the actual cycloidal function is slightly shifted in relation to the model function. The rotor, working according to this function, will reach the maximum blade pitch angle earlier than if it operated on the basis of the other function. Such a difference may result from the imperfect calibration of the cycloidal adjustment, slight inaccuracies in the production process of the mechanical adjustment, or backlash created during the operation of the fan.

3. Numerical model

On the basis of the described cyclorotor model, a geometry was developed on the basis of which a numerical model was created in Ansys CFX. It was decided to shorten the inlet, which significantly accelerated the calculations. This change did not affect the quality of the calculations. Figure 5 shows the geometry of the numerical model with its individual elements.

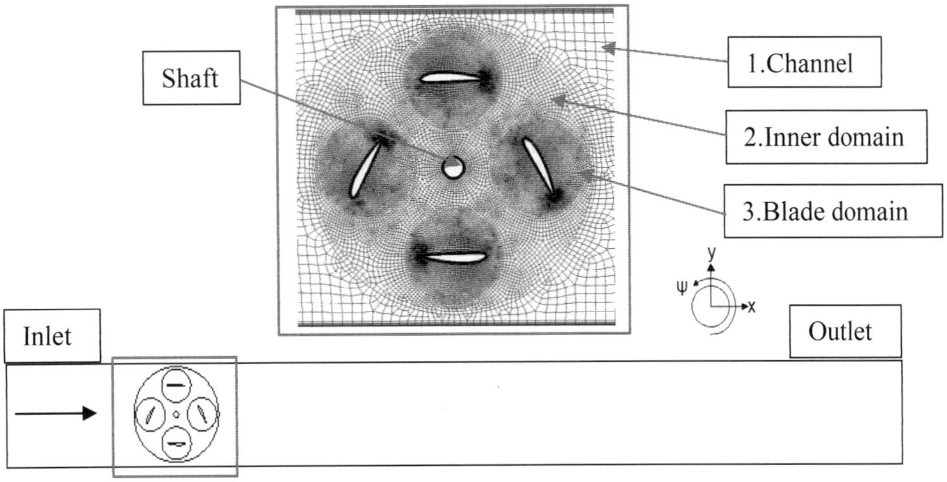

Figure 5. Schematic drawing of geometry and sample meshes

The geometry used in the numerical model was split into 3 parts. The first part was a shortened tunnel, the second was the inner domain rotating inside the tunnel, and the third were 4 blade domains. As in the case of the CRF working in the test stand, the geometry includes the shaft. The numerical model used has a certain thickness along the Z axis, while it modeled the 2D flow phenomena. A mesh study was performed to check the mesh density and the effect of the time step. The sizes of the studied meshes were:100k (Mesh 1), 200k (Mesh 2), 400k (Mesh 3) and 600k elements (Mesh 4).

For the rotor shaft, blades and the tunnel wall, the mesh was refined in order to obtain the boundary layer. The condition of symmetry was set for the side walls of the tunnel. On the inlet, the initial enviromental parameters were set. An opening condition was set at the outlet. The y^+ parameter was approximately 1, which allowed the use of the SST turbulence model. The model was chosen because of its popularity in similar studies and the advantage of combining the features of the k-ε and k-ω models. Due to the solver used and the nature of the analyzed fan, it turned out to be necessary to write a script in Ansys CEL that would permit changing the angles of the blades during the rotation of the rotor.

Based on the measurements, the course of the actual cycloidal function was determined for each of the blades. In the numerical model, for each of the four blades with azimuth position ψ=0;1/2π; π;3/2π, the initial blade inclination angle θ was set. The cycloidal function, which can be described as *θ=f(ψ)*, was then transformed into the angle increment function *Δθ=f(ψ)*, where Δθ is the angular difference between the next and the current position of the blade during rotation through the angle of ψ. The angle of increment, Δθ, can therefore be described by the relationship:

$$\Delta\theta = (\theta_{i+1} - \theta_i) \tag{2}$$

For each blade, the function *Δθ=f(ψ)* was approximated by the sum of sines in the Matlab software.

A short analysis was carried out in which the described function was introduced to the CFX program in the form of a sixth-degree polynomial, however, due to the level of the polynomial, satisfactory results were not obtained. This could be due to the necessity to introduce much more precise parameters for this type of interpolation. It was important that the start and end values of each function were the same. The discrepancies will result in the fact, that after the blade rotation by 2π and the passage through the initial azimuthal position of the blade, the angle θ for the next rotation of rotor will additionally change by a value resulting from the divergence of functions, which resulted in a complete change of the cycloidal function, and in extreme cases, after several dozen rotation caused the blade to rotate a significant angle. An example plot showing *Δθ=f(ψ)* for an amplitude θ_0=40° is shown in Figure 6.

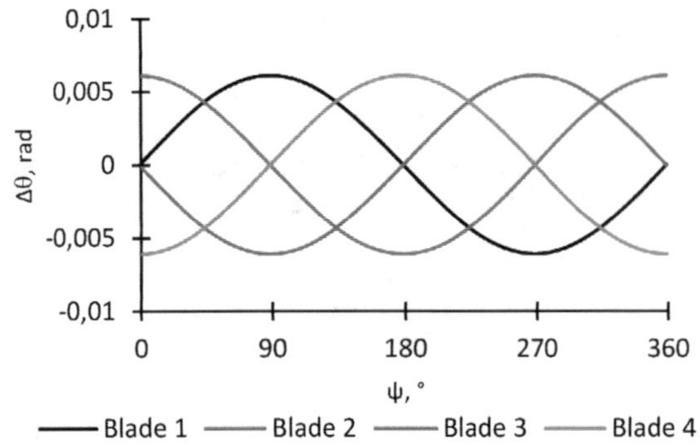

Figure 6. Graph showing the $\Delta\theta=f(\psi)$ function for amplitude $\theta_0=40°$

An analysis of the influence of the mesh density was performed using afromentioned meshes. The basic analysis was carried out for a time steps of 360 (the blade changes the angle every 1°). The fan inlet and outlet velocity profile were used as a comparative criterion. The best results were obtained for Mesh 2. Further mesh densification for such a time step did not affect the results, i.e the velocity profile were the same at the inlet and outlet of the fan. Rejecting Mesh 1 due to different profile runs, tests were carried out for a higher time step of 720 (the blade changes the angle every 0.5°). Due to the calculation time and insignificant differences in the velocity profiles, the densest mesh was rejected, and Mesh 2 was chosen as the optimal one. For the selected mesh, calculations were carried out for the time step equal to 1440 (the blade changes the angle every 0.25°) and it was found that for the selected mesh, a further increase in the time step did not affect the course of the velocity profiles. On this basis, Mesh 2 and time step 720 were selected for further calculations.Thanks to the application of the LDA measuring technique, measurements were carried out on an experimental stand. For this purpose, it was necessary to develop two measurement grids: inlet and outlet (Figure 7). The inlet mesh was located 2R (twice the rotor radius) from the rotor center, while the second grid was rectangular in shape and was located 2R from the rotor center by a length of 4R (ended at 6R).

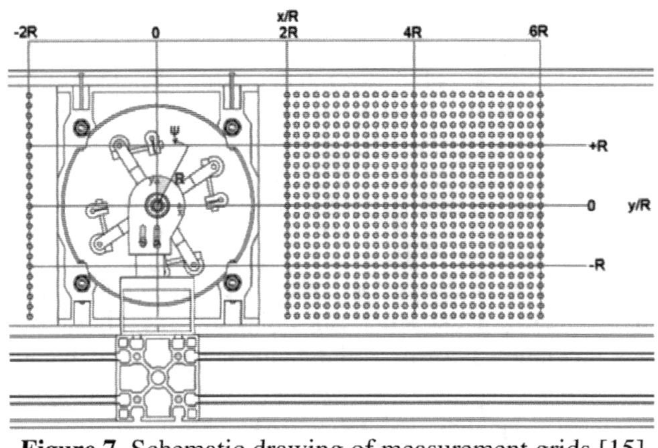

Figure 7. Schematic drawing of measurement grids [15]

Due to the structure of the rotor and the entire LDA system, the boundary layer inside the channel was not measured. Measurements were made for two variants of the fan with a cycloidal rotor, equipped with CLARK Y and NACA 0012 blades. Each fan variant worked at a speed of 750rpm for and amplitude of 40°. Table 1 and Table 2 shows the velocity profiles in relation to the channel height. To better understand the CRF flow, the absolute velocity was divided into c_x (axis) and c_y (radial) components along with the marked standard deviation determined by the LDA.

Table 1. Velocity profiles for CLARK Y fan working with θ_0=40° and 750rpm

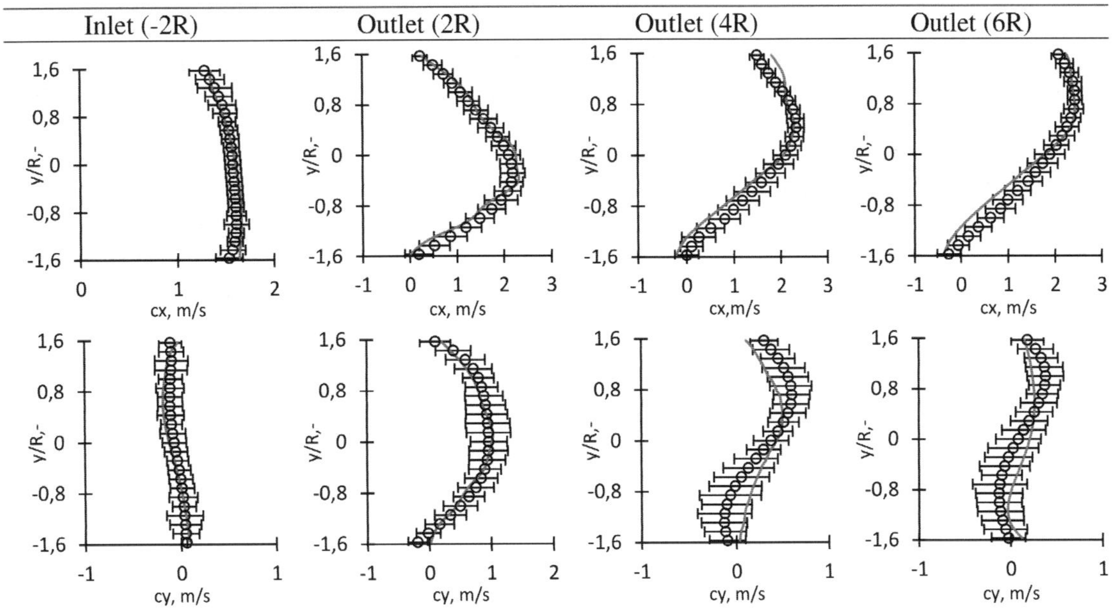

Table 1 shows the fan inlet and outlet velocity profiles for the variant working with CLARK Y profile blades, θ_0=40° and 750rpm. When analyzing the inlet plane, reduced velocity in the upper part of channel in c_x component can be noticed. This phenomena is probably caused by the impact of the rotor blades, caused by a change in the angle of the fan blades.The shape of the profile shows that the rotor absorbs the flow mostly at the bottom of the channel. Directly on the outlet, the flow head is slightly below the rotor axis. By analyzing the c_y component, it can be stated, that the fan ejects the jet from the bottom of its rotor, and the relatively high velocities suggest that there will be a further curvature of the flow towards the top of the tunnel. Moving away from the rotor, the flow is directed towards the top wall. The shape of the c_y velocity profile suggests further flow curvature, but with a lower intensity than in the transverse plane. The head of the flow in the last plane (6R) is almost in the upper axis of the rotor blade, and the course of the radial component suggests further, slight curvature of the stream. Moving away from the rotor, the flow turns towards the upper wall, which creates stagnation areas in the lower part of the channel. It can be concluded that a satisfactory agreement was obtained between the CFD model data and the experiment.

Table 2. Velocity profiles of c_x and c_y for NACA 0012 fan working with θ_0=40° and 750rpm

Inlet (-2R)	Outlet (2R)	Outlet (4R)	Outlet (6R)

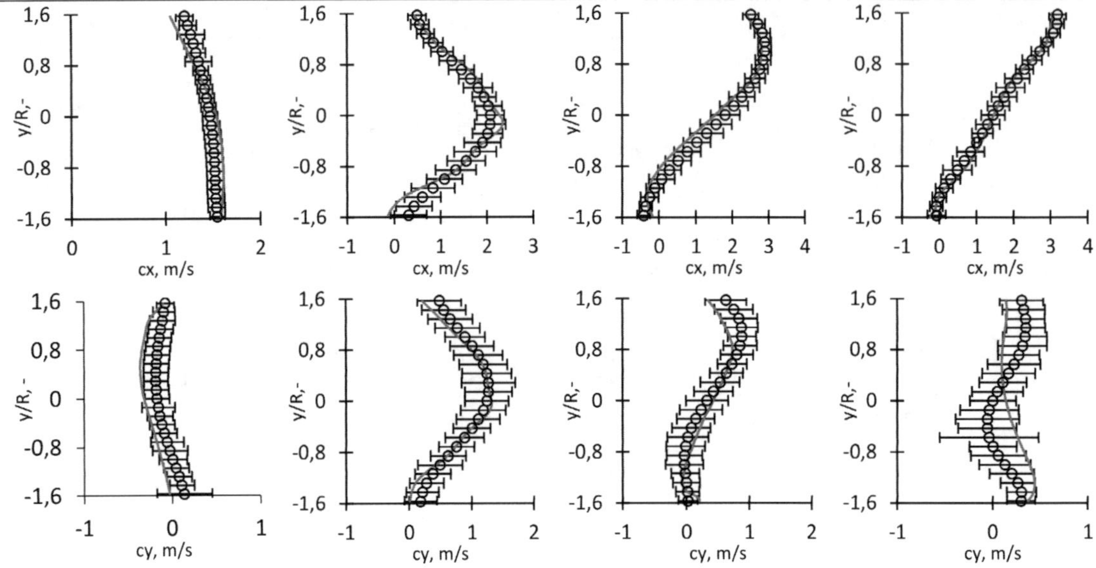

To show the difference between the profiles used, calculations and measurements were made for a fan equipped with symmetrical NACA 0012 blades, operating in the same conditions (Table 2). The usage of symmetrical blades resulted in a slightly greater reduction in speed in the upper part of the impeller inlet channel. The reason for this phenomenon, despite having the same amplitude, may be the greater angle of attack of the NACA 0012 blade compared to the CLARK Y blade. This behavior is also indicated by the c_y component, which is more bent from the rotor axis. Directly at the outlet, the velocity profile takes the shape and values close to the asymmetric fan variant, with the difference that its face is slightly closer to the rotor axis. As in the analogous case, the component c_y suggests that for further planes there will be a curvature of the flow towards the upper wall of the tunnel. Moving away from the rotor, it can be seen that the rotor equipped with symmetrical blades, operating at the same amplitude, curves the flow much more severely, which is best seen in the last measurement plane, since the c_x component velocity profile is almost a straight line with an invisible flow head. This suggests that the flow is probably traveling along the top of the tunnel.

4. Conclusions

This article presents the concept of a fan with a cycloidal rotor and briefly describes the principle of its operation. Due to the limitations of the Ansys CFX software, it was necessary to write a script in Ansys CEL that allowed the change of the blade pitch angles based on the demanded function. The short analysis of four mesh of different densities was performed, for which the analysis of the influence of the time step was carried out. For the numerical model selected in this way, numerical calculations were carried out for two variants of CRF operating with NACA 0012 or CLARK Y blade profiles. Each variant worked for a constant rotational speed and same amplitude. CFD results were compared with the data obtained from measurements using the LDA technique. On this basis, the following conclusions were obtained:

-a satisfactory validation of CFD results with experimental data was obtained. Small differences visible between the determined speed profiles may result from small differences between the real cycloidal function and that implemented in CFD, inaccuracies arising during the production, construction, and calibration of the CRF, clearances arising during machine operation and leaks resulting from the complex structure of the regulation system.

-comparing the blade profiles used, it can be stated that for the described amplitudes and rotational speed, the use of a symmetrical profile causes a slight reduction of the flow stream, as well as a greater flow curvature. This characteristic of the flow, may be caused by the greater angles of attack of the NACA 0012 profile, which is likely to result in greater flow curvature and increased losses.

Acknowledgment
The presented results were performed within the Research Statutory Funds of the Silesian University of Technology.

References

[1] Boirum C, Post S 2009 Review of Hitosric and Modern Cyclogyro Design *45th AIAA/ASME/SAE/ASEE Jopint Propulsion Conference,* Denver, USA.
[2] Benedict M, Ramasamy M, Chopra I and Leishman JG 2009 Experiments on the optimization of the MAV-scale cycloidal rotor characteristics towards improving their aerodynamic performance *Proceedings of the International Specialists Meeting on Unmanned Rotorcraft.*
[3] Ficken N L and Dickerson M C 1969 Experimental Performance and steering characteristics of cycloidal propellers. Washington: *Nava Ship Research and Development Center, Hydromechanics Laboratory Research and Development*, Report **2983**.
[4] Siegel S 2014 Wave radiation of a cycloidal wave energy converter *Applied Ocean Research* **49**(1)
[5] Xisto C, Pascoa J and Trancossi M 2016 Geometrical Parameters Influencing the Aerodynamic Efficiency of a Small-Scale Self-Pitch High Solidity VAWT. *Journal of Solar Energy Engineering* **138**(3)
[6] Sachse H 1926 Kirsten - Boeing Propeller *Technical Notes* **351**
[7] Wheatley J B 1933 Simplified Aerodynamic Analysis of the Cyclogyro Rotating - Wing System *Technical Notes* **467**
[8] Wheatley J B and Windler R 1935 Wind - Tunnel Testes of a Cyclogiro Rotor *Technical Notes* **528**
[9] Andrisani A, Angeli D and Dumas A 2016 Optimal pitching schedules for a cycloidal rotor in hovering *Aircraft Engineering and Aerospace Technology* **88**(5)
[10] Xisto C, Leger Monteiro AJ, Pascoa J, Gagnon L , Masarati P and Angeli D 2016 Parametric Analysis of a Large-Scale Cycloidal Rotor in Hovering Conditions *Journal of Aerospace Engineering* **30**(1)
[11] Xisto C, Pascoa J, Abdollahzadeh M, Leger Monteiro AJ, Masarati P and Gagnon L 2016 PECyT-Plasma enhanced cycloidal thruster *50th AIAA/ASME/SAE/ASEE Joint Propulsion Conference.*
[12] Gagnon L, Morandini M, Quaranta G, Masarati P, Xisto C and Pascoa J 2018 Aeroelastic Analysis of a Cycloidal Rotor Under Various Operating Conditions *Journal of Aircraft* **55**(4)
[13] Adams Z, Benedict M, Hrishikeshavan V and Chopra I 2013 Design, Development, and Flight Test of a Small-Scale Cyclogyro UAV Utilizing a Novel Cam-Based Passive Blade Pitching Mechanism *International Journal of Micro Air Vehicles* **5**(2)
[14] Staśko T, Tomala M, Majkut M, Nawrat K and Smołka K 2022 Influence of Geometrical Parameters on the Shape of the Cycloidal Function Curve of a Fan with a Cycloidal Rotor *Energies* **15**(7)
[15] Staśko T, Majkut M, Dykas S and Smołka K 2021 An attempt to measure cycloidal rotor fan in a rectangular duct *MATEC Web of Conferences* **345**, 00029
[16] Staśko T, Majkut M, Dykas S and Smołka K 2021 Selection of a numerical model to predict the flow in a fan with a cycloidal rotor *Archives of Thermodynamics* **42**(4).

Journal of Physics: Conference Series **2367** (2022) 012006 doi:10.1088/1742-6596/2367/1/012006

Statistical analysis of flow velocity field inside random packed beds

Maciej Marek

Faculty of Mechanical Engineering and Computer Science, Czestochowa University of Technology, Czestochowa, Poland

E-mail: maciej.marek@pcz.pl

Abstract. Immersed boundary method is used for simulation of fluid flow through random packed beds of various particles (cylinders and rings) obtained in a separate numerical simulation of the packing process. Due to complexities of the random packing structure the flow velocity is also highly complex despite the laminar regime. The statistical characteristics of the flow field are examined in detail and compared between various packings – not only probability distribution of flow velocity but also novel statistical measures that provide useful information about the flow applicable to simplified models of flow in similar structures. The measures are based on the statistics of the flow deflection angle i.e. the angle between the local flow velocity vector and the inlet velocity direction. The proposed two-dimensional map allows for assessment of the probability distribution of the flow deflection angle for different values of the velocity magnitude.

1. Introduction

Random packings of particles of various shapes (e.g. spheres, cylinders or rings) are employed in many branches of chemical engineering [1, 2]. They are typically used as a filler in chemical reactors in order to increase the contact surface between reactants flowing in the inter-particle space (pores). Recent developments in experimental and numerical methods allow for examination of the flow field at the particle level (e.g. PRDNS – Particle Resolved Direct Numerical Simulation [3, 4, 5]). With all advantages of DNS modelling, however, one must accept also the disadvantages: high computational cost (limiting the size of the computational domains), the need for microscale representation of the geometry and meshing complexity. Also, the information provided by such studies is often too detailed and needs to be reduced to statistical properties of the flow field – typically mean profiles of axial velocity or probability density function (pdf) of velocity components [6]. There are some attempts to identify a general formula for pdfs of the flow velocity in porous media of various types [7, 8]. Some of the studies focus not only on the velocity field but also on the statistical characteristics of trajectories of particles transported in media of complex structure [9].

Simplified models, omitting the details of the packed bed geometry and aiming at direct calculation of the averaged flow characteristic, are still popular (see [10, 11, 12] for example)but they usually require some input from experiment or more detailed simulations for tuning of the model constants or setting up the procedures mimicking the influence of the packed bed on the fluid flow [13, 14, 15].

Due to complexity of the random packing, the structure of the flow is highly complicated. Thus, it is of practical importance to provide a reduced description of the velocity field that can be used in simplified flow models. The aim of the present study is to use PRDNS simulations of fluid flow in random packings of Raschig rings and full cylinders for calculation of certain statistical measures of the flow velocity field delivered in detail by the PRDNS model. This work proposes a novel type of analysis in which a two-dimensional map that represents the joint probability distribution of the flow deflection angle and the flow velocity magnitude can be obtained.

2. Methods

2.1. Flow configuration

The flow configuration considered in the present work is shown in figure 1a. The random packing of equilateral particles is obtained using the method described in [4]. The height H_p and diameter D_p of a particle are equal to 0.025m. In the case of Raschig rings the inner diameter is $D_i = 0.02$m. The particles are packed in a cylindrical container (column) of diameter $D_c = 0.3$m. In order to avoid the bottom effect [16], only a section of the generated bed of height $L = 0.3$m is taken for further analysis. The height of the whole domain is equal to $H_c = 0.42$m including two empty buffer zones $L_{b,in} = L_{b,out} = 0.06$m below and above the packed bed for separation from the inlet and outlet plane, respectively. The fluid (air) is introduced into the domain through the top plane with a uniform velocity profile. The inlet velocity v_{in} is set to relatively low value of 0.05 m/s to ensure the laminarity of the flow ($Re_p \approx 80$). The properties of the fluid are $\rho = 1.15\text{kg/m}^3$ (density), $\mu = 1.85 \times 10^{-5}$Pa·s (dynamic viscosity).

The no-slip boundary condition is imposed on all solid surfaces, including the column's wall and the particles. The pressure on the outflow plane is set to 0 (Dirichlet boundary condition).

The statistics of the flow velocity field are gathered in the steady-flow conditions only in the interior of the packed bed between planes P1 and P2 located one particle diameter below the top of the bed and above the bed base, respectively.

2.2. PRDNS model

The PRDNS model has two main components – one is used for the generation of the geometry of the packed bed (see [4, 17]); the other one is a finite volume flow solver designed for a structured, staggered computational grid (see [4, 5]).

The random structure of the packing is obtained by sequential deposition of particles (rings, full cylinders) into a cylindrical container. The model does not try to capture the mechanics of the process in detail, still, it was shown that the final structures have correct orientation distribution of particles and void fraction (both global and local) [17, 16].

Although the solver is based on a structured grid, the simulation of the flow in a complex geometry can be performed due to implementation of the Immersed Boundary Method (IBM) [18, 19]. Direct forcing and a modified projection method allows for solution of the Navier-Stokes equations for an incompressible flow without additional explicit terms:

$$\rho\frac{\partial \mathbf{U}}{\partial t} + \rho\mathbf{U}\cdot\nabla\mathbf{U} = -\nabla p + \mu\nabla^2\mathbf{U}\ , \qquad \nabla\cdot\mathbf{U} = 0 \tag{1}$$

where $\mathbf{U} = (u, v, w)$ – flow velocity and p – pressure. The no-slip boundary condition is employed in the prediction and the correction step of the projection method [20, 21] in which the flow velocity near the solid boundaries is damped to 0 (see the details in [4]). Sample results concerning the simulation of the flow in a random packing of Raschig rings are shown in figure 1b.

The resolution of the computational grid for each case was set to 160 × 224 × 160 cells. The computational domain, encompassing the physical domain (the column with the packing), was

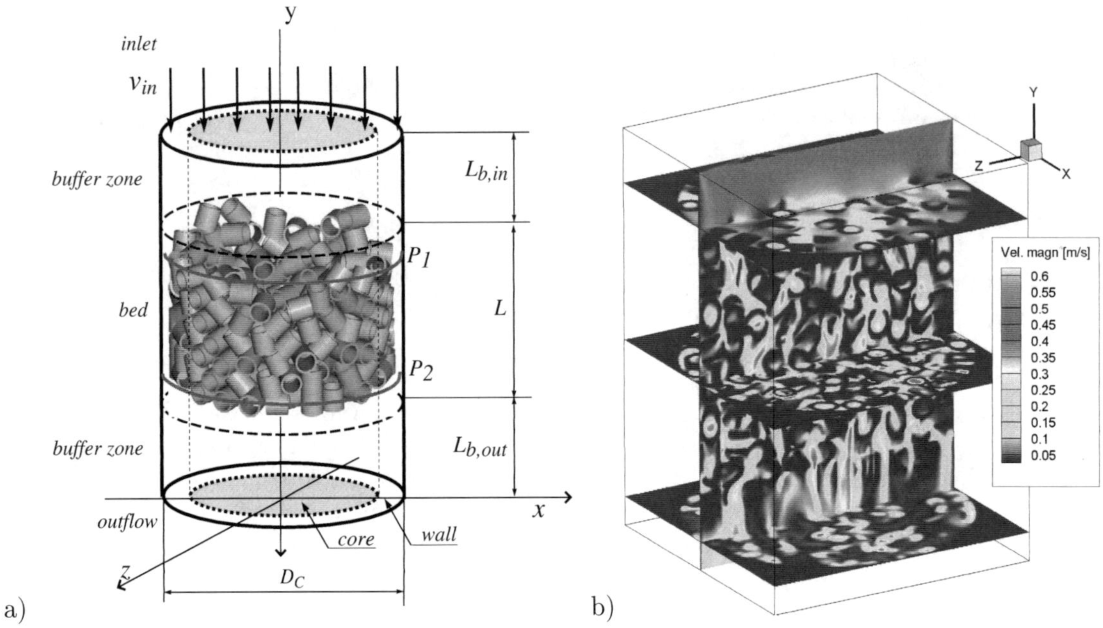

Figure 1. Flow configuration (a) and sample results of PRDNS calculations for flow velocity magnitude in three cross-sections (packed bed of rings) (b).

a rectangular box of size $0.3m \times 0.42m \times 0.3m$ in respective dimensions (x, y, z). According to the guidelines in [4], for the examined flow regime the grid resolution should be sufficient to resolve the flow within the random packing. It is confirmed further by the results shown in figure 2d.

The void fraction ϵ (global porosity) of the examined random packings, defined as the ratio of the volume of the pore spaces to the total volume of the packed bed, was equal to $\epsilon = 0.4$ for the cylindrical particles and $\epsilon = 0.78$ for the rings.

2.3. Statistical measures of the flow

Beside the typical statistical measures encountered in analyses of different kinds of flows, like probability density functions (pdf) of various velocity components, this work introduces a novel type of map which takes into account both flow velocity magnitude and the deflection angle. The deflection angle θ is defined as the angle between the flow velocity vector and the axis of the column (coinciding with y axis of the coordinate system and the inlet velocity direction) and can be calculated as:

$$\theta = \arccos \frac{v}{|\mathbf{U}|} \tag{2}$$

i.e. it is equal 0 when the flow is parallel to the column's axis and equal 180° when it is exactly opposite to the inlet direction (this is possible due to complex structure of pore space in the packed bed).

The pdf of the deflection angle has been previously examined in the literature, see for instance [15] where later it has been employed in a simplified CFD model of flow in random packed bed. Here, pdfs of this type are shown further in figure 3.

However, a measure which takes into account only the deflection angle has a disadvantage. In calculations of frequency of different values of the deflection angle, small velocities and large velocities are included with the same weight. In [15] this problem was resolved by obtaining pdfs of a quantity defined as the product of the deflection angle and the flow velocity magnitude. The present work proposes a more direct approach – a two-dimensional map in which frequency of deflection angles for a given velocity magnitude is examined. Thus, the joint pdf is defined with respect to two variables $p(\theta, |\mathbf{U}|)$.

3. Results

The simulations have been performed for two types of random packed beds: with Raschig rings and cylindrical particles. After the steady-state has been reached, the pdfs of stream-wise velocity component v, the deflection angle θ and the joint pdf $p(\theta, |\mathbf{U}|)$ were calculated in the interior of the packed bed (between planes P1 and P2). The data regarding the flow velocity were gathered only at the void space at least one grid cell away from the solid surfaces in three zones: global – meaning the whole void space; core – the cells not further away from the column's axis than 0.8 of the column's radius R_c; wall region – complementary to the core, close to the column's surface.

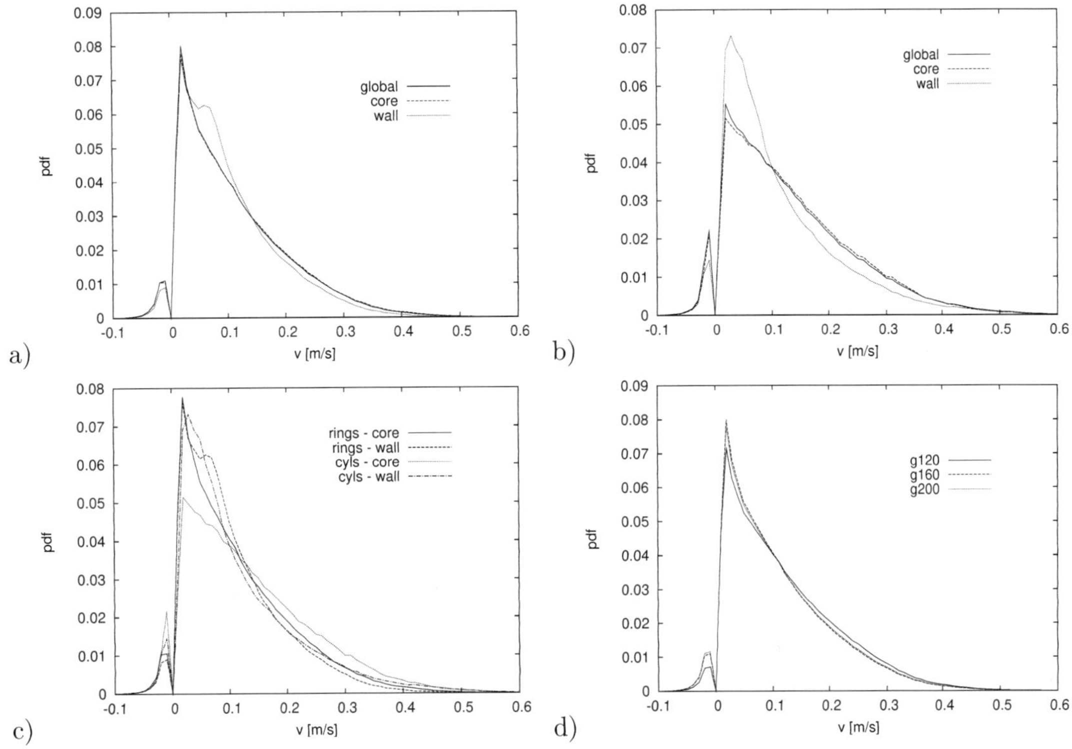

Figure 2. The pdfs of the stream-wise velocity component v for: a) rings, b) cylinders, c) both particle types (core and wall region only), d) grid dependency test (rings, global), .

The results concerning the pdf of the stream-wise velocity component v (parallel to the column's axis) are shown in figure 2. One may observe that the maximum probability occurs for values close to 0, which is a consequence of the no-slip boundary condition and the fact that a significant part of the domain is dominated by solid walls. In comparison to the positive values of v (having the same direction as the inlet velocity), the distribution tail related to negative values is considerably smaller, still, its presence signifies reversed flow (opposite to the inlet velocity) in some regions of the packed bed. For both types of particles, there is not much difference between the global pdfs and the ones from the core zone. Thus, the contribution from the wall zone to the global pdf is not significant. However, the pdfs in the wall zone differ to some extent from the ones in the core region (figure 2a) – the difference is particularly visible in the case of cylindrical particles (figure 2b). The reason for this phenomenon is much lower void fraction near the column's wall, as the wall introduces special ordering of particles and the limit on packing density. The void fraction of a packed bed of full cylinders is mostly governed by packing density (rings have a significant void space by themselves), so this phenomenon has the largest impact on such packings.

The comparison of pdfs for both types of packings is shown in figure 2c. As can be seen, the largest reversed flow occurs in packings of cylinders in the core region, the smallest – for rings in the wall region. Also, larger flow velocities are more probable for cylinders in the core of the packing than for rings. This is again related to the smaller void fraction of packed bed of full cylinders (mean flow velocity can be estimated as v_{in}/ϵ).

The influence of the selected grid density on the pdfs has been examined by performing simulations on three grids: $120 \times 168 \times 120$ (denoted as g200), $160 \times 224 \times 160$ (g160) and $200 \times 280 \times 200$ (g200). Figure 2d shows the results for the packing of rings and the global zone. As can be seen, the difference between the results for grids g160 and g200 is not significant which proves that g160 is adequate for further analysis.

As Matyka et al. suggested in [7], the flow velocity in porous media can be well described by the exponential distribution of type:

$$p(v) = A\exp(-Bv^{\gamma}) \tag{3}$$

The fitting of this distribution to the positive tail of $p(v)$ gives the following values of parameters: A=0.083, B=9.6, $\gamma = 1.1$ (for rings, global) and A=0.054, B=11, $\gamma = 1.5$ (for cylinders, global). The uncertainty of the parameters in less than 3%, which suggests that law (3) can be used to approximate the pdf of the stream-wise component in the considered random packings.

The pdfs of deflection angle θ are shown in figure 3. They are all unimodal functions with the maximum value between 20 and 30 degrees. Also here, the contribution of the wall zone to the global pdf is not significant – the global distributions almost coincide with the distributions from the core zone. However, if the wall zone is analysed separately, the distribution is more concentrated near lower angles due to increased void space in that region. The pdfs for the packing of rings display an interesting sudden change of slope near 90 degrees (figure 3a) not visible in the case of cylinders (figure 3b). For the latter case, the distributions are smooth over the whole range of angles and one may notice increased probability of large values of θ ($> 100°$) related to regions with pronounced reversed flow. The comparison of all pdfs is shown in figure 3c. The largest deflection angles correspond to the core region of the packing of rings. This is may be the result of fluid flow travelling through the inner void spaces of rings which are often oriented nearly perpendicularly to the column's axis.

The analysis shown in figure 4 takes into account not only the deflection angle but also the magnitude of the flow velocity (joint pdf $p(\theta, |\mathbf{U}|)$). As could be expected, the higher the flow velocity the more the probability distribution is concentrated near lower values of the deflection angle. In particular, the reversed flow is observed mostly with velocities comparable to the inlet velocity. The width of the distribution decreases faster with increasing velocity for the packings of rings. For cylindrical particles, velocities as high as order of magnitude larger than the inlet velocity can be observed but they are deflected mostly by angles close to 20°.

In contrast to $p(\theta)$, the joint pdf for packings of cylinders is visibly bimodal with the second local maximum occuring near $120 - 130°$. This bimodal character is less pronounced in the wall region where larger void fraction does not favour reversed flow and increased flow velocity (see figure 4f).

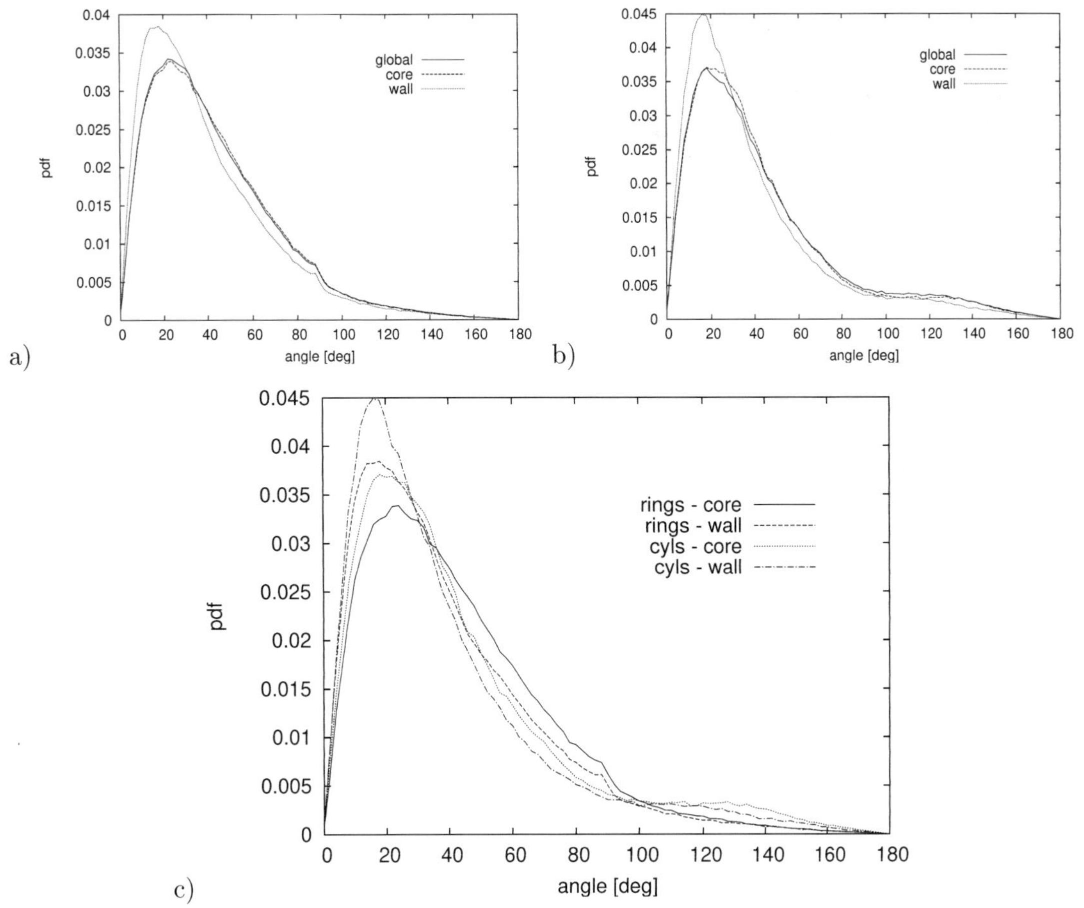

Figure 3. The pdfs of the deflection angle θ for various packings and bed regions: a) rings, b) cylinders, c) both particle types (core and wall region only).

4. Conclusions

The PRDNS simulations performed for steady-state fluid flows in random packed beds of rings and cylinders allowed for statistical analysis of flow velocity field in the laminar flow regime inside the packing structure in various zones – including the core and the wall region.

The pdf $p(\theta)$ of the flow deflection angle is a unimodal function with a local maximum close to $20 - 30°$. For the packings or rings, the distribution falls quickly for $\theta > 90°$, for full cylinders – it reaches plateau up to $130°$, which is related to certain amount of reversed flow occurring particularly in the core region.

The 2D maps visualising the joint pdf $p(\theta, |\mathbf{U}|)$ allow for direct assessment of contribution of various flow velocities to given deflection angles. For different types of packings they can be qualitatively different – in the case of cylinders the joint pdf is a bimodal function with the second, minor maximum corresponding to the reversed flow.

The pdfs obtained in PRDNS simulations can be used for further analysis of fluid flow related phenomena occurring in chemical reactors or in simplified fluid flow models in which the random packed bed is not treated explicitly at the particle level but as homogeneous medium with certain averaged characteristics.

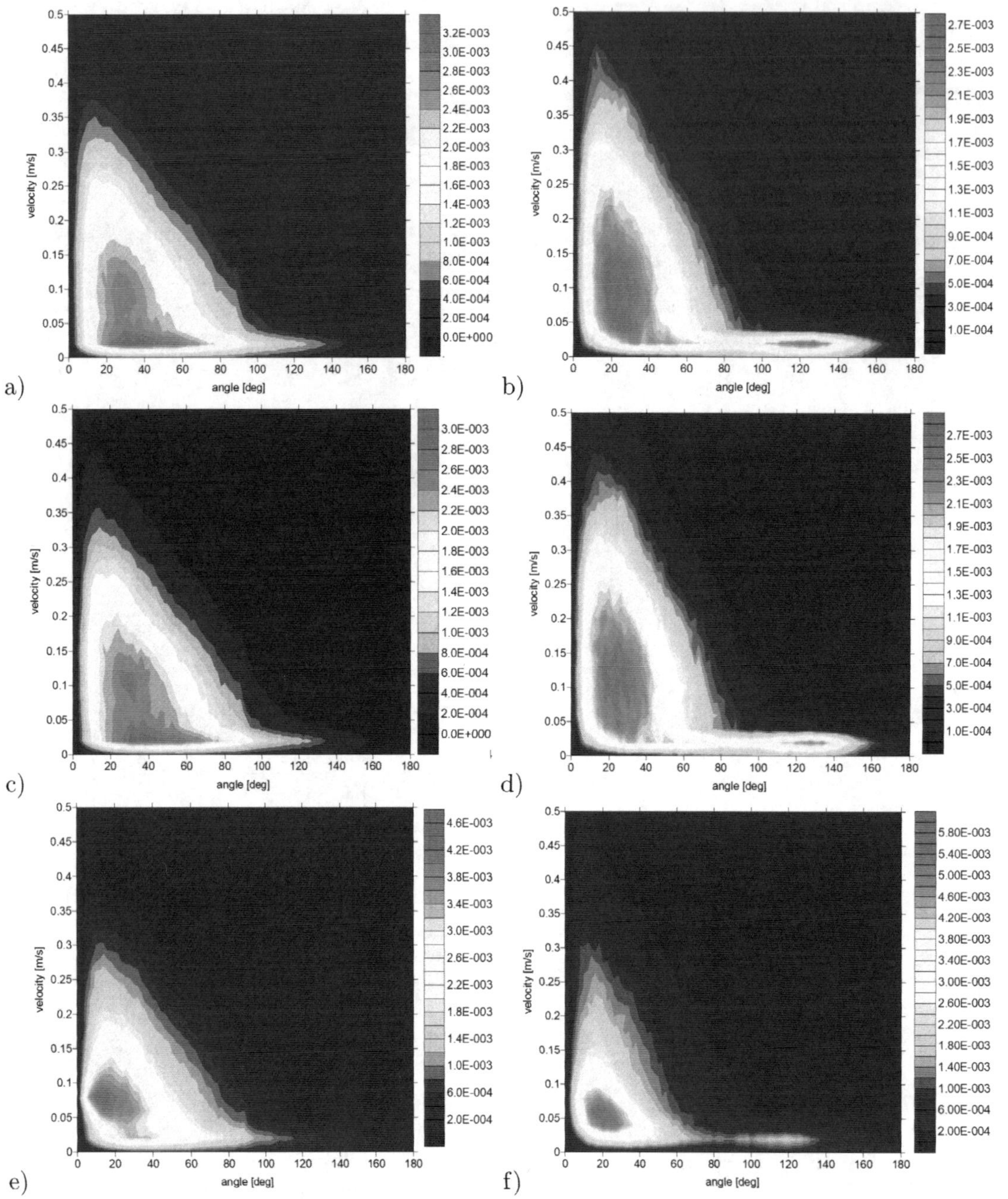

Figure 4. The joint pdfs $p(\theta, |\mathbf{U}|)$ for various packings and bed regions: a) rings, global; b) cylinders, global; c) rings, core; d) cylinders, core; e) rings, wall; f) cylinders, wall.

Acknowledgements

The investigation was supported by National Agency for Academic Exchange under Grant no. PPI/APM/2019/1/00062. The author would like to thank for cooperation to Pedro M. Faia and Maria da Graça Rasteiro from University of Coimbra (Portugal).

Journal of Physics: Conference Series **2367** (2022) 012006 doi:10.1088/1742-6596/2367/1/012006

References

[1] Billet R 1995 *Packed towers* (VCH Verlagsgesellschaft mbH, Weinheim)
[2] Maćkowiak J 2010 *Fluid dynamics of packed columns* (Springer-Verlag, Berlin)
[3] Nijemeisland M, Dixon A G and Stitt E H 2004 *Chemical Engineering Science* **59** 5185–5191
[4] Marek M 2017 *Chemical Engineering Science* **161** 382–393
[5] Marek M 2019 *Chemical Engineering Science* **197** 296–305
[6] Sheikh B and Qiu T 2018 *Comp. Fluids* **171** 15–28
[7] Matyka M, Golembiewski J and Koza Z 2016 *Phys. Rev. E* **93** 013110
[8] Nguyen V T and Papavassiliou D V 2021 *Industrial & Engineering Chemistry Research* **60** 13979–13990
[9] Monga R, Brenner O, Meyer D W and Jenny P 2022 *Advances in Water Resources* **161** 104106
[10] Niegodajew P and Asendrych D 2016 *Applied Mathematical Modelling* **40 (23-24)** 10222–10237
[11] Asendrych D, Niegodajew P and Drobniak S 2013 *Chemical and Process Engineering* **34(2)** 269–282
[12] Yang X, Zheng Z C, Winecki S and Eckels S 2013 *Applied Mathematical Modelling* **37** 9052–9062
[13] Bazmi M, Hashemabadi S and Bayat M 2012 *International Communications in Heat and Mass Transfer* **39** 736–743
[14] Iliuta I, Hamidipour M, Schweich D and Larachi F 2012 *Chemical Engineering Science* **73** 299–313
[15] Niegodajew P, Asendrych D, Marek M and Drobniak S 2014 *Journal of Physics: Conference Series* **530** 012053
[16] Niegodajew P and Marek M 2016 *Powder Technology* **297** 193–201
[17] Marek M 2013 *Chemical and Process Engineering* **34(3)** 347–359
[18] Iaccarino G and Verzicco R 2003 *Appl. Mech. Rev.* **56(3)** 331–347
[19] Mittal R and Iaccarino G 2005 *Annu. Rev. Fluid. Mech* **37** 239–261
[20] Chorin A J 1968 *Mathematics of Computation* **22(104)** 745–762
[21] Tryggvason G, Scardovelli R and Zaleski S 2011 *Direct numerical simulations of gas-liquid multiphase flows* (Cambridge University Press, Cambridge)

Application of high accuracy Penalized Vortex in Cell method for the high Reynolds number turbomachinery flows.

Dominik Błoński[1], Henryk Kudela[1] and Katarzyna Strzelecka[1]

[1]Wrocław University of Technology, The Faculty of Mechanical and Power Engineering , wyb. Stanisława Wyspiańskiego 27, 50-370 Wrocław, Poland

E-mail: dominik.blonski@pwr.edu.pl

Abstract. This paper presents the high-order Penalized Vortex in Cell method applied to flow past an airfoil with a cavity and a T-106A turbine cascade. The first part of the paper treats about the parallel environment and algorithm performance where times for each operation are measured. Then further validation on the flow past cylinder at Re=9500 was shown, and flow past an airfoil simulation results were extended. For unbounded flow with stream-function correction based on Biot-Savart law, velocity profiles at the boundary were shown. A volumetric imbalance was determined. In the second part, formulation of pVIC method for periodic flows was discussed, and an example of calculation with a turbine cascade with a moving bar wake generator was demonstrated.

1. Introduction

Complex vortex structures occurring in the flows through the turbomachinery blade systems of and mutual interactions between neighboring stages are responsible for most energy losses. To prevent them, the precise identification of the vortex structures is essential. The negative effect of vortices can be eliminated by a thorough analysis of the flow field and optimization of the flow path. This approach allows the efficiency and performance of rotating machinery to be improved. The leading cause of vortical flows are connected with boundary layer separation, the existence of endwalls, blade tip clearance, low rotor solidity, trailing edge wakes, and interaction with the vortices that already exist in the flow. Eddies can be classified as primary and secondary flows. It can be assumed that the primary flows are similar to those occurring on the mean diameter. Secondary flows occur near the vicinity of endwalls.

This work presents continuation of work on the application of the vortex particle method and its adaptation to simulate turbomachinery flows [1]. The vortex particle method, with the aid of parallel computations and the use of high-order compact schemes, is characterized by high robustness and improved accuracy. This allows for precise tracking of the vorticity field when flowing past bodies with high Reynolds numbers. In previous paper, the in-house code, utilizing Penalized Vortex in Cell method, was used to determine the effect of the vortex trapping cavity on the pressure side of the airfoil on the behavior of the nearby flow field. The cavity is designed to reduce the recirculation zone's negative impact and limit the intensity of the vortex structures in this zone. This paper is focused on the extension of results presented in [1], where some fine details regarding the numerical implementation and performance evaluation

Journal of Physics: Conference Series **2367** (2022) 012007 doi:10.1088/1742-6596/2367/1/012007

are performed. Additionally, the investigation included the periodic flow case of the T106A low-pressure, high-solidity turbine cascade, where body motion simulation is tested.

2. Penalized Vortex in Cell method

In the computational domain, the presence of irregular solids is usually troublesome. The algorithm used in the paper is based on a very effective immersed boundary method. The computational domain is treated as a porous medium with areas of different porosity. The fluid is assumed to be perfectly permeable (zero porosity), and the solid is completely impermeable. Due to this approach, the mesh on does not have to be adjusted to the solid body. The main advantage of the penalized Vortex in Cell method is that for the realization of the no-slip boundary condition, no change in systems of linear equations has to be made. In the algorithm, this is done using the velocity correction, which rely on simple algebraic operations. This approach allows simulating the fluid motion by solving the Navier-Stokes-Brinkman equation and requires adding one more step to the entire algorithm. In classical Vortex in Cell method, which is known to be a hybrid method, viscous spitting is used, and it allows for solving incompressible vorticity evolution equation in two steps. Advection is solved in the Lagrangian particle framework, and a continuous vorticity field is approximated by the system of the vortex particles. The diffusion is solved on the Eulerian grid. The presence of a rigid body is taken into account by the penalization method [2, 3].

Viscous splitting is replaced by three-step splitting. The idea of penalization was presented by Jean-Paul Caltagiron [4] and Philipp Angot [5]. In order to enforce no-slip wall condtion, generation of appropriate value of vorticity on the solid body wall and adjusting the mesh to the shape of the body is replaced velocity field correction method. Angot showed that utilizing the velocity penalization method, the obtained solution convergences to the solution of the Navier-Stokes equation with appropriate boundary conditions formulation. The convergence analysis for three dimensions was presented by Cottet, [6]. Navier-Stokes-Brinkman equation in primivite variables is formulated as

$$\frac{\partial \mathbf{u}}{\partial t} + (\mathbf{u} \cdot \nabla)\mathbf{u} = -\frac{1}{\rho}\nabla p + \nu \nabla^2 \mathbf{u} + \lambda \chi \mathbf{u}, \tag{1}$$

$$\nabla \cdot \mathbf{u} = 0. \tag{2}$$

This equation can be transformed to the Helmholtz equations that govern the evolution of the vorticity:

$$\frac{\partial \omega}{\partial t} + (\nabla \omega)\cdot \mathbf{u} = \nu \nabla^2 \omega + \nabla \times (\chi \lambda \mathbf{u}), \tag{3}$$

$$\nabla^2 \psi = -\omega, \qquad \mathbf{u} = (u, v) = (\frac{\partial \psi}{\partial y}, -\frac{\partial \psi}{\partial x}), \tag{4}$$

where χ – characteristic function, λ – penalization coefficient ($\lambda = 10^8$), ψ – stream function, ω – vorticity, $\mathbf{u}$ – velocity vector. The definition of the fluid and solid region is done by assigning a proper value of characteristic function χ, which is equal to 1 in solids and 0 in fluids. Solid body porous flow properties are defined by a non-dimensional penalization parameter $\lambda = \mu\psi H/\rho k u_\infty$, where k is the intrinsic permeability, μ the viscosity, ψ porosity, H is body height, ρ density and u_∞ is free-stream velocity.

Preparation of the grid requires a solid body geometry created in CAD software and exported as a stereolithography file (STL). The algorithm that define solid and fluid domain is designed to find the distance between each grid point and the triangle of the surface shell. The details of algorithm for characteristic funtion designation is presented in [1].

When characteristic function is known continuous penalization of a velocity equation can be solved. The Euler scheme was used to discretize the equation, (5). The implicit scheme is unconditionally stable, and the corrected velocity field is obtained by a simple expression. Due to penalization the change of the velocity field with time is expressed by the equation by equation (6).

$$\frac{\partial \mathbf{u}}{\partial t} = \lambda \chi \mathbf{u} \tag{5}$$

$$\mathbf{u}^{n+1} = \mathbf{u}^n - \frac{\lambda \chi \Delta t \mathbf{u}}{1 + \lambda \chi \Delta t} \tag{6}$$

The velocity field correction is made to account for the presence of a solid. When this value is known the vorticity correction is done. The change in the vorticity field is expressed by the equation (7). The vorticity correction is done using equation (8). The entire algorithm of the "Vortex in Cell" method consists of 10 points, tab. 1. Points 2-10 are repeated at each time step.

$$\frac{\partial \omega}{\partial t} = \nabla \times (\lambda \chi \mathbf{u}) \tag{7}$$

$$\omega^{n+1} = \omega^n - \nabla \times \frac{\lambda \chi \Delta t \mathbf{u}}{1 + \lambda \chi \Delta t} \tag{8}$$

Table 1: The Penalized Vortex in Cell method algorithm.

No.	Equation	Description	Method
1	-	Initialization	CAD file, *Point2trimesh* function
2	$\nabla \psi = -\omega$	Poisson equation	High order compact
3	$\mathbf{u} = (\partial_y \psi, -\partial_x \psi)$	Velocity evaluation	High order compact
4	$\partial_t \mathbf{u} = \lambda \chi \mathbf{u}$	Velocity Penalization	Euler implicit
5	$\partial_t \omega = \nabla \times (\lambda \chi \mathbf{u})$	Vorticity correction	High order compact
6	$\mathbf{F} = \frac{d}{dt} \int_{\Omega_s} \mathbf{u} - \mathbf{u}_{\Omega_s} \, d\mathbf{x}$	Force evaluation	Trapezoid rule, Euler explicit
7	$\frac{d\mathbf{x_p}}{dt} = \mathbf{u}(\mathbf{x_p})$	Advection	Runge-Kutta 4
8	$\omega_i = \sum_{p=1}^{N} \alpha_p(\mathbf{x_p}) \varphi\left(\frac{\mathbf{x} - \mathbf{x_p}}{h}\right) \frac{1}{h^2}$	Remeshing	$Z_2(\mathbf{x})$ kardial kernel
9	$\partial_t \omega = \nu \Delta \omega$	Diffusion	Crank-Nicholson, ADI, HOC
10	$\psi = \psi_\infty + \psi_\omega$	Stream func. correction	Biot-Savart law, Cubic spline

3. Method validation and computational performance

In recent paper [1] the order convergence of the method and grid independence test was presented. It was proven that in order to achieve convergence, approximately 10 grid cells are needed across boundary layer thickness. In this paper, the calculations for the flow past impulsively started cylinder for $Re = 9500$ were repeated for the finest mesh. The duration of the calculations was extended to dimensionless time equal to $\tau = 6$. The results of the current work are compared against the results of Gillis [7]. Both methods are second-order and use numerical schemes of increased accuracy. Gillis uses a more advanced method of penalization that uses an accurate determination of the derivative at the boundary of the solid. The results presented by Gillis can be considered a reference to the long-term solver behavior test. In figure 2, the change in the drag coefficient corresponds very well with the reference results. The accuracy is maintained up to $\tau = 6$, and it can be seen that further mesh refinement might be needed to fully compensate for the differences between the two solutions. Figure 1 presents the vorticity field for the developed flow at $\tau = 5$. The comparison of the results obtained in the present work (gray background) and Gillis' work (white background, [7]) shows that the similarity between the two solutions is substantial. The position and shape of the smallest vortex structures are the same.

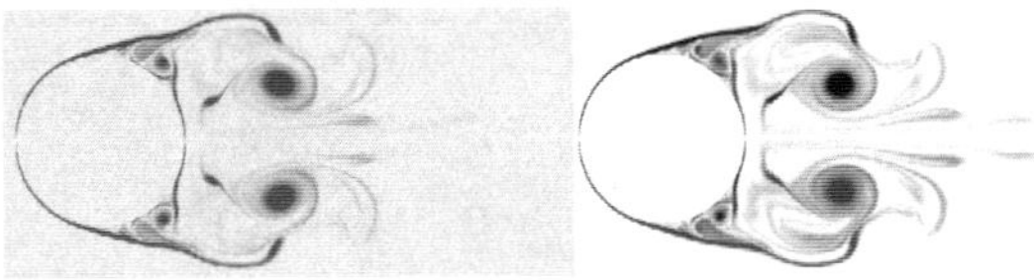

Figure 1: Vorticity distribution for the flow over the cylinder at $Re = 9500$, $\tau = 5$.

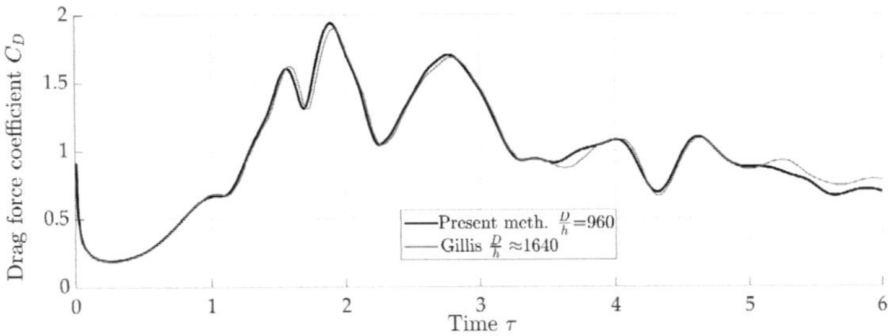

Figure 2: Drag coefficient change in time for flow past cylinder at $Re = 9500$.

An important feature of a good computational algorithm is the appropriate balancing of the computational time for each step, figure 3. A performance test was performed to determine the robustness of the in-house code. The computational domain was divided into 288 subdomains. The decomposition was done in a Cartesian grid topology with $N_x \times N_y = 96 \times 48$ divisions. Each computational process was solving a local problem on a 200×200 mesh. The global problem consisted of $(96\,200) \times (48\,200) = 184\,320\,000$ mesh elements. The test showed that most of the algorithm's time is occupied by the Poisson equation. Reducing the accuracy of this step to the second-order with the 5-point central scheme leads to an approximately fourfold reduction in the time of solving a system of linear equations. Evaluation of velocity and solving the diffusion equation are performed very effectively, although the tridiagonal system of linear equations are solved twice for each equation. The penalization, particle advection, and remeshing steps are performed with similar efficiency.

4. Flow past airfoil with vortex trapping cavity

For calculation, the authors tested 3 geometries for various angles of attack. Each geometry base on a Risø B1-18 airfoil and optimized cavities. It was shown that the instantaneous vorticity distribution influences significantly both the lift and drag force. In all investigated cases in [1], cavity of sharp triangular shape on the upper surface of an airfoil lead to an increase of lift coefficient due to additional circulation in cavity. The presence of the cavity deteriorates the C_L/C_D ratio for $\alpha = 3°$ and $\alpha = 6°$. For $\alpha = 9°$ sharp edged cavity outperforms the smooth profile – C_L/C_D ratio is higher for an airfoil with triangular cavity than for the standard Risø airfoil. The drag coefficient is higher for a profile with a cavity in respect to the smooth profile in all cases and it increases with the increase of the angle of attack. Additionally, it should be noticed that the optimal angle of attack moved toward higher values.

Apart from the analysis of the velocity field and the vorticity distribution, it is worth

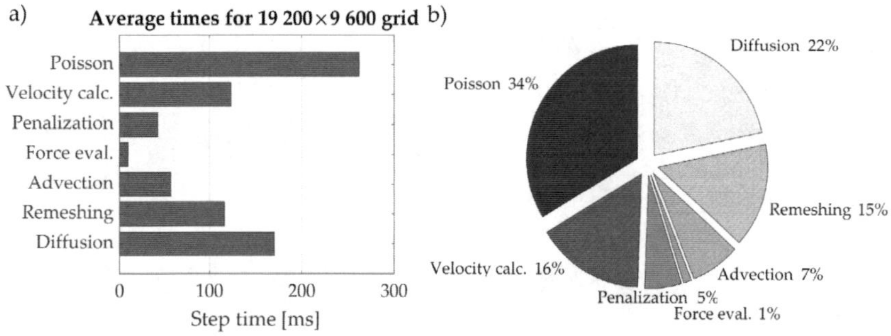

Figure 3: Profiling the solver performance based on the average times: (a) absolute times, (b) percentage

supplementing the results with some details of the boundary condition correction for the Poisson equation. Correction is based on the Biot-Savart law, and it allows to modify the values of stream-function according to the vorticity distribution in the domain. The approach is intented to improve results for the unbounded domain of finite dimensions. This is useful when the solver is running on a structural grid where the grid element size is constant and there are difficulties with fulfilling far-field boundary condition. Velocity profiles located on the boundaries of the computational domain were prepared for $\tau = 20$. The vorticity field for this time is consistent with the figure 4.

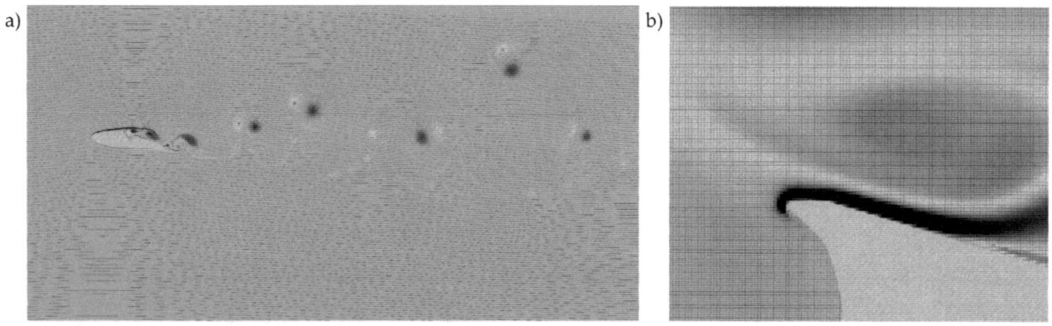

Figure 4: (a) Computational domain, (b) details of the numerical grid.

Figure 5 presents the distribution of the normal velocity component. The presence of solid has an evident impact on the inlet and velocity profiles indicate that the vorticity in the domain strongly affects all boundaries of the computational domain. There is a section of reduced velocity where its value is corrected by approximately 5% in relation to the undisturbed flow. The top and bottom boundaries indicate that the streamlines are deflected by the presence of the airfoil. The inflow of fluid occurs at the upper boundary, while at the lower one, the fluid leaves the domain. The outlet is characterized by an area influenced by the wake created behind the profile. The correction of the boundary condition for the Poisson equation based on the vorticity distribution inside the domain is a significant modification in the mathematical formulation of the problem. For the sake of further validation, the volume balance was determined and after calculating the integrals of stream-function on all boundaries, the total unbalance is at the level of 0.048%. The error has got several sources. The first is the interpolation used at the boundary. Every 25th node was used for interpolation, which translated into the use of 309 interpolation nodes to obtain the correct inlet profile. Since at the boundary the function is smooth, it can be suspected that increased error is caused by the vorticity that has already left the domain and is no longer taken into consideration. Only mesh elements in the computational domain are used to calculate the corrected value of the stream function.

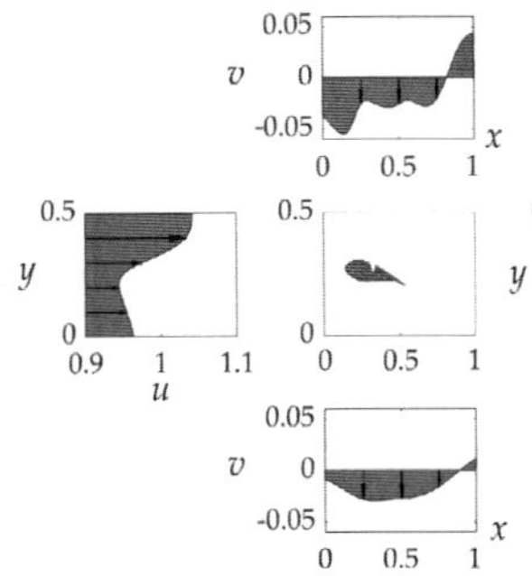

Figure 5: Velocity profiles at domain boundaries when stream function correction is applied.

5. Simulation of flow past turbine cascade

Simulation of flows through cascades and rotor stages is a challenge in the design and analysis of turbomachinery. In the vortex particle method, the periodic boundary condition for the Poisson equation might be problematic. Appropriate modification of the system of linear equations allows for defining the volumetric flow and fluid velocity angle at the boundary. In most cases, a uniform, perpendicular to the boundary flow is used, in which the Dirichlet boundary condition in the form of a first order polynomial is utilized. Adding periodicity results in stream function being discontinuous. Another problem with turbomachinery flows is that defining inflow at a certain angle requires defining the tangent velocity with a Neumann boundary condition and this prevents from using Dirichlet boundary.

The quantity $\delta\psi$ will define the amount of media flowing through the boundary. Applying the classical boundary condition to periodicity and taking into account the flow at angle, it can be written that

$$\begin{aligned} \psi_{i,n_y+1} &= \psi_{i,0} + \delta\psi, \\ \psi_{i,-1} &= \psi_{i,n_y} - \delta\psi, \\ \partial_x \psi_{0,j} &= -v_{inlet}, \\ \partial_x \psi_{n_x,j} &= v_{outlet}. \end{aligned} \tag{9}$$

A formulation of the Poisson equation with two periodic boundaries, inflow and outflow at an angle (defines v velocity component) and specified volumetric flow (defines u velocity component) can be put in the matrix form $A\mathbf{x} = \mathbf{b} + \mathbf{p}$, where $\mathbf{p}$ is a vector responsible for handling periodic boundary condition discontinuity and enforcing flow rate with the use of $\delta\psi$ term. The matrix A will be singular unless one value of the numerical grid is explicitly defined. This is done by modyfing a single row of matrix A. The value of the stream function (typically $\psi_i = 0$) is pinned to the vector of the right-hand side of the Poisson equation.

For the simulation of wake and turbine cascade interaction, the computational domain was adjusted to reproduce the experiment conducted by Stieger [8, 9]. Characteristic dimensions and flow velocities have been normalized, figure 6. The Reynolds number was defined by modifying the viscosity and is equal to $Re = 1.6 \cdot 10^5$. The exit velocity was used to determine the Re number. The inlet and outlet angle has been set constant and are equal to $\alpha_1 = 37.7°$ and $\alpha_2 = 63.2°$. The top and bottom boundaries are defined using a periodic boundary condition.

The test was carried out for flow coefficient $\phi = 1.2$, that is defined as a ratio of an axial to bar velocity. Stieger performed his experiment for $\phi = 0.83$.

Each computational process was assigned to perform the calculation on a small part of the mesh of dimensions $n_x \times n_y = 256 \times 256$. Numerical grid consist of $N_x \times N_y = 160 \times 32 = 5120$ subdomains which resulted in a mesh size of 335 544 320 elements. Each physical core of the processor was got 4 subdomains (1 physical core = 4 computational processes).

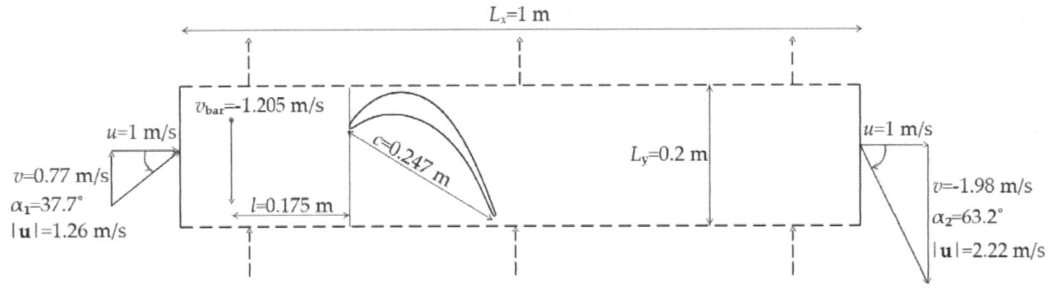

Figure 6: T106A turbine cascade simulation setup.

With the penalization method, where the equations of fluid motion have been transformed to fluid motion in a porous bed with variable permeability, simulation of solid body motion within the computational domain is easy to implement. If the time step is sufficiently small, the motion of the solid can be simulated by modifying the permeability of the bed as a function of time. It is required to change the value of the characteristic function χ and maintain the continuity of the transition between the media. A moving bar wake generator can be represented as supergaussian

$$\chi(\mathbf{x}, t) = A \exp\left(- \left(\frac{(x - ut)^2}{2\sigma_x^2} \right)^{P_x} - \left(\frac{(y - vt)^2}{2\sigma_y^2} \right)^{P_y} \right), \tag{10}$$

where: A - amplitude of the function ($A = 1$), x, y - position in space, u - horizontal component of velocity, $v = u_{bar}$ - vertical component of velocity (equal to the speed of vertical movement of the cylinder), P_x, P_y - slope of Gaussian functions, σ_x, σ_y – Gaussian spread. For more complex bodies simplified approach might be not valid and characteristic function χ advection eqation should be solved [10].

The loss of stability of the boundary layer, figure 7 (a), is a hydrodynamic phenomenon worth paying attention to. After an impulsively started flow, the thickness of the boundary layer increases with time, and at some point, it detaches from the airfoil. The presence of vortices generated by a moving bar works as a trigger and accelerates the occurrence of boundary layer separation.

The comparison with the two-dimensional simulation of the flow over the cylinder with the Reynolds number $Re = 10^5$ (Rossi [11]) shows that the size of the finest eddies in the wake area is approximately the same. For comparison, it was assumed that the size of the vortices is calculated in relation to the characteristic dimension - the diameter of the cylinder and the chord of the profile, respectively.

At the end of the simulation, the flow is entirely dominated by the presence of randomly moving vortex pairs. Vortices have a strong tendency to aggregate into bigger structures. This phenomenon is not observed in three-dimensional flows. By following the evolution of the vorticity evolution, on the concave part of the cascade passage, vortices have problems with leaving this very low-velocity zone. In the advanced phase of the flow, this results in their local build-up, and over time, they diffuse or merge with the positive-vorticity boundary layer. It

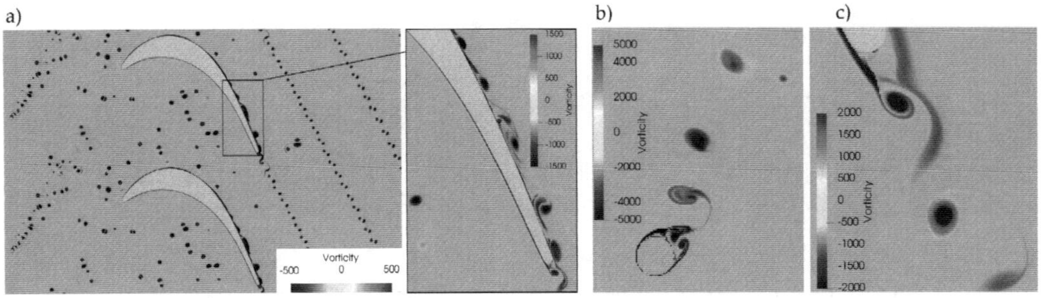

Figure 7: Flow past T106A turbine cascade, (a) boundary layer separation, (b) bar wake, (c) trailing edge wake.

was also noticed that vortex structures often collide with each other due to chaotic motion, so the destabilization of the bar von Karman vortex street is observed quite often. Figure 7 (b) and (c) show the comparison of bar and trailing edge wake. It can be seen that both vortex structures are similiar. The results presented above show that this method is promising, robust, and technically, it allows for complex problems in turbomachinery to investigated. However, if calculations are done without any turbulence model, formulation of the problem should be done in three dimensions. Two-dimensional simulation is not able to capture the nature of real advection dominated flow.

6. Summary

In the paper extension of the results of Penalized Vortex in Cell method is presented. Authors show solver performance profiling and conservation of volumetric balance in the unbounded flow when stream function correction is applied. By developing periodic boundary conditions and performing calculations for the T106A low-pressure turbine cascade, it has been shown that the Penalized Vortex in Cell method is an effective tool for solving nonstationary problems related to turbomachinery. Thanks to the simple formulation of equations, it is possible to perform simulations of moving bodies, research rotor-stator interactions or investigate the interactions of turbulent vortex structures with the turbine blades.

Acknowledgments

Calculations have been made using resources provided by Wroclaw Centre for Net- working and Supercomputing (http://wcss.pl), grant No. 444/2021.

References

[1] Błoński D, Strzelecka K and Kudela H 2021 *Energies* **14** ISSN 1996-1073
[2] Mimeau C, Gallizio F, Cottet G H and Mortazavi I 2015 *International Journal for Numerical Methods in Fluids* **79**
[3] Bost C, Cottet G H and Maitre E 2009 *arXiv*
[4] Caltagirone J P 1994 *Comptes Rendus de l'Académie des Sciences. Série II* **318**
[5] Angot P, Bruneau C H and Fabrie P 1999 *Numerische Mathematik* **81** 497–520
[6] Bost C, Cottet G H and Maitre E 2010 *SIAM Journal on Numerical Analysis* **48** 1313–1337
[7] Gillis T, Marichal Y, Winckelmans G and Chatelain P 2019 *Journal of Computational Physics* **394** 700–718
[8] Stieger R and Hodson H 2004 *Journal of Turbomachinery* **127**
[9] Stieger R 2005 *The Effects of Wakes on Separating Boundary Layers in Low Pressure Turbines* (University of Cambridge)
[10] Coquerelle M and Cottet G H 2008 *Journal of Computational Physics* **227** 9121–9137
[11] Rossi E, Colagrossi A, Bouscasse B and Graziani G 2015 *Comm. in Comp. Phys.* **18** 351–379

Journal of Physics: Conference Series **2367** (2022) 012008 doi:10.1088/1742-6596/2367/1/012008

Conservative weakly compressible smoothed particle hydrodynamics applied to flows with high vorticity

Adam Kajzer

Institute of Fluid-Flow Machinery, Polish Academy of Sciences,
Fiszera 14, 80-231 Gdańsk, Poland

E-mail: akajzer@imp.gda.pl

Abstract. The Smoothed Particle Hydrodynamics (SPH) method in the so-called weakly compressible variant (which mimics the incompressibility conditions) is studied in the context of its resolving power of vortical flows. It is well known that the conservative formulations of SPH have serious problems to provide reasonably accurate solutions even in simple flow cases. This deficiency is additionally emphasized by the fact that conservative SPH formulations are not numerically convergent. We investigate and discuss chosen techniques to improve the results; yet, the convergence issue remains. Maintaining the conservative properties (often presented in the literature as the SPH biggest advantage) requires procedures which are in contradiction with the accuracy improvements. Some myths about SPH are discussed and denied.

1. Introduction

Smoothed Particle Hydrodynamics is a Lagrangian computational approach which has been proposed in the late 1970s as a numerical method for astrophysical hydrodynamics, see [1] for a review. This class of flows mainly involve the compressible inviscid gas dynamics problems. Moreover, large density variations are common there. Therefore, the mathematical modelling in the Lagrangian frame is advantageous in such cases since "resolution follows the mass" [2]. For the astrophysical applications it is crucial for the numerical method to provide discrete solutions which possess physically relevant features: conservation of mass, linear and angular momentum, total energy and Galilean invariance. All of these are assured by the proper SPH formulations.

On the other hand, the demand of physical consistency leads to numerical problems, e.g. SPH is not convergent in the usual meaning. The main reasons are: (i) the Lagrangian nature itself which introduces the so-called particle noise; (ii) the necessity of introducing two discrete length scales: the particle spacing Δr_0 and smoothing length h which act in opposite w.r.t. the resolving power; (iii) the form of the discrete governing equations providing the conservation.

The SPH method has been used for other flow classes, e.g. environmental and engineering applications [1]. However, for the reasons discussed in this paper, it seems that its only reasonable use relates to the violent free surafce flows (where fine flow structures are less relevant) as Lagrangian methods easily track material fronts.

In this work we analyse the application of SPH for nearly incompressible flows, using the weakly compressible (WC) formulation that consists in assuming the finite (yet high enough) speed of sound and relating the pressure to the density by an equation of state (EOS). We focus on simple laminar flows involving high vorticity in periodic two-dimensional domains.

We examine the basic SPH formulation and chosen improvements. Even in the simple cases considered and even applying modern sophisticated SPH techniques it is difficult to obtain accurate solutions. Also, results of quality comparable to those obtained by the Eulerian, fixed mesh based methods are not achievable with SPH at reasonable cost.

2. Weakly compressible Smoothed Particle Hydrodynamics

As mentioned earlier, SPH uses the Lagrangian description of the fluid medium. The mathematical model expresses the conservation of mass and linear momentum:

$$\frac{d\rho}{dt} = -\rho \nabla \cdot \mathbf{u}, \tag{1}$$

$$\frac{d\mathbf{u}}{dt} = \frac{1}{\rho} \nabla \cdot (-p\mathbf{I} + \tau) \tag{2}$$

and the EOS (here we use the barotropic isothermal relation) $p = (\rho - \rho_0)c_s^2 + p_0$. In the above equations ρ is the local density, ρ_0 is the fluid density at rest (material property), $\mathbf{u}$ is the velocity vector, p is the pressure and $\tau = \mu(\nabla\mathbf{u} + (\nabla\mathbf{u})^T - (\frac{2}{3}\nabla \cdot \mathbf{u})\mathbf{I})$ is the tensor of the viscous stresses ($\mathbf{I}$ is the unit tensor). In the WCSPH literature the viscous stresses are often defined as for truly incompressible flow, i.e. $\mu/\rho_0\nabla^2\mathbf{u}$. This choice is computationally less demanding but it requires verification against the setting of the Mach number in particular flow cases. The speed of sound c_s is chosen such that during the simulation the Mach number $\mathrm{Ma} = |\mathbf{u}|/c_s < 0.1$ that results in density variations lower than 1% which is a commonly agreed incompressibility limit. In the EOS, p_0 denotes the background pressure that assures the positivity of pressure in the whole domain. The negative pressure causes particle clumping since in this case it is no longer a repulsive force. The freedom in choice of p_0 is limited since high values may lead to numerical problems; some details will be given later.

The basics of SPH discretistation consist in the integral approximations of flow fields. For a generic variable f one has:

$$f(\mathbf{r}) = \int_\Omega f(\mathbf{r}')\delta(\mathbf{r} - \mathbf{r}')d\mathbf{r}', \tag{3}$$

where δ is the Dirac delta function, Ω is the flow domain and $\mathbf{r} = (x, y)$ (in 2D). The above relation is exact but useless in practice. One obtains a useful approximation by replacing δ function with some other function $W^h : \Omega \mapsto R$ that involves a parameter h (smoothing length) and such that $W^h \to \delta$ when $h \to 0$. The functions W^h, called later as the kernel functions, should also: (i) have a compact support (for analytical and computational reasons); (ii) be even functions of $|\mathbf{r}|$ (symmetry is required to provide conservation of physical quantities); (iii) be normalised; (iv) be non-negative; (v) be, possibly (but not neccessarily), continuously differentiable. The parameter h is proportional to the kernel support radius. The approximation of f is defined by:

$$f(\mathbf{r}) \approx \int_\Omega f(\mathbf{r}')W^h(\mathbf{r} - \mathbf{r}')d\mathbf{r}'. \tag{4}$$

The compact support allows one to make a following approximation of the derivatives (below we only give example of derivative w.r.t. x direction):

$$\partial_x f(\mathbf{r}) = \int_\Omega f(\mathbf{r}')\partial_x W^h(\mathbf{r} - \mathbf{r}')d\mathbf{r}'. \tag{5}$$

The above equation reveals one of the "advantages" of SPH: computing the spatial differential operators does not require the computation of the derivatives of the field itself but involves only the derivative of the kernel function. This makes the method conceptually simple in the context of algorithmics. However, more accurate (in specific situations only) least-square-like procedures to approximate the gradient were elaborated [3, 4]; those involve the geometrical configuration of the discretisation points (particles) and require special care to assure the exact conservation. Using the Taylor expansion one can show that the above approximations are 2nd-order accurate.

In SPH one uses particles (which, to some extent, can be interpreted as the fluid elements) to discretise the space. The integral operators in Equations (4) and (5) are replaced by the summation over the particles:

$$f(\mathbf{r}_i) \approx \sum_j f(\mathbf{r}_j) W^h(\mathbf{r}_i - \mathbf{r}_j) V_j. \tag{6}$$

The compact support allows one to make a following approximation of the derivatives, cf. Equation (5):

$$\partial_x f(\mathbf{r}_i) \approx \sum_j f(\mathbf{r}_j) \partial_x W^h(\mathbf{r}_i - \mathbf{r}_j) V_j \tag{7}$$

where $\mathbf{r}_i$ and $\mathbf{r}_j$ are the positions of particles i and j and V_j is the volume of particle j (its definition will be discussed later). For brevity we will use the shorthand notation $W^h(\mathbf{r}_i - \mathbf{r}_j) = W^h_{ij}$. Notice that the above formulae realise the method of rectangles which is sensitive to the distribution of the quadrature points; SPH introduces two sources of errors: due to the kernel approximation and due to the approximate integration.

The discretisation of the flow governing equations is not straightforward, one can not simply apply the discrete approximations (6) and (7) since it would lead to lack of symmetry and in consequence non-conservative method. The conservative model can be obtained in various ways, e.g. using the Lagrange functional for the discrete particle system. A few forms of the SPH equation of motion (expressing the conservation of linear momentum) are known in the literature; one of the most common is:

$$\frac{d\mathbf{u}_i}{dt} = \sum_j m_j \left(-\left(\frac{p_i}{\rho_i^2} + \frac{p_j}{\rho_j^2} \right) \mathbf{I} + \Pi_{ij} \right) \cdot \nabla W^h_{ij} \tag{8}$$

where Π_{ij} corresponds to the properly defined discrete viscous term (possibly including the artificial viscosity). Notice that the equation above is antisymmetric (when proper discretisation of the visocus term is provided) w.r.t. the indices i and j. Therefore, the particles i and j act on each other with the same force with the opposite sign and hence obey the Newton third law which leads to conservation of momentum. The mass conservation equation is discretised in the form

$$\frac{d\rho_i}{dt} = \sum_j m_j (\mathbf{u}_i - \mathbf{u}_j) \cdot \nabla W^h_{ij} \tag{9}$$

which is variationally consistent with Equation (8). Another possibility to obtain the density field is to use its definition ($\rho = m/V$) and the SPH approximation:

$$\rho_i = \sum_j \rho_j \frac{m_j}{\rho_j} W^h_{ij} = \sum_j m_j W^h_{ij}. \tag{10}$$

Above we used the natural definition of particle volume $V_i = m_i/\rho_i$. Other choices are possible and are not neutral for the accuracy of the method (improvement is available), see [5].

Finally, the particles are advected using

$$\frac{d\mathbf{r}_i}{dt} = \mathbf{u}_i. \tag{11}$$

2.1. Considered SPH variants

Below we present two approaches to the SPH method. The first one is more classical (denoted later by clSPH): we utilise Equations (10), (8) and (11). We found that if instead of Equation (10) we use (9) the results are barely distinguishable. Being free from numerical or artificial viscosity the clSPH conserves the angular momentum exactly.

The second one is a variant of the so-called Riemann or Godunov SPH (GSPH). In GSPH one computes the particle interactions utilising a solution (preferably approximate) of the local inter-particle Riemann problem. The equations of mass and momentum conservation (its inviscid part) are rewritten as [6]:

$$\frac{d\rho_i}{dt} = 2\sum_j m_j(\mathbf{u}_i - \mathbf{u}^*_{ij}) \cdot \nabla W^h_{ij}, \tag{12}$$

$$\frac{d\mathbf{u}_i}{dt} = -\sum_j m_j p^*_{ij} \left(\frac{1}{\rho_i^2} + \frac{1}{\rho_j^2}\right) \mathbf{I} \cdot \nabla W^h_{ij}, \tag{13}$$

where $\mathbf{u}^*_{ij}$ and p^*_{ij} are solutions of the Riemann problem between particles i and j. If instead of the Riemann solution one takes the arithmetic average state, the formulation is equivalent to the standard one. As the approximate Riemann solver we use the Riemann solver of Roe, see [6]. To additionally reduce the numerical diffusion we use the 2nd-order reconstruction of the left and right Riemann states of the primitive variables and we apply the monotonised central limiter with the limiting constant 1.9. The GSPH conserves the angular momentum up to gradient approximation errors [7].

The viscous term is discretised as follows:

$$\Pi_{ij} = \frac{\tau_i}{\rho_i^2} + \frac{\tau_j}{\rho_j^2}. \tag{14}$$

2.2. Some numerical details

As the kernel function we utilise the Wendland fifth order polynomial, a popular choice for SPH applications. The Wendland functions have the attractive property of non-negative coefficients in the Fourier expansion. This allows to avoid the artificial particle clumping (which dramatically reduces the SPH accuracy), without introducing a high background pressure. Alternatively, one can use spline polynomial kernels, however they do not possess the anti-clumping property and therefore one has to assure the background pressure to be high enough (numerical diffusion increase) to prevent the particles clustering. On the other hand, the Wendland functions require a higher number (than spline functions) of particles inside the kernel support to obtain accurate results. Moreover, the spline provides a better partition of unity approximation (see next section for details). So, as usual in the numerical methods, the robustness vs. accuracy dilemma appears.

The derivatives of velocity components and the pressure (required by the viscous stress and the Riemann states reconstruction) are computed as follows (only the example in the x direction and u velocity component is given):

$$(\partial_x u)_i = \sum_j \frac{m_j}{\rho_j}(u_j - u_i)\partial_x W^h_{ij}. \tag{15}$$

For the time integration we use the 1st-order semi-implicit Euler method, see [8]. We point here that when SPH is used for inviscid flows and when exact energy conservation is important, one often uses 2nd-order symplectic methods. These methods are not straightforward to be applied to viscous flows simultaneously keeping 2nd-order accuracy. We, however, do not care about these issues since we consider viscous flows at relatively low Reynolds numbers for which the physical dissipation rate of the kinetic energy is much higher than the numerical one stemming from a particular time integration method. Moreover, the spatial discretisation errors dominate so there is no need to apply the time integration method of at least the same order as the spatial discretisation (which in the case of SPH is of 2nd-order only theoretically). Additionally, single stage time integration allows easy handling of periodic domains without introducing ghost particles: particles are simply shifted form one side of the domain to the other. This is not straightforward in, e.g. chosen variants of the Runge-Kutta methods where subsequent stages depend on the solution at different previous stages.

3. Chosen problems of SPH

3.1. Initial particle distribution

It is known (however, surprisingly often not taken into account) that the initial particle distribution plays a very important (or even crucial) role for the simulation quality and accuracy. The exact advection offered by SPH is also its major drawback: if the particles are initially placed in some regular manner (e.g. Cartesian lattice), the fact that they track the Lagrangian trajectories leads to the creation of void regions in the domain, destroying the accuracy or even leading to unphysical results. Moreover, the regular arrangement of the particles does not realise the physically correct state of minimal energy: the particles tending to this state may modify solution in unphysical way, see Figure 1. Therefore in our simulations we relax the particles initial positions according to the method given in [9]. Other methods are also available, see e.g. [5], but the one we have chosen has the advantage of simplicity while it offers similar results and efficiency as the more sophisticated methods. Notice that the fact that the SPH simulation requires this special treatment is in contradiction with the SPH "myth" that it does not require meshing: the traditional mesh is indeed not required but a considerable computational time has to be spent on the particles relaxation even in simple periodic domains (in complex domains with free surfaces or fluid-fluid interfaces some human-active preprocessing would be required).

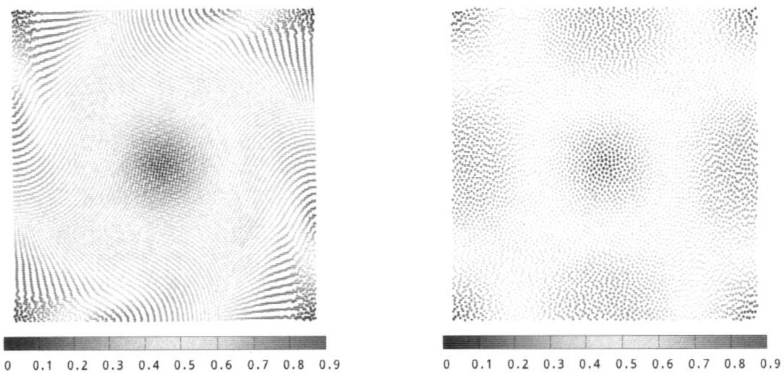

Figure 1. The velocity field at $t^+ = 0.25$ in the TGV case with uniform (left panel) and relaxed (right panel) initial particles positions. Only a quarter of the domain is shown. The colour scale is normalised by U_0 (see Section 4. for details).

3.2. Particle noise

The standard conservative SPH approximations are not able to exactly restore even the constant functions. All trials to obtain a method which is exact for the constant functions break the symmetry, see [7]. Even if the particle positions are initially relaxed, during the simulation it may happen that the particles are in some irregular arrangement in a region of size of order h (the lower is h the more probable is such a situation). This is somehow analogical to degenerated mesh cells (e.g., with very sharp corners). This degrades the accuracy and leads to significant dispersive errors (numerical noise). A way (in fact the only way [7]) to overcome this problem is to increase $h/\Delta r_0$. However this degrades the effective resolution and introduces numerical diffusion.

3.3. Low accuracy, lack of convergence

SPH is not a convergent method in the classical meaning. To decrease the errors one has to simultaneously reduce h, reduce Δr_0, and increase $h/\Delta r_0$. Clearly these conditions are in contradiction. Reduction of h increases the rectangle rule errors, reduction of Δr_0 with constant h does not increase the spatial resolution of the simulation. Therefore the achievable SPH convergence order is very low (far below 1) and is obtained at very high computational cost. Also for the convergence one has to give up the exact conservation by applying non-symmetric discretisation.

3.4. Initial condition for density

Since we use WCSPH for nearly incompressible flows the initial conditions for the problems of interest involve the velocity field and the pressure. However, the pressure field in the WC formulation uniquely depends on the density field (and vice versa). The correct density field can be found by using the inverse of EOS. Notice, however, that the density summation formula, Equation (10), should be also fulfilled for consistency (at least in clSPH). This means that for a given relaxed particle arrangement one has to set non-uniform particle masses to realise the proper density field, which in turn requires the introduction of a purely geometrical definition of particle volume; we use $V_i(t = 0) = (\sum_j W_{ij}^h)^{-1}$. The problem of initial density field is surprisingly often omitted in the SPH literature and usually the equal mass particles are used. Also, authors often set the initial density field to be uniform, which, obviously, in some cases leads to incorrect solution.

3.5. What do the above issues mean for WCSPH?

All the deficiencies mentioned above are especially harmful in the weakly compressible modelling. In the WC approach, the relevant pressure variations stem from very low density variations. Therefore, care is needed to decrease the numerical errors much below the level of physically relevant density variations (less than 1%). This is difficult to obtain at reasonable cost since reduction of the errors (partition of unity, truncation errors) requires the use of high numbers of neighbors for each particle ($\mathcal{O}(10^2)$ in 2D). The GSPH approach allows to decrease the numerical noise at a reasonable number of neighbors but, on the other hand, as a Godunov type method, it is not very well suited for WC modelling due to the disparity of the speed of sound and convective velocity scale (excessive diffusion appears unless sophisticated or tuned Riemann solver is used).

SPH with its elementary concept was invented for astrophysical simulations where large density gradients appear. Therefore, as we discuss above, it is not well suited for nearly incompressible flows. One can argue that free surface flows exhibit some similarities to astrophysics: large density gradients appear at the free surface. On the other hand, the dynamically relevant part of the domain is filled by liquid, a nearly incompressible medium.

Our choice of the SPH variants is done by analogy to the Eulerian, mesh based methods. The clSPH corresponds to the numerical (also artificial) diffusion-free methods based on the centred fluxes, often used for the numerical simulation of flows supporting fine vortical structures. In such methods one reduces the dispersive errors by increasing the order of the scheme using large stencils which correspond to large kernel support. However, in SPH the large stencil does not mean increased formal order of accuracy. The GSPH variant corresponds to the high-resolution upwind type methods and reduces the dispersive errors by introduction of numerical diffusion. Another approach is the introduction of artificial viscosity and mass diffusion term (δ-SPH, [10], not considered in this work); this approach however requires tuning of numerical parameters.

4. Results

We have chosen two flow cases to test the accuracy and resolving power of the considered SPH approaches. These are: the well-known case of Taylor-Green vortex (TGV) at Reynolds number

Re=100 and the periodic shear layer (PSL) at Re = 500, see [11]. Both cases are two-dimensional, the domain is periodic box $L_x \times L_y = 1 \times 1$. The simulations are conducted up to dimensionless time $t^+ = tU_0/L_x = 1$, where $U_0 = 1$ is a reference velocity. The required Reynolds numbers are obtained by setting proper viscosity of the fluid. The reference density ρ_0 is set to 1. The background pressure $p_0 = \rho_0 U_0^2$ was found to be high enough to assure the positive pressure in the whole domain during the simulations. For the TGV case the simulations were performed using four spatial resolutions: $h = L_x/N$ with $N = 100$ and 200, combined with $h/\Delta r_0 = 1.5$ and 2. For the PSL we used $N = 150$ and 300, combined with $h/\Delta r_0 = 1.5$ and 2. The reference data are taken as the solution of Equations (1) and (2) (rewritten in Eulerian frame) obtained with a centred (non-diffusive) 2nd-order finite volume (FV) method; for TGV case there exists the analytical solution in the truly incompressible case which is not the exact (but very close) solution in WC case. The reference data were computed using 120×120 grid for TGV and 160×160 grid for PSL.

In Figure 2 chosen results of the TGV case simulation are presented. As can be seen the clSPH performs poorly: significant dispersive errors are visible in the histories of the maximal velocity and pressure extremal values. Moreover, increasing N and $h/\Delta r_0$ does not improve the reults. GSPH performs better. However, although it introduces significant numerical diffusion it is still not free from the dispersive errors in the pressure (density) field; such a situation can be easily avoided in the mesh based methods. Nevertheless, the increase of N and $h/\Delta r_0$ brings some improvement of the results. Notice that discrepancy in the pressure extremal values occurs from the very beginning of the simulation; it is the consequence of poor partition of unity resulting from the relaxation process with low $h/\Delta r_0$.

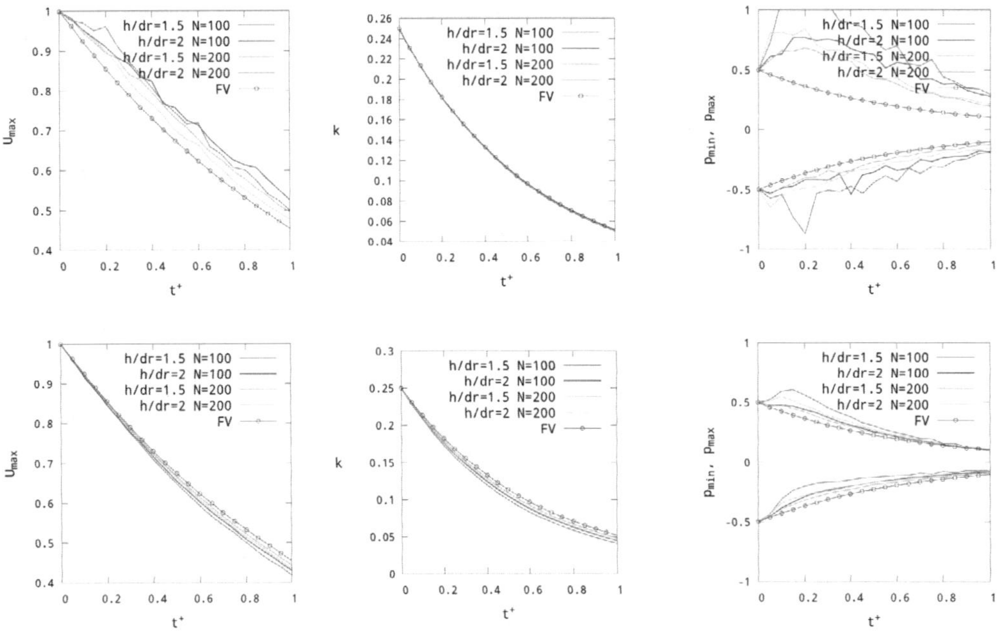

Figure 2. Simulation of TGV using clSPH (top panels) and GSPH (bottom panels): the histories of the maximal velocity magnitude, total kinetic energy and the pressure extremal values (reduced by p_0) are shown from left to right.

In Fig. 3 we present the results of PSL simulation. The vorticity fields are noisy and, as expected, the noise weakens with the increase of resolution. Unforunately, although the Reynolds number is not high and the flow is laminar, even at high spatial resolutions the simulations are underresolved as seen in the histories of the kinetic energy and enstrophy.

A remark on the computational cost should be made: using our in-house SPH and FV codes executed on the eight core CPU the GSPH simulation of PSL case at the highest considered resolution takes more than three hours (without the relaxation process) while the FV based simulation takes less than one minute.

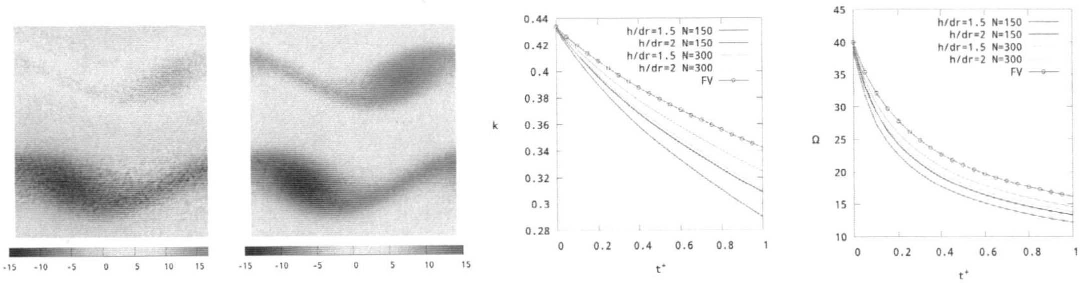

Figure 3. Simulation of PSL. From left to right: the vorticity field at $t^+ = 1$ with $h/\Delta r_0 = 1.5, N = 300$; the same with $h/\Delta r_0 = 2$; the normalised kinetic energy and enstrophy.

5. Summary

None of the SPH variants we studied allows to solve the simple, low-Re flows at reasonable cost. The reason is the general SPH problem of low accuracy and particle noise, which is even more serious for nearly incompressible flows where density variations are low. Thus, the use of SPH seems to be limited to cases where the fine flow structures are negligible (e.g., violent free surface flows) and where the angular momentum conservation and Galilean invariance are crucial.

A question arises: can we do any better? The answer is positive: the weighted particle methods introduced by Vila [12, 13] and developed [9, 7, 14] offer the required conservative properties and formal 2nd-order convergence. They use the flux form of the equations in the Lagrangian frame and least-square-like approximations, making the conservation and convergence rather independent of each other and being only slightly more complex than SPH.

Acknowledgments

The Author would like to thank Prof. Jacek Pozorski for useful discussions. This research was funded by NCN, grant number 2018/29/N/ST8/00267.

References

[1] Monaghan J J 2012 *Ann. Rev. Fl. Mech.* **44** 323–346
[2] Price D J 2012 *J. Comp. Phys.* **231** 759–794
[3] Bonet J and Lok T-S L 1999 *Comput. Methods Appl. Mech. Engrg* **180** 97–115
[4] García-Senz D, Cabezón R M and Escartín J A 2012 *Astron. & Astroph.* **538** A9
[5] Rosswog S 2015 *Mon. Not. R. Astron. Soc.* **448** 3628–3664
[6] Puri K and Ramachandran P 2014 *J. Comp. Phys.* **270** 432–458
[7] Hopkins P F 2015 *Mon. Not. R. Astron. Soc.* **450** 53–110
[8] Violeau D 2012 *Fluid Mechanics and the SPH Method* (Oxford: Oxford University Press)
[9] Gaburov E and Nitadori K 2011 *Mon. Not. R. Astron. Soc.* **414** 129–154
[10] Sun P N, Colagrossi A, Marrone S, Antuono M and Zhang A M 2018 *Comput. Phys. Comm.* **224** 63–80
[11] Minion M L and Brown D L 1997 *J. Comp. Phys.* **138** 734–765
[12] Vila J P 1999 *Math. Mod. and Meth. App. Sci.* **9 (2)** 161–209
[13] Lanson N and Vila J P 2008 *SIAM J. Numer. Anal.* **46 (4)** 1912–1934
[14] Panuelos J, Wadsley J and Kevlahan N 2020 *J. Comp. Phys.* **414** 109454

Journal of Physics: Conference Series **2367** (2022) 012009 doi:10.1088/1742-6596/2367/1/012009

A parametric analysis of concentration losses in an anode of a solid oxide fuel cell

T Prokop[1], G Brus[1] and J Szmyd[1]

[1] AGH University of Science and Technology, Department of Fundamental Research in Energy Engineering, al. Mickiewicza 30, 30-059 Kraków, Poland

E-mail: prokopt@agh.edu.pl

Abstract. The purpose of this paper is to discuss the loss of voltage in a Solid Oxide Fuel Cell (SOFC) as a result of the irreversible diffusive transport of gases through the microscopic pores of its fuel electrode. The voltage loss is estimated by solving the diffusive-convective equations, and current conduction equations in a computational domain based on data from Focused Ion Beam Scanning Electron Microscopy. The simulation includes both the binary diffusion and the free-molecular diffusion phenomena by employing Cylindrical Pore Interpolation Model. Butler-Volmer model is used to compute the reaction rate. Ion and electron current conduction is based on empirical relationships reported in the literature. The total losses are decomposed to identify the contribution of the mass transport irreversibilities. The parametric study results form a concave surface, showing non-linear relationship between electrode thickness, and diffusion-related voltage losses. The optimal active layer thickness is estimated.

1. Introduction

Fuel cells are devices capable of generating electrical energy by directly converting the chemical energy of fuels. Solid Oxide Fuel Cells (SOFCs) are a class of fuel cells where a ceramic material is used as the electrolyte. The advantages of SOFCs include: high energy conversion efficiency, good capacity for cogeneration due to high operational temperature, and wide variety of available fuels. SOFC electrodes are composed of a porous, ceramic-metal composite consisting of phases with particular transport functions. In a typical SOFC anode, gaseous reagents diffuse via open pores, oxygen ions are conducted through ceramic material (i.e. Yttrium-Stabilized Zirconia), and the half-reaction electrons are collected through connected grains of metal, such as Nickel. The complexity of SOFC microstructure is an important research challenge, which affects the efficiency and reliability of SOFC systems. Over the course of the last decade, research groups have used Focused Ion Beam Scanning Electron Microscopy (FIB-SEM) data to analyze the three-dimensional transport phenomena in SOFC electrodes [1, 2]. Nevertheless, this approach has not been very common, as most SOFC models discussed in the literature use averaged parameters (such as phase volume fraction and tortuosity) as the input. Furthermore, it has been noted that some aspects of the microscopic mass transfer have been neglected in common SOFC electrode models [3]. On the other hand, the research presented by Bao et al. [4] demonstrated that the models, which assume dilute, constant-pressure solution perform adequately for a mixture of $H_2 - H_2O - Ar$, provided that a correction is made to include the Knudsen diffusion. To achieve these results, Bao et al. [4] compared experimentally measured concentration overpotential to twelve values obtained by simulation. The twelve variants involved different

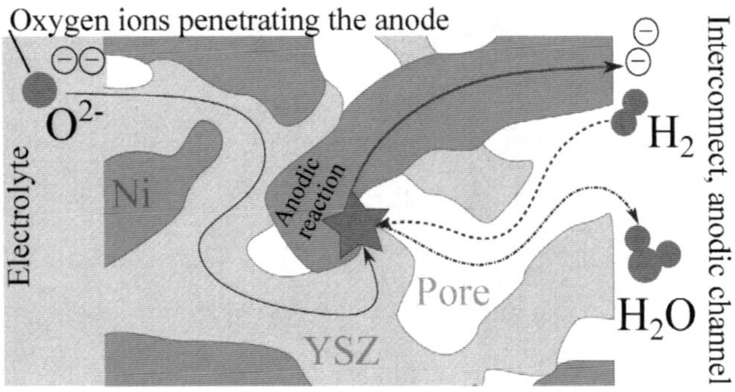

Figure 1. Transport phenomena in a porous SOFC anode.

models (the Stefan-Maxwell Model, the Fick's Model and the Dusty Gas Model), different methods of calculating the flux ratio (stoichometry-based and Graham Law based), as well as inclusion of the Knudsen diffusivity. The stoichometry-based Stefan-Maxwell model (SMM) with diffusivity coefficient accounting for Knudsen phenomena performed at similar error to the error of the stoichometry-based Dusty Gas Model (DGM) with no constant pressure assumption. More research on mass transfer of diffusion in SOFC electrodes is required.

In our study, we focus on solving mass and electric charge transfer equations in a non-continuous, three-dimensional computational domain based on the data from Focused Ion Beam Scanning Electron Microscopy. The drop of available voltage is then decomposed to predict the concentration overpotential, and provide insight into the irreversibilities related to the diffusion of gaseous reagents. A parametric study is performed to analyze the anode performance at different thicknesses and operation temperatures.

2. Methodology

The mathematical model is constructed using the potential-flux relationships for the transport of gas reagents, ions and electrons in a computational domain composed of multiple, irregularly distributed phases with specific transport properties — either electron conduction, or ion conduction, or gas diffusion (see figure 1). The reaction rate is modeled only in proximity to the Triple Phase Boundaries. The distribution of the computational domain phases is based on a digital three-dimensional material reconstruction done using Focused Ion Beam Scanning Electron Microscopy (discussed to a greater extent in a previous publication) [11]. Special attention was given to the pore-scale mass transfer. Since the specific diameter of the pores in an SOFC anode is measured at $\sim 0.1\,\mu\text{m} - 1\,\mu\text{m}$, locally linearized Cylindrical Pore Interpolation Model (CPIM) by Young and Todd [5] was used to account for the gas flow in the transitional regime between the limits of the continuum, and the free-molecular diffusion (see figure 2). In CPIM, the pores are approximated as cylindrical micro-channels, for which the momentum equation can be solved analytically (as a more general variant of the Poiseuille flow). This approach circumvents the need to measure sample permeability, or to compute it by solving the Navier-Stokes equations numerically. The model encompasses both Fick's and Knudsen diffusion phenomena, as well as the slip arising due to different masses of the gas particles. If ideal gas mixture behavior is assumed, the model derived by Young and Todd [5] can be written in the following form:

$$\begin{cases} \nabla X_j = \frac{RT}{p} \sum\limits_{j \neq k}^{n} \left[D_k^{-1} X_j \mathrm{J}_k - D_j^{-1} X_k \mathrm{J}_j \right] \\ \nabla p = -K \sum\limits_{j \neq k}^{n} M_k^{1/2} \mathrm{J}_k \ , \end{cases} \tag{1}$$

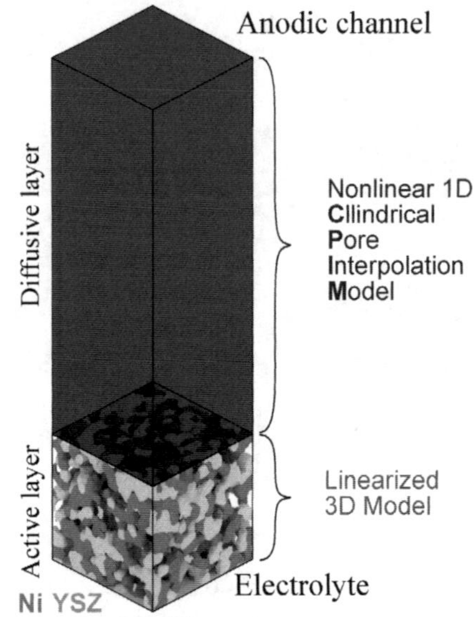

Figure 2. Visualization of the computational domain with non-continuous distribution of micro-channels in the active layer, and linearization of the porous electrode diffusion model in the diffusive layer.

where $X_j = p_j/p$ is the molar fraction of substrate j of n, p_j (Pa) is the partial pressure of substrate j, p (Pa) is the total pressure, J_j ($\mathrm{mol\,s^{-1}}$) is its molar flux, $D_{j,}$ ($\mathrm{m^{-2}s}$) is the diffusivity of mixture jk, interpolated from continuous (Fick's) an non-continuous (Knudsen's) regime diffusivities by means of the Bosanquet formula:

$$D_j^{-1} = D_{jk,\mathrm{Fick}}^{-1} + D_{j\mathrm{K}}^{-1} , \tag{2}$$

where D_{jk} ($\mathrm{m^{-2}s}$) and $D_{i\mathrm{K}}$ ($\mathrm{m^{-2}s}$) are, respectively, the binary (continuous-flow) diffusivity of the mixture jk and the Knudsen (free-molecular) diffusivity of the species j. It is worth noting that $D_j \neq D_k$, and that these values are not substitutes for the diffusivity coefficient in an application of Fick's Laws. M_j is the specific molar mass of species j, K is the pressure drop equation coefficient, obtained from the following relationship:

$$K = \left\{ \left[\frac{3}{4r} \left(\frac{\pi RT}{2} \right)^{1/2} \right]^{-1} + \left[\frac{8RT\mu}{p} \frac{1}{\sum\limits_{\substack{j=1 \\ i\neq jk}} M_k^{1/2} \mathrm{J}_k} \right]^{-1} \right\}^{-1} , \tag{3}$$

where μ ($\mathrm{kg\,m^{-1}\,s^{-1}}$) is the dynamic viscosity of the gas mixture and r is the local pore diameter. It can be computed using Wilke formula [6, 7]:

$$\mu = \sum_{j=1} \frac{X_j \mu_j}{\sum\limits_{k=1} X_k \kappa_{jk}} , \tag{4}$$

in which the term κ_{jk} is obtained from:

$$\kappa_{jk} = \frac{[1 + (\mu_j/\mu_k)^{(1/2)} (M_k/M_j)^{(1/4)}]^2}{[8(1 + M_k/M_j)]^{1/2}} . \tag{5}$$

The specific viscosity μ_j of substrate j, at given operation conditions may be obtained from thermodynamic tables such as those provided by the National Institute of Standards and Technology of America [8].

In the proposed model, the Equation set (1) is applied in the diffusive layer of the electrode. However, due to high computational cost, it is prudent to propose a linearized mass conservation equation, which can be employed locally in the active layer. If the local pressure gradient is neglected, and the fluxes are reciprocal ($\mathrm{J}_j = \mathrm{J}_k$), the molar fraction formula in equation set (1) reduces to an expression analogous to the Fick's law. One of the requirements for that assumption is that the chemical reaction occurring in the electrode does not produce a pressure gradient (so one mole of substrates is replaced by one mole of products). Such is the case in an anodic SOFC reaction, where hydrogen is oxidized and water vapor is produced. Equations (6) describe the relationship between the charge transfer rate and the partial pressure of the gaseous reagents:

$$\nabla \cdot \left\{ D_{jk} \frac{1}{RT} \nabla p_j \right\} = \begin{cases} \dfrac{i}{2F}, & j = \mathrm{H_2}, \quad k = \mathrm{H_2O} \\ -\dfrac{i}{2F}, & j = \mathrm{H_2O}, \quad k = \mathrm{H_2} \end{cases} , \tag{6}$$

where i ($\mathrm{A\,m^{-3}}$) is current density, $p_{\mathrm{H_2}}$ and $p_{\mathrm{H_2O}}$ (Pa) are partial pressures of hydrogen and water vapor, D is the diffusion coefficient ($\mathrm{m^2\ s^{-1}}$), R is the universal gas constant ($\mathrm{J\,mol^{-1}\,K^{-1}}$), T (K) is temperature, and F is Faraday's constant ($\mathrm{C\ mol^{-1}}$). Since the binary gas composition is assumed, a singular diffusion coefficient suffices for both gaseous reagents. Assuming binary mixture, and approximating negligible pressure gradient due to slip, Young and Todd's model may be used to obtain a locally linear interpolation between Knudsen and Fick diffusivities, yielding the following coefficient:

$$D_i = D_j = \left[\frac{p_i}{p} \left(\frac{1}{D_{ij}} + \frac{1}{D_{i\mathrm{K}}} \right) + \frac{p_j}{p} \left(\frac{1}{D_{ij}} + \frac{1}{D_{j\mathrm{K}}} \right) \right]^{-1} . \tag{7}$$

The Fick binary diffusion coefficient can be obtained from the Fuller-Schettler-Giddings model [10]:

$$D_{ij} = D_{\mathrm{Fick}} = 10^2 \frac{T^{1.75} \left(\dfrac{1}{M_i} + \dfrac{1}{M_j} \right)}{p \left(v_i^{1/3} + v_j^{1/3} \right)} , \tag{8}$$

and the Knudsen diffusion coefficient can be derived from the relationship (9) [9]:

$$D_{i\mathrm{K}} = D_{\mathrm{Knudsen}} = \frac{d_{\mathrm{pore}}}{3} \sqrt{\frac{8RT}{\pi M_i}} , \tag{9}$$

where p is the absolute pressure in pascals, M_n is the molar mass of species i, v_n is its specific, cross sectional constant in the Fuller-Schettler-Giddings correlation [10] d_{pore} (m) is pore diameter.

The mathematical model for the active layer accounts not only for the gas mixture, but also for the oxygen ions and the electrons transported via the ceramic and the metallic phases of the fuel electrode. Coupled with Butler-Volmer model for local reaction rates, the aforementioned relationships form a system of Poisson differential equations:

$$\begin{cases} i = \nabla \cdot (\sigma_{\text{el}} \nabla \phi_{\text{el}}) \\ -i = \nabla \cdot (\sigma_{\text{ion}} \nabla \phi_{\text{ion}}) \\ \dfrac{i}{2F} = \nabla \cdot \left(\dfrac{D_{\text{H}_2}}{RT} \nabla p_{\text{H}_2} \right) \\ -\dfrac{i}{2F} = \nabla \cdot \left(\dfrac{D_{\text{H}_2\text{O}}}{RT} \nabla p_{\text{H}_2\text{O}} \right) \\ i = i_0 l_{\text{tpb}} \cdot \left[\dfrac{\bar{p}_{\text{H}_2}}{p^{\text{bulk}}_{\text{H}_2}} \exp\left(\dfrac{\alpha_{\text{an,fwd}} F}{RT} \eta \right) - \dfrac{\bar{p}_{\text{H}_2\text{O}}}{p^{\text{bulk}}_{\text{H}_2\text{O}}} \exp\left(-\dfrac{\alpha_{\text{an,bcw}} F}{RT} \eta \right) \right] \\ \eta = \phi_{\text{el}} - \phi_{\text{ion}} - \dfrac{RT}{2F} \ln \left(\dfrac{p^{\text{bulk}}_{\text{H}_2}}{\bar{p}_{\text{H}_2}} \dfrac{\bar{p}_{\text{H}_2\text{O}}}{p^{\text{bulk}}_{\text{H}_2\text{O}}} \right), \end{cases} \tag{10}$$

where ϕ (V) is the electric potential, σ (S m^{-1}), α is the charge transfer coefficient, $\bar{p}$ (Pa) is active layer-averaged partial pressure, and p_{bulk} (Pa) is partial pressure at boundary. Outside of a 20 µm – 40 µm thick reactive layer, reaction current is neglected, and the phase electric potentials are treated as constants. The computational domain was modeled after real phase distribution in a sample of an SOFC anode, investigated by means of Focused-Ion-Beam Scanning Electron Microscopy (FIB-SEM). Discretization is performed according to the Finite-Volume Method (FVM). The solver is based on the Successive Over-Relaxation (SOR) method. Model validation, as well as information on the microstructures used in the study are available in another paper [11].

3. Results

Solving the equation set (10) allowed computation of the overpotentials generated on the microstructure of an SOFC anode. Figure 3 displays the results of a parametric analysis of the voltage loss η (V) with respect to temperature T, current density, and anode thickness. It can be seen that as the current density increases, the results form a concave surface for the considered range. For thin electrodes, as the thickness is increased, the overpotential becomes lower. This is related to an increasing number of available reaction sites at which fuel oxidation may occur. However, at around 20 µm, a local minimum emerges, and additional electrode thickness results in more losses, largely due to increasing the length of diffusion pathways. The optimal thickness is lower for higher temperatures. The parametric study was extended with focus on the voltage loss occurring specifically due to the gas diffusion irreversibilities (concentration loss). The concentration losses for various values of temperature, electrode thickness, and current density are presented in figure 4. As expected, additional thickness results in greater losses related to the gas diffusion. The curvature appears to be non-convex, with inflections around the ($\sim$ 20 µm – 40 µm) mark. This can be attributed to a simultaneous increase of active layer thickness (and thus - reaction site availability) for a thin electrode, as well as irregularities of the microstructure. An increase in temperature leads to higher concentration losses.

4. Conclusions

In this research, a numerical model taking into consideration the local distributions of microstructure features was used to investigate an SOFC anode. The electrode voltage losses due to mass transport irreversibilities were computed for a range of electrode thicknesses, current densities, and temperatures. The non-monotonous effect of increasing the electrode's thickness on voltage losses was determined to be the result of simultaneous decrease of polarization due to better reaction site availability, and greater resistance to gas diffusion. The following optima were found for the investigated electrode: 33 µm at 923 K, 27 µm at 973 K, 18 µm at 1023 K.

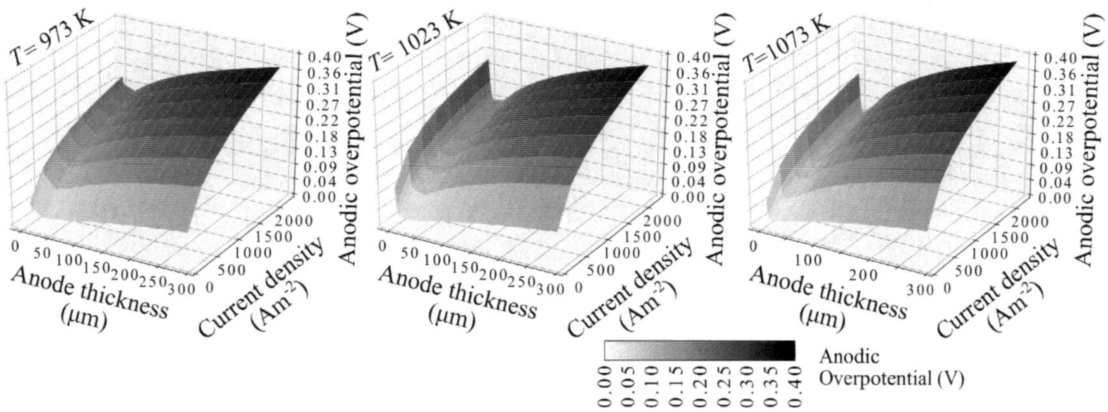

Figure 3. Parametric study of the anodic polarization at different electrode thicknesses and temperatures. Simulation parameters: p = 100,000 Pa, $p_{H_2,b}$ = 60,000 Pa.

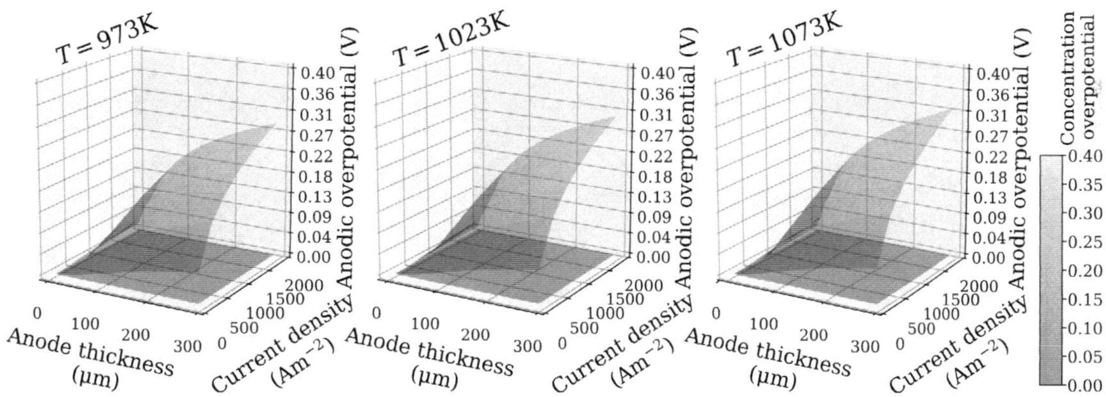

Figure 4. Anodic voltage losses due to gas transport irreversibilities at different electrode thicknesses and temperatures. Simulation parameters: Total pressure: 100 000 Pa, Hydrogen partial pressure at the anodic channel boundary: 60 000 Pa.

Therefore, it can be concluded that the optimal active layer thickness is not solely related to the microstructural parameters, but also to the conditions at which the cell operates.

Acknowledgments

Numerical code developed during National Science Centre's grant (SONATA-10, Grant No. UMO-2015/19/D/ST8/00839) was used. Further work partly supported by program "Excellence Initiative – Research University" for the AGH University of Science and Technology. This research was supported in part by PL-Grid Infrastructure. Avizo software by Thermo Fisher Scientific and Matplotlib Python library were used.

References

[1] Kanno D, Shikazono N, Takagi N, Matsuzaki K and Kasagi N 2011 Evaluation of SOFC anode polarization simulation using three-dimensional microstructures reconstructed by FIB tomography *Electrochim. Acta* **56** pp 4015–21

[2] Kishimoto M, Iwai, H, Saito M and Yoshida H 2011 Quantitative evaluation of solid oxide fuel cell porous anode microstructure based on focused ion beam and scanning electron microscope technique and prediction of anode overpotentials *J. Power Sour.* **196** pp 4555–63

[3] Bertei A and Nicolella C 2015 Common inconsistencies in modeling gas transport in porous electrodes: The dusty-gas model and the Fick law *J. Power Sour.* **279** pp 133–7
[4] Bao C, Jiang Z and Zhang X 2016 Modeling mass transfer in solid oxide fuel cell anode: I. Comparison between Fickian, Stefan-Maxwell and dusty-gas models *J. Power Sour.* **310** pp 32–40
[5] Young JB and Todd B 2005 Modelling of multi-component gas flows in capillaries and porous solids *Int J Heat Mass Transf.* **48** pp 5338–53
[6] Wilke CR 1950 A viscosity equation for gas mixtures *J. Chem. Phys.* **18** pp 517–19
[7] Poling B, Prausnitz J and O'Connell J 2000 *The Properties of Gases and Liquids 5E* (New York: McGraw-Hill Education)
[8] Linstrom P and Mallard W 2020 *NIST Chemistry Webbook: NIST Standard Reference Database Number 69* (Gaithersburg: NIST) url: https://doi.org/10.18434/T4D303
[9] Welty J, Wicks C and Wilson R 2019 *Fundamentals of Momentum, Heat, and Mass Transfer, 7th Edition* (Hoboken: Wiley)
[10] Fuller EN, Schettler PD and Giddings JC 1966 A new method for prediction of binary gas-phase diffusion coefficients *Industrial and Engineering Chemistry* **58** pp 18–27
[11] Prokop T, Berent K, Mozdzierz M, Szmyd J and Brus G 2019 A Three-Dimensional Microstructure-Scale Simulation of a Solid Oxide Fuel Cell Anode – The Analysis of Stack Performance Enhancement After a Long-Term Operation *Energies* **12** p 4784

Journal of Physics: Conference Series **2367** (2022) 012010 doi:10.1088/1742-6596/2367/1/012010

CFD analysis of a multi-rotor flying robot for air pollution inspection

G Suchanek[1] and R Filipek[1]

[1]AGH University of Science and Technology, Department of Power Systems and Environmental Protection Facilities, Faculty of Mechanical Engineering and Robotics, 30 Mickiewicza Av., 30-059 Krakow, Poland

E-mail: suchanek@agh.edu.pl

Abstract. The aim of the work was to create a CFD model of the flow generated around the drone to estimate the impact of field parameters on the results of actual measurements from PM sensors that are positioned differently in relation to the propellers. The model created with the use of the ANSYS Fluent software allowed one to determine the criterion of their sufficient distance. The robots with four, six and eight rotors were analyzed. For these, the turbulence intensity, velocity and pressure distributions were determined. The paper also presents the results of PM measurements carried out under field conditions using two sensors mounted on the hexacopter robot.

1. Introduction

The polluted air, in particular the high concentration of airborne particles, is one of the main environmental threats to human health. It is responsible for, i.e., cancer of the respiratory system, the prevalence of allergies and increases the relative risk of premature death [1]. There are also less obvious costs, such as environmental costs for, e.g., decaying building facades. Sources of this type of pollution are, but are not limited to: wildfires [2], low chimney emissions, heat power plants [3], or other processes where low-quality solid fuels or waste are incinerated. Therefore, there is a constant need to look for and develop solutions that would make it possible to locate the sources of this type of pollution. A dedicated multirotor flying robot equipped with a particle concentration measurement system can be used to autonomously locate sources of air pollution. However, a robot of this type significantly disturbs the ambient flow of gases and particulate matter.

The use of multi-rotor robots to measure particulate matter (PM) has been described, among others, in [4], where measurements were carried out in the city during periods of heavy traffic with the use of a quadcopter robot and a simple low-cost recorder with particulate matter sensors. Another example should be mentioned [5], where a mobile robot cart with a low-cost sensor was developed to map spatial PM distributions over time. However, the use of a wheeled robot is a great simplification and limitation. In [6], the possibilities of using the low-cost sensor system were analyzed to obtain reliable data. The paper [7] presents an innovative approach, where the novelty lies in the fact that the robot can mitigate (only NO_2 for now) the detected air pollution. Moreover, the effectiveness of this solution is limited by the loading capacity of the robot. The influence of rotors on the reliability of the measurement was also investigated [8]. However, no actual units were presented to measure air pollutants. In the work [9], similar experiments were carried out in both the wind tunnel and the open air, using a quadcopter robot, an array of low-cost sensors and research-grade instruments. The results have shown that a concentration of $PM2.5$ increases by as much as 400% when comparing the measurements taken before and after the propellers were turned on. The subject of the investigation is also the influence of meteorological factors such as wind and precipitation [10] on the change in PM concentration. Another example may be [11] where low-cost sensors have been evaluated at high altitudes in mountains, where there is lower atmospheric pressure. However, in these there is no answer to the question of how a local change in the ambient conditions caused by the robot affects the measured particles.

On the basis of the above-mentioned examples, it is possible to indicate some research gaps related, but not limited to: analyzing only one robot configuration or the arbitrarily selected location of the measuring system. Therefore, the purpose of the work was to estimate the magnitude of aerodynamic phenomena on an example of robots with four, six and eight rotors. To achieve this goal, it was necessary to determine velocity, turbulence and pressure distributions for a single, two (one above the other), four and six rotor configurations. These distributions were obtained with Ansys Fluent software, widely used to solve academic problems. Finally, particle matter measurements were performed under field conditions to verify whether the results obtained can be explained on the basis of the prepared numeric model.

2. Computational models

As mentioned earlier, the subjects of the analysis were configurations with one, two, four and ultimately six rotors. For these, the velocity, pressure field distributions and turbulent intensity were determined. These configurations are the result of the three analyzed multirotor type quad, hexa, and octocopter robots.

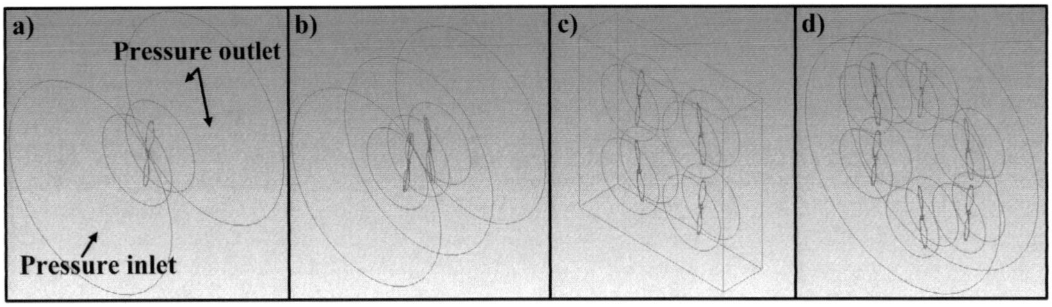

Figure 1. Computational domains: one (a), two (b), four (c), six (d) rotors configurations.

First, a single-rotor reference analysis was performed in a large zone (Figure 1a). Two propellers used in the described robots were analyzed: a smaller 1045 type (diameter of 10 inches and 4.5 inch pitch) and a larger of type 1238 (diameter of 12 inches and 3.8 inch pitch). These conditions corresponded to the conditions during the tests on a dynamometer station. The computational domain for these cases has the shape of a cylinder with dimensions: for a 1045 propeller with a diameter of 0.7 m and height of 0.5 m, for larger one with a diameter of 0.8 m and height 0.6 m. The geometric model of the rotors (propellers) was made in CATIA engineering software based on 3D scans. With various types of propellers with dimensions previously given, it was chosen to scan those that generated the largest thrust, as measured at the dynamometer station. Similarly, on the basis of the measurements, it was possible to determine the rotation of the propeller of the robot in the hover state, which amounted to: 4029 rpm (quad), 4598 rpm (hexa) and 4185 rpm (octo), and such a value was also adopted for simulations.

The first robot analyzed is an octocopter-type robot. It is in a redundant configuration, which means that it has four arms and for each arm there are two drive sets, a top and a bottom one. The total weight of this robot is 4.3 kg and is equipped with a 1238 propeller. Due to the large external dimensions and the high computing complexity that ensues, it was decided to perform the analysis for a counter-rotating configuration of the two propellers, as shown in Figure 1 b. For this purpose, it was necessary to create two dynamic zones over ourselves, similar to the shape of a cylinder with a diameter of 0.36 m and a height of 0.12 m. Dynamic zones rotate at identical rotational speeds but in opposite directions. The surrounding is the static zone, which is cylinder-shaped with a diameter of 0.8 m and height 0.6 m. The next robot analyzed is a quadcopter robot equipped with four rotors, one on each arm. This robot is equipped with 1045 propellers and weighs 1.35 kg. The prepared geometric model (Figure 1c) includes four cylinder-shaped rotary zones with a diameter of 0.3 m and a height of 0.1 m. The robot environment is a rectangular static zone with dimensions of 0.75 m × 0.75 m × 0.2 m. The construction of a hexacopter robot (Figure 1d) is very similar. It is equipped with the same type of propeller, so the same dynamic zones have been adopted. Additional rotors were taken into account, and in this case the static zone has the shape of a cylinder with a diameter of 0.5 m and a height of 0.2 m. The robot weighs 2.54 kg. In the cases where the configuration of four or six rotors was modeled, the prepared model included only the closest robot environment. This is because the measuring system arm cannot be too far from the center of gravity of the robot because it would severely degrade the stability of the robot in the air.

The next stage of the work was to build computational models in ANSYS Fluent [12]. This fluid flow modeling software solves partial differential equations resulting from the conservation of mass and momentum. Additional transport equations must be added if the flow is turbulent. The mass conservation equation is written in a generalized form as follows

$$\frac{\partial \rho}{\partial t} + \nabla \cdot (\rho \vec{v}) = S_m \tag{1}$$

where ρ is the density of the fluid, $\vec{v}$ is the velocity, and S_m is the mass source. The principle of momentum conservation for an inertial reference frame is described [13] by

$$\frac{\partial}{\partial t}(\rho \vec{v}) + \nabla \cdot (\rho \vec{v}\vec{v}) = -\nabla p + \nabla \cdot (\bar{\bar{\tau}}) + \rho \vec{g} + \vec{F} \tag{2}$$

where p is the static pressure, $\bar{\bar{\tau}}$ is the stress tensor, as well as $\rho \vec{g}$ and $\vec{F}$ are the gravitational body force and external body forces, respectively. The stress tensor $\bar{\bar{\tau}}$ is given by

$$\bar{\bar{\tau}} = \mu \left[\left(\nabla \vec{v} + \nabla \vec{v}^T \right) - \frac{2}{3} \nabla \cdot \vec{v} I \right] \tag{3}$$

where μ is the molecular viscosity, I is the unit tensor, and the second term on the right-hand side is the effect of volume dilation. A direct numerical solution to these equations is known as the DNS (Direct Navier-Stokes) method, but it is impractical due to the very high demand for computing power. In practice, as in these studies, other methods, such as the Reynolds-averaged Navier-Stokes (RANS) method, are widely used. In Reynolds averaging, the solution variables in the instantaneous (exact) Navier–Stokes equations are decomposed into the mean (spatial–averaged or time–averaged) and fluctuating components. For velocity, pressure and other scalar quantity components it is written as

$$u_i = \bar{u}_i + u_i', \quad \phi = \bar{\phi} + \phi' \tag{4}$$

where $\bar{u}_i$ and u_i' are the mean and fluctuating velocity components ($i = 1, 2, 3$) as well as $\bar{\phi}$ and ϕ' denote analogously scalar parameters such as pressure, energy or species concentration. The boundary conditions in the model were assumed as shown in Figure 1a, i.e. the upper surfaces of the static zones as the pressure-inlet and the lower and side surfaces as the pressure-outlet. Both the inflow and outflow conditions are based on the known pressure value at the boundary. On these surfaces, the pressure value was assumed to be zero with the reference value of 1 atm. For the modeling of turbulence, a standard $k - \omega$ SST model [14] was used and all calculations were performed for steady state.

For the different rotor configurations analyzed, it was necessary to use the multiple reference frame (MRF) approach. The MRF model is a steady-state approximation. In that approach, specific rotational speeds are assigned to each rotational zone and the flow in each of them is solved using the equations of the moving coordinate system. A local coordinate system transformation is performed on the internal contact surfaces of the domains (interfaces) between the moving and stationary zones. This allows the flow parameters in one zone to affect the flows at the boundary of an adjacent one. This MRF approach is very computationally efficient, but it is planned to use the sliding mesh (SM) approach [15] in future work, as it is highly accurate for high rotational speeds. In the SM approach, the rotary domain mesh slide relative to the stationary one along the interface surface.

The finite volume method (FVM) based on a nonstructural mesh was used because of the high complexity of the propeller geometry. For example, for the smaller of the propellers, it was defined that the element size in the rotary zone does not exceed 0.003 m, and in a static zone 0.01 m. Furthermore, to consider the aerodynamic phenomena modeled near the rotor as correct, it is necessary to ensure that the dimensionless wall distance y^+ on the entire rotor surface is less than 1. To obtain an y^+ at a specified level, a boundary layer was modeled. This goal was achieved through a significant local refinement of the mesh (figure 2c) on the surface of the wall (rotor). For this purpose, the following parameters were specified: the first layer height of 0.006 mm, 8 boundary layers and 1.2 inflation ratio. The assumed values were the result of a compromise between the required computing power and the accuracy and usability of the model. Further refinement of the mesh did not significantly improve the solution. Figure 2a shows the determined parameters of the force and torque of the rotor, while figure 2b shows the maximum values of the flow velocity in the sections analyzed marked as A, B and C. These sections are located 0.05 m, 0.1 m, and 0.2 m, respectively, below the surface of the rotor. All of these parameters were plotted as functions of a different number of elements. The criterion adopted here is that for the next step in computational mesh refinement, their difference from the previous step changes by less than 10%. In the final case, the difference in the rotor parameters was about 1% and in the case of the speed about 8%. This allowed us to conclude that the grid independence of the solution was obtained.

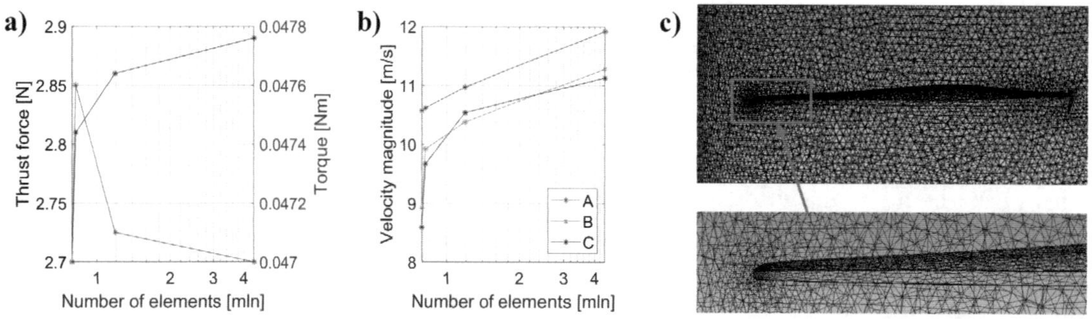

Figure 2. Parameters determined during the examination of the independence of the solution from the computational mesh: a) thrust force and torque, b) velocities, c) mesh — close-up on the rotor surface.

3. CFD simulation results

Figure 3 shows longitudinal sections of the velocity field, axes, as well as A, B, and C marking the cross sections for a single 12 inch wide propeller, which are located 0.05 m, 0.1 m and 0.2 m below the rotor, respectively. In the case of a configuration of two rotors above each other, these distances were defined below the bottom rotor. For these sections, data from the ANSYS Fluent program were collected and then, on this basis, the envelopes of the velocity distribution and turbulent intensity were determined using scripts prepared in Matlab software.

The turbulence intensity parameter is by definition expressed as a percentage and is given as the quotient of the RMS value of the turbulent velocity fluctuations and the mean velocity of

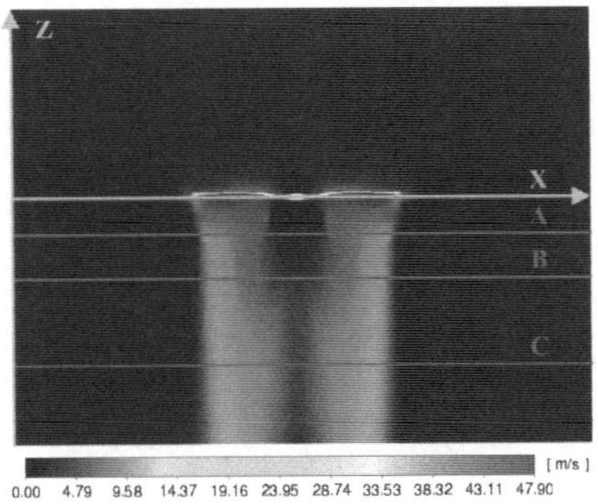

Figure 3. Distribution of the flow velocity magnitude in the cross-section for a single (S) 12-inch rotor with marked perpendicular measurement cross-sections (A,B,C).

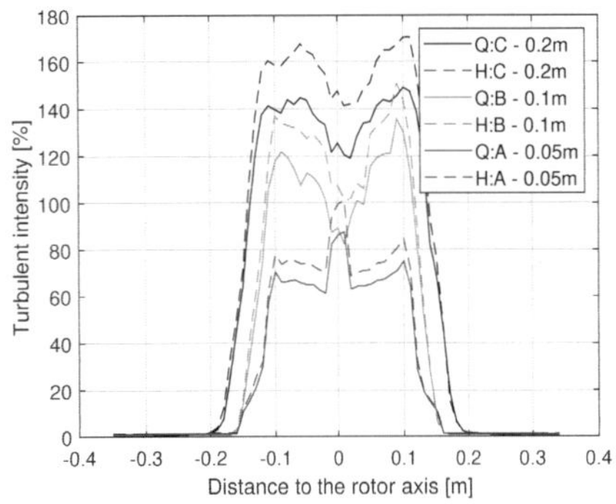

Figure 4. Turbulent intensity in selected cross-sections for the single rotor at rotational speed corresponding to four (Q) and six (H) rotors robots in the hover state.

the flow at the specified point. This value allowed us to compare different configurations and, on its basis, to determine how far the input of the measuring system should be located. Initially, a suitable place was considered, where the intensity level was less than 2 %. Table 1 lists the minimum distances to the rotor axis d_{min} and the flow velocities for those $v(d_{min})$ for which the intensity is lower than the adopted threshold. Values *10 inch Q* and *10 inch H* were determined from 10 inch single propeller models for the operating conditions of: Q - quad-, H - hexacopter. The values *12 inch S* and *12 inch D* denote the 12 inch single (S) and dual (D) propeller configurations for the operating conditions of the octocopter, respectively. From the table 1 it

Table 1. Parameters at analyzed sections for turbulent intensity $< 2\%$.

Section	Parameter	10 inch Q	10 inch H	12 inch S	12 inch D
A	d_{min} $[m]$	0.150	0.150	0.175	0.177
	$v(d_{min})$ $[\frac{m}{s}]$	0.80	0.91	0.80	0.62
B	d_{min} $[m]$	0.153	0.153	0.144	0.174
	$v(d_{min})$ $[\frac{m}{s}]$	0.49	0.55	0.64	0.50
C	d_{min} $[m]$	0.189	0.192	0.175	0.227
	$v(d_{min})$ $[\frac{m}{s}]$	0.27	0.32	0.36	0.32

can be seen that the minimum distance d_{min} is increasing for a configuration with more rotors. It should be noted that this distance is always greater than the radii of the propellers, which are 0.127 m and 0.15 m, respectively. Similarly, the flow speed increases, but for the 2 % threshold adopted it never exceed 1 $\frac{m}{s}$ in all cases. Regarding the distances d_{min}, the differences between sections A and B are small for each configuration. This is due to the fact that for cross section A (which roughly corresponds to the position of the robot arm), the intensity increases rapidly to the level of 10 %, which can be seen in Figure 4. When comparing the configurations of the octocopter robot, it should also be taken into account that the distances for a dual configuration are defined below the bottom rotor, which are 0.11 m, 0.16 m and 0.26 m, respectively, under the axis of the robot arm. In Figures 5 and 6 the fluid flow velocity streamlines are shown for four and six rotor configurations, respectively.

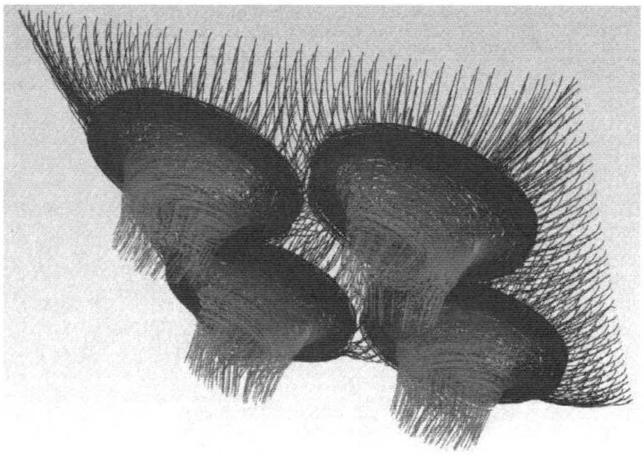

Figure 5. Fluid flow velocity streamlines for four rotors configuration CFD model.

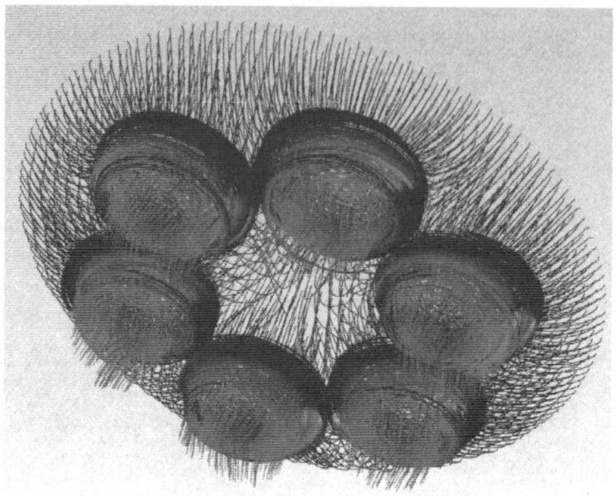

Figure 6. Fluid flow velocity streamlines for six rotors configuration CFD model.

4. Prototype design and field experiments

A six-rotor robot was selected for field work because it had sufficient lifting capacity to work with the sensor mounted on the arm while generating less disturbance than an eight-rotor robot. The robot is equipped with six sets of: BLDC motors, a 30 A ESCs and a 10 inch propellers.

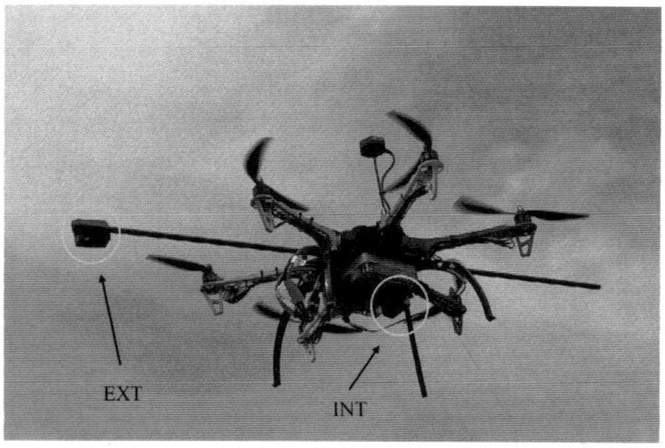

Figure 7. The hexacopter robot in the air.

The maximum lifting capacity of this robot is equal to about 5.16 kg, and it has been equipped with batteries that, according to estimates, should provide about 25 minutes of flight time. The robot was equipped with a symmetric external arm. Its length in relation to the robot's COG was 0.5 m. The first PM sensor was placed on the arm in front of the robot and the second PM sensor (both PMS5003 type) was installed under the robot. Such a solution was chosen because the sensor mass was not large and the direct assembly on the arm did not affect its measurement properties. The Plantower PMS5003 particulate matter sensor has the following (selected) measurement properties: range of measurement 0.3-1.0, 1.0-2.5, 2.5-10.0 μm, resolution of 1 $\frac{\mu m}{m3}$, single response time < 1 s, and physical size of 50 mm × 38 mm × 21 mm. Figure 7 shows the location of both sensors. The constructed control and measurement system has the ability to collect data from both sensors simultaneously. The following nomenclature will be used from this point on: the internal (INT) sensor is the sensor located directly below the robot and the external (EXT) sensor is the one located on the arm. Therefore, the horizontal position of the internal sensor is the same as that of the robot. It is located 0.1 m vertically under the propeller axis. Tracking the position of the external sensor was easy because the robot received a constant orientation (forward toward the north) while taking measurements. Therefore, the position of this sensor could be considered as a simple offset. The external sensor is extended 0.5 m horizontally between the axes of the adjacent propellers, while vertically it is 0.05 m below the propeller axis.

The most important aspect of these tests was to check if there is a relationship of some sort between the measured particle concentrations. For this purpose, Pearson's correlations between measured concentrations for five factors measured or determined, such as robot position (X, Y), flight time, flight direction, and distance from pollution sources, were examined. In addition to these, there were still three not measured factors, such as the intensity of the pollution source and the speed and direction of the wind. During the tests, the concentration of particles PM_{10} was analyzed because they covered the vicinity of the particles counted by the sensor.

The results from two flights with and without the source of pollution are shown in Figures 8 and 9 are containing the analysis of the PM_{10} concentration measurements on both sensors, visualized as histograms and scatter plots with the Perason correlation coefficient determined. The PM sensors used have a variable sampling time, so it was necessary to resample incoming data to a uniform frequency of 1Hz.

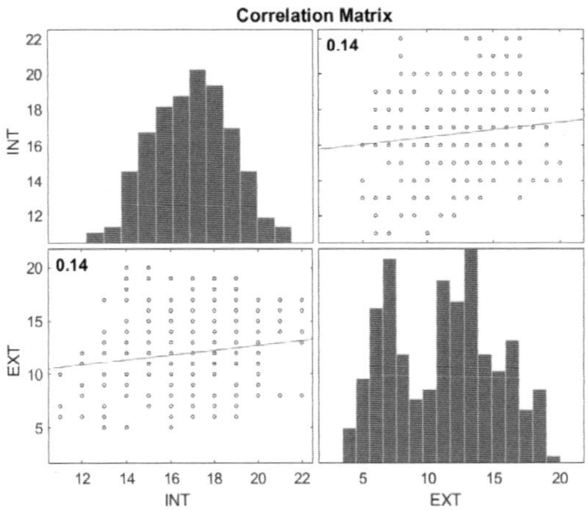

Figure 8. Histogram and correlation of measured PM_{10} without pollution source.

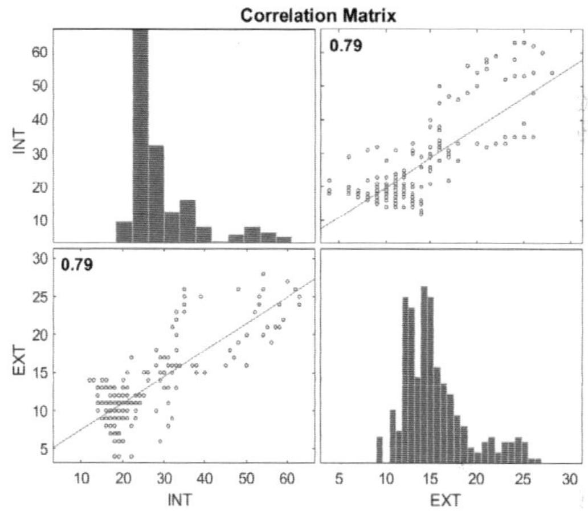

Figure 9. Histogram and correlation of measured PM_{10} with pollution source.

On the basis of the presented data, it can be seen that there is a significant correlation between the measured concentrations in the flight with the source present. The concentration measured on the internal sensor was higher than that measured on the external sensor. For conditions without a source, this correlation does not occur because the concentration of ambient air pollution is fairly constant, so the two averages are compared.

5. Conclusions

In summary, an analysis was performed for the rotor configurations of the robots of the greatest practical importance: quad-, hexa- and octocopter. For those, five cases were modeled and analyzed in this work: two single propellers (10 and 12 inches), counter-rotating pair, four and six rotor configurations. Next, on that basis, it was possible to determine the criterion of a sufficient distance of the PM sensor from the propellers. Then PM measurements were made using the constructed hexacopter robot with the simultaneous use of two sensors: on the extended arm and under the robot. With the data from the field experiments, it is possible to check whether the differences in the measurements can be explained on the basis of the prepared numerical CFD model. For this purpose, the conditions that occur at the sensors' locations were listed with the mean measured concentrations and presented in table 2. The following symbols were used: R - reference measurement for the flight without S1 and S2 - averages for the flight with the pollution source present. These means of S1 and S2 were determined as averages calculated taking samples below and above the maximum value recorded on the external sensor for the reference measurement, respectively. In the case of an internal sensor, the air speed in

Table 2. Summarized results from the computational model and experiment.

	CFD				Experiment		
Sensor location	Velocity [m/s] magnitude	Pressure [Pa] static	dynamic	Turbulent [%] intensity	PM_{10} [$\mu g/m^3$] R	S1	S2
INT	0.8723	-0.4729	0.4733	4.46	16.62	21.24	47.17
EXT	0.1768	-0.0198	0.0198	0.89	11.95	11.05	23.37

its surroundings is about five times greater, which may cause a greater accumulation of particles near it. In summary, it should be stated that CFD can explain the existence of differences in measured concentrations at the two analyzed points, providing information on how to design the air pollutant measurement system. The linear dependence between the concentrations measured in Figure 9 was given by equation $PM_{10,INT} = 1.79PM_{10,EXT} + 1.94$. If such a relationship would be generally correct, as the work carried out shows, then only the sensor located under the robot could be used to calculate the concentrations on the arm without the need for it. However, a general affirmation of the validity of this relationship requires investigation of the impact of changes in pressure and flow speed on the ambient concentration of particles. Plans for such work are planned for the future as well as for comparing the true concentration with the data reported by both sensors. To achieve this goal, it is necessary to perform reference concentration measurements using a robot without rotating propellers. Such measurements can be made by mounting a robot on a long stick in the field or using a dust chamber. Finally, the numerical model also gives the possibility to analyze different configurations of robots. However, the level of disturbance generated by a robot with a larger number of rotors is greater than that with a smaller number of rotors, with the benefit of greater lifting capacity.

Acknowledgments
This research was supported in part by PLGrid Infrastructure.

References
[1] Maciejewska K 2020 *Air Quality, Atmosphere & Health* **13** 659–72
[2] Holder A L *et al.* 2020 *Sensors* **20** ISSN 1424-8220
[3] Cichowicz R and Dobrzański M 2021 *Energies* **14** ISSN 1996-1073
[4] Mayuga G P *et al.* 2018 Airborne particulate matter monitoring using uavs for smart cities and urban areas *TENCON 2018 - 2018 IEEE Region 10 Conference* pp 1398–1402
[5] Cashikar A *et al.* 2019 *Journal of Environmental Engineering* **145** 04019057
[6] Narayana M V *et al.* 2022 *Sensors* **22** ISSN 1424-8220
[7] Rohi G *et al.* 2020 *Heliyon* **6** e03252 ISSN 2405-8440
[8] Koziar Y *et al.* 2019 Quadrotor design for outdoor air quality monitoring *2019 IEEE 39th International Conference on Electronics and Nanotechnology (ELNANO)* pp 736–739
[9] Hedworth H A *et al.* 2021 *Journal of Aerosol Science* **152** 105702 ISSN 0021-8502
[10] Zhang B *et al.* 2018 *Meteorology and Atmospheric Physics* **130** 383–92
[11] Li H *et al.* 2020 *Atmosphere* **11** ISSN 2073-4433
[12] Ansoft Inc. 2020 *Ansys Fluent Theory Guide. Release 2020 R1*
[13] Batchelor G K 1967 *Cambridge Univ. Press. Cambridge, England*
[14] Romik D and Czajka I 2022 *Energies* **15** ISSN 1996-1073
[15] Franzke R *et al.* 2019 *Energies* **12** ISSN 1996-1073

Journal of Physics: Conference Series **2367** (2022) 012011 doi:10.1088/1742-6596/2367/1/012011

Numerical analysis of the influence of wall roughness on the turbulent boundary layer separation

P Kaminski[1] and A Tyliszczak[1]

[1] Czestochowa University of Technology, Faculty of Computers Science and Mechanical Engineering, Al. Armii Krajowej 21, 42-201 Czestochowa, Poland

E-mail: `piotr.kaminski@pcz.pl`, `artur.tyliszczak@pcz.pl`

Abstract. Passive control of turbulent boundary layer (TBL) separation in a channel with adverse pressure gradient (APG) condition towards the outlet is investigated using Reynolds-averaged Navier-Stokes (RANS) methods and Large Eddy simulations (LES) with the help of commercial ANSYS software. Isothermal and incompressible flow in the near-wall area of the flat plate as well as various configurations of wavy wall is studied. It is shown that the amplitude and period of the waviness significantly influence the near wall region and point of separation.

1. Introduction

The boundary layer separation is present in almost all situations in which there is a flow around objects. This phenomena is usually caused by the adverse pressure gradient ($\frac{dp}{dx} > 0$) and occurs when the wall shear stresses are equal to zero ($\tau_w = 0$). The turbulent boundary layer separation can often cause significant energy losses due to the increase of aerodynamic drag and the loss of lift force, an example of which can be the stall effect. The work [1] has shown that the bigger the magnitude of the adverse pressure gradient, the more significant reduction of the wall shear stresses occurs. This result is an earlier and more pronounced separation. The commonness of this phenomenon makes the research and development of passive flow control methods an extremely important task. The added benefit of those is the fact that in comparison to the active flow control methods the passive approach does not need an external energy source. This makes them favorable in terms of the energy efficiency, which in modern days is a crucial topic. Among the passive flow control techniques there are, for example: vortex generators [2], dimples, grooves, baseball ball stitches [3] and the wavy wall which is a very recent discovery [4, 5] and which is the subject of this work. Recent literature has shown that by applying specific geometry modifications one can achieve the increase of wall shear stresses, which in result move the point of separation downstream. In the work [6] it has been shown that using waviness of a certain amplitude and length the 13% increase in the wall shear stresses was obtained. Also, the increase of near-wall flow velocity has been observed. In the work [7], a case of sinusoidal waviness has been examined. It has been shown that the tested wall topology provides an increase in wall shear stresses, which further confirms the results obtained in [6]. While the usage of wall waviness can be beneficial to moving the point of separation downstream, some cases can drastically result in the opposite effect [8]. In the present work we focus on the numerical analysis of the near-wall area of the flow in the vicinity of flat and wavy wall topologies which are considered as wall roughness. A wide range of the waviness amplitudes (A) and number of periods (T) and their influence on the point of separation has been examined. The research consisted of three stages: RANS simulations for the flat plate, various combinations of wavy wall parameters and the LES simulations.

2. Numerical modelling

The test geometry of this work as well as the boundary conditions are based on the experimental set-up and data from [6]. The original geometry is presented in figure 1.

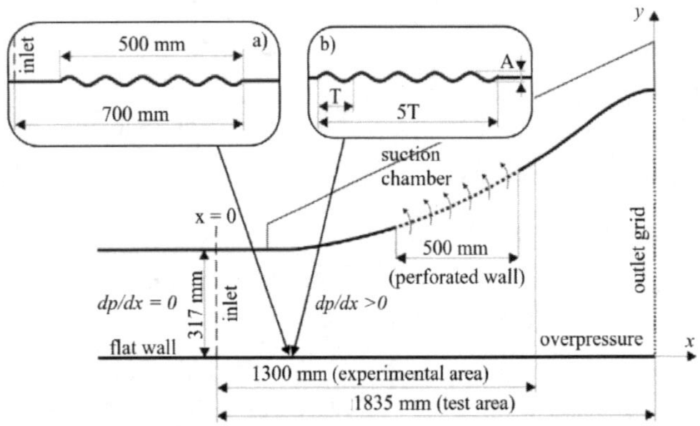

Figure 1: Geometry based on [6], a) wavy wall shape, b) wavy wall parameters: A - amplitude, T- period.

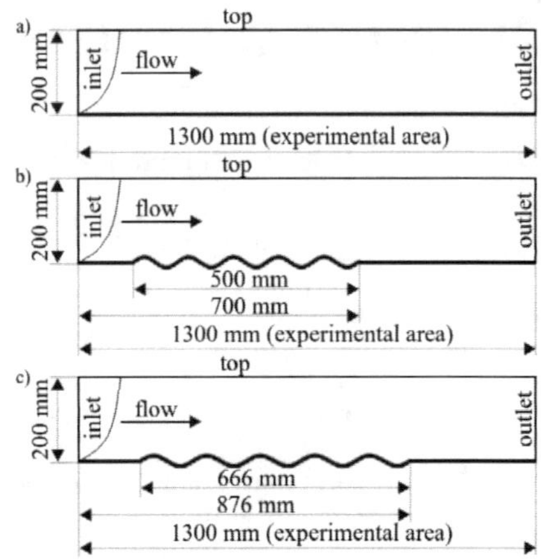

Figure 2: Computational domains: a) flat plate; wavy wall for the momentum thickness based Re_θ: b) 4900, c) 14600.

It is important to notice that the channel is not only a diffuser which induces the APG, but it also is equipped with the suction chamber which makes the gradient even more pronounced. Figure 2 presents the geometries defined in ANSYS which are based on the original experimental stand seen in figure 1. In the work [6] the following profiles have been measured:

- velocity magnitude U and fluctuations u' for the $inlet$ boundary at $x = 0\ mm$
- $\overline{u}$ and $\overline{v}$ (x and y velocity components) profiles for the top boundary at a constant height of $y = 200\ mm$

The obtained data has been implemented in the ANSYS Fluent simulations via the user defined functions (UDF) on the $inlet$ and top boundaries. This allowed to convert a diffuser channel into a rectangular computational domain.

Two types of wavy walls have been examined depending on the Reynolds number. For the $Re_\theta = \frac{U\theta}{\nu} = 4900$ (U - free stream velocity, θ - momentum thickness, ν - kinematic viscosity) the waviness starts at a distance $l_{in} = 200\ mm$ from the inlet (figure 2.b) and is defined by the equation 1:

$$y(x) = A \cdot sin\left(\frac{(x - l_{in}) \cdot 2\pi}{100}\right) \tag{1}$$

where the amplitude A is defined as:

$$A = ((0.0000222 \cdot (x - l_{in})^2) - 0.0049893 \cdot (x - l_{in}) + 9.8370899) \tag{2}$$

For $Re_\theta = 14600$ the waviness starts at a distance $l_{in} = 210\ mm$ from the inlet (figure 2.c) and is defined by the equation 3:

$$y(x) = A \cdot sin((x - l_{in}) \cdot 0.0075075 \cdot 2\pi) \tag{3}$$

where the amplitude A is defined as:

$$A = (0.00000366x^2 + 0.00061432x + 3.3515721) \tag{4}$$

2.1. RANS turbulence model

A wide range of simulations have been conducted in order to find a RANS turbulence model, which allows for proper agreement with the experimental data from [6]. The tests have shown that, for the flat plate, the standard $k - \epsilon$ with a standard wall function was the most suitable. Some models which were tested and deemed inappropriate were, among others: $k - \omega$ (standard, GEKO, BSL, SST), Reynolds Stress Model, Transition-SST, Spalart Allmaras. None of these have allowed for accurate representation of experimental data from [6].

The RANS simulations utilizing the standard k-epsilon model have been ran as 2D after verification with 3D solution which was identical to the 2D results. Conducting two-dimensional analysis allowed to greatly reduce computational time.

Because the k-epsilon model does not allow to introduce the velocity fluctuations profile, it was necessary to convert it into the turbulent kinetic energy k and dissipation rate $epsilon$ profiles. The kinetic energy is defined as the half of the square sum of velocity fluctuation components. Because only the 'x'-component was known, it was assumed that:

$$k = \frac{3}{2} \cdot (u')^2 \tag{5}$$

The energy dissipation rate:

$$\epsilon = \frac{C_\mu^{3/4} \cdot k^{3/2}}{l} \tag{6}$$

where: $C_\mu = 0.09$ - the $k - \epsilon$ model constant, $l = 0.005$ - mixing length - the value was obtained during the tests.

The $k - \epsilon$ model constant C_μ is brought to the power of $3/4$ (equation 6) instead of 1 because Fluent uses length scale definition based on the mixing length instead of the turbulence length. The standard wall function requires a mesh that is of $y^+ \geq 30$, which results in the height of first near wall element being $\Delta y = 2.6mm$. The meshes used for the simulations are presented in the figure 3.

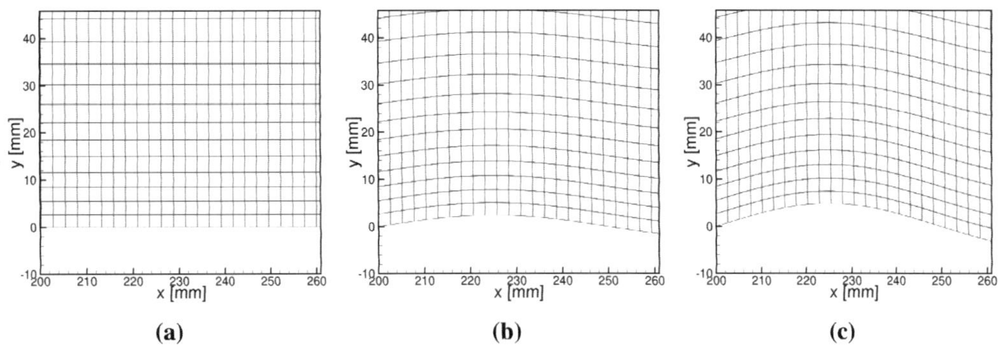

Figure 3: Computational meshes for the RANS: a) flat channel; wavy wall amplitudes: b) 0.25A, c) 0.5A.

During tests, for each turbulence model a number of meshes were analyzed as well as finer and coarse cases. This allowed to verify whether any improvement can be obtained. For the standard $k - \epsilon$ model with the standard wall function a higher density mesh provided significant decrease in the solution accuracy. Therefore, the mesh presented in figure 3 was used for all of the RANS simulation cases.

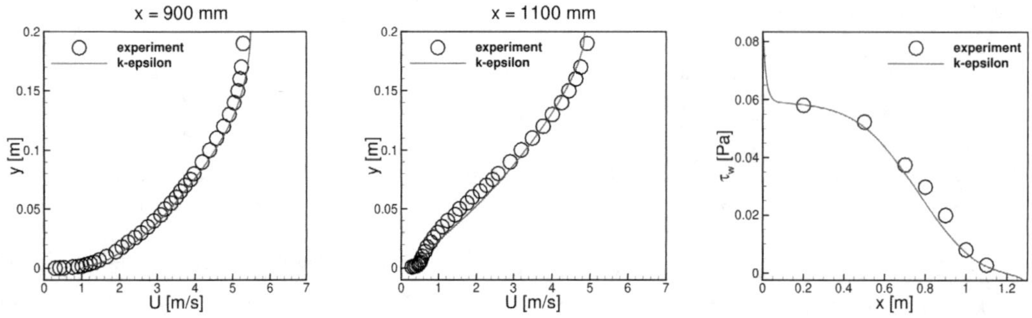

(a) (b) (c)

Figure 4: RANS solution comparison with the experimental data from [6] for the Re$_\theta$=4900 flat channel case: a) x=900 mm, b) x=1100 mm, c) wall shear stresses.

The figure 4 presents the comparison of the results obtained with the standard $k - \epsilon$ model on the mesh seen in figure 3.a with the experimental data for the flat plate and the momentum thickness based Reynolds number 4900. It can be observed that the model provides decent accuracy both in predicting the velocity magnitude as well as the wall shear stresses.

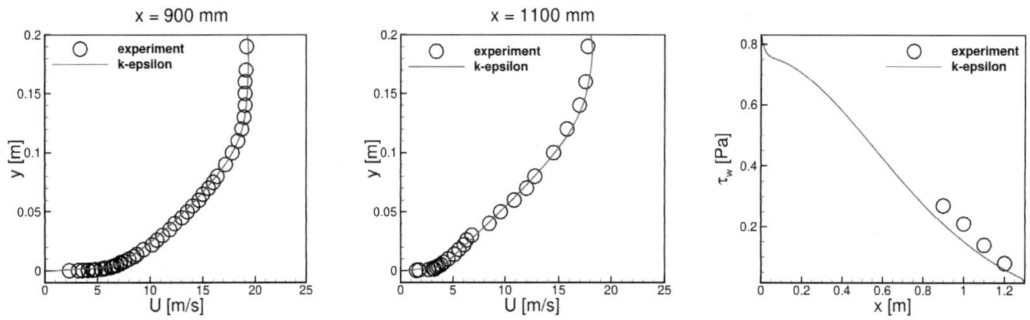

(a) (b) (c)

Figure 5: RANS solution comparison with the experimental data from [6] for the Re$_\theta$=14600 flat channel case: a) x=900 mm, b) x=1100 mm, c) wall shear stresses.

The results for the momentum thickness based Reynolds number 14600 are shown in figure 5. It can be seen that the model also provides a very good accuracy in predicting the velocity magnitude. There is a small disagreement when comparing the wall shear stresses with the experimental data, which could indicate that the standard $k - \epsilon$ model does not allow for a proper near-wall velocity gradient resolution. Despite this fact, we have chosen this model for the further analysis. It is due to providing the most accurate results among other RANS in the present study.

2.2. LES

The LES simulations have been carried out in Fluent using WALE subgrid model and utilized a 2nd order implicit time integration method, a bounded central differencing discretization of the momentum equations, a second order discretization for the pressure gradient. The pressure-velocity coupling was achieved applying the SIMPLE algorithm.

The x and y plane of 3D domains for LES computations are presented in figure 2. The domain was homogenous in the perpendicular (z) direction and its depth was $z = 240\ mm$ with periodic walls boundary condition.

The dimensions of the first near-wall element of the mesh were: $\Delta x = 1.25\ mm$, $\Delta y = 0.25\ mm$, $\Delta z = 1.25\ mm$. The mesh was uniform in the x and z direction.

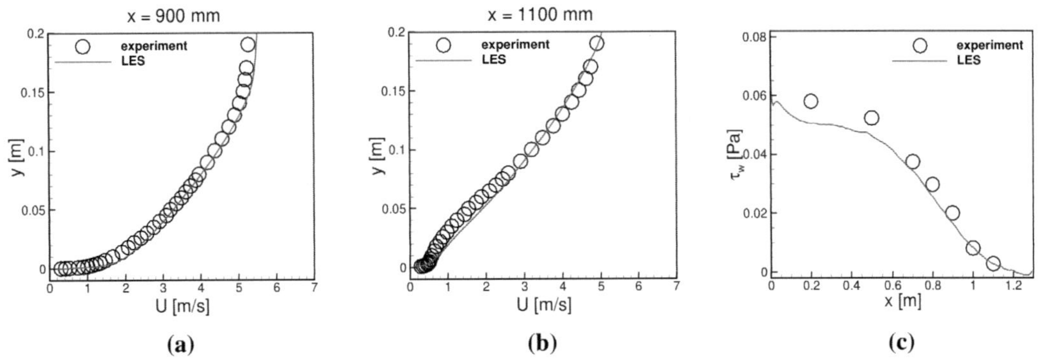

Figure 6: LES solution comparison with the experimental data from [6] for the Re$_\theta$=4900 flat channel case: a) x=900 mm, b) x=1100 mm, c) wall shear stresses.

Figure 6 presents the comparison of the LES simulation results with the experimental data [6]. The presented data is a time-averaged solution collected from 10 cross-sectional slices of the domain after 32 flow-through times of data sampling. The results from the slices have then been spatially averaged to obtain improved convergence. While the LES, in this case, does not provide as accurate wall shear stress prediction when compared to the previous RANS results, it provides much better representation of the real effect induced by the presence of the waviness. Further studies shown that RANS $k-\epsilon$ model is not applicable to the optimization of the wavy wall. Therefore, the LES was chosen as a default tool. Figure 7 presents a sample of instantaneous velocity contours for the Re$_\theta$=4900, flat plate case obtained from the LES.

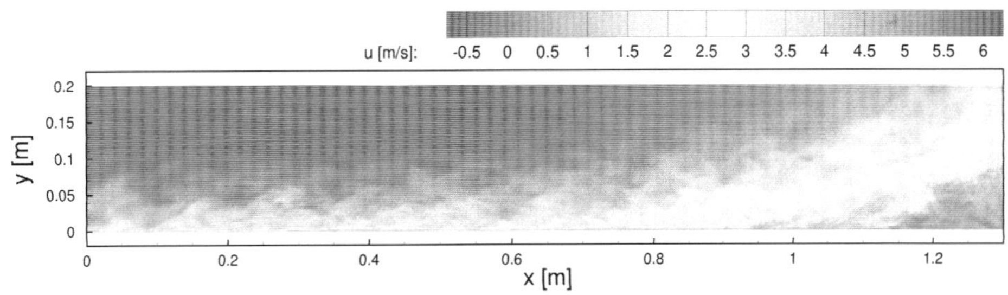

Figure 7: LES - instantaneous streamwise velocity contours for the Re$_\theta$=4900.

3. Results

Having the computational model, which provides satisfactory agreement with the experimental data for the flat channel, the analysis for the wavy wall was conducted. The table 1 presents the simulation cases. For each of the Reynolds numbers 8 different wavy wall amplitudes and 3 numbers of periods were analyzed. This gives 24 cases for each inlet flow velocity. For the LES, only five cases for the Re$_\theta$=4900 have been considered in this work. Future studies will extend the number of cases and wavy wall parameters.

Table 1: Wavy wall parameters.

Model	Re_θ	Wavy wall amplitude scale	Number of wavy wall periods
RANS	4900	0A, 0.125A, 0.25A, ..., 1.0A	4T, 5T, 6T
	14600	0A, 0.25A, 0.5A, ..., 2.0A	4T, 5T, 6T
LES	4900	0A, 0.125A, 0.25A, ..., 0.5A	5T

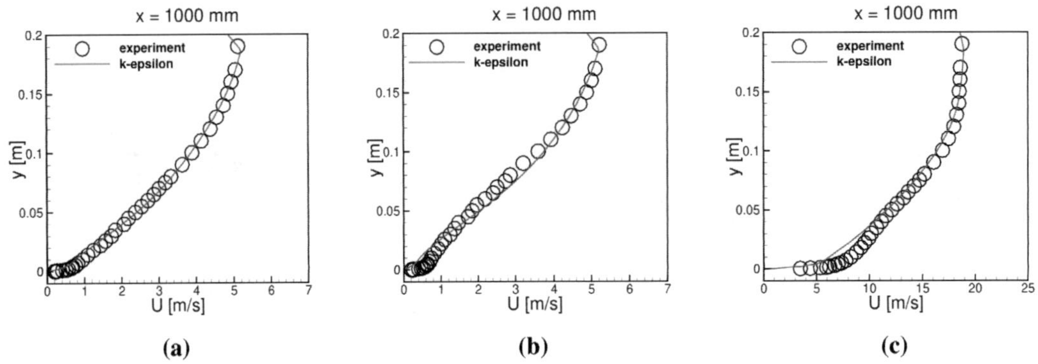

Figure 8: RANS solution comparison with the experimental data from [6] for Re_θ=4900: a) amplitude 0.25, b) amplitude 0.5A; Re_θ=14600: c) amplitude 1.0A.

Figure 8 presents the comparison of the simulation results for the momentum thickness based Reynolds numbers 4900 and 14600 with the experimental data. It can be observed that the standard $k - \epsilon$ model provides decent agreement with the experimental data. For Re_θ=14600, the simulation data sligthly diverges for $y < 0.03\ m$ normal to the wall.

3.1. RANS solution

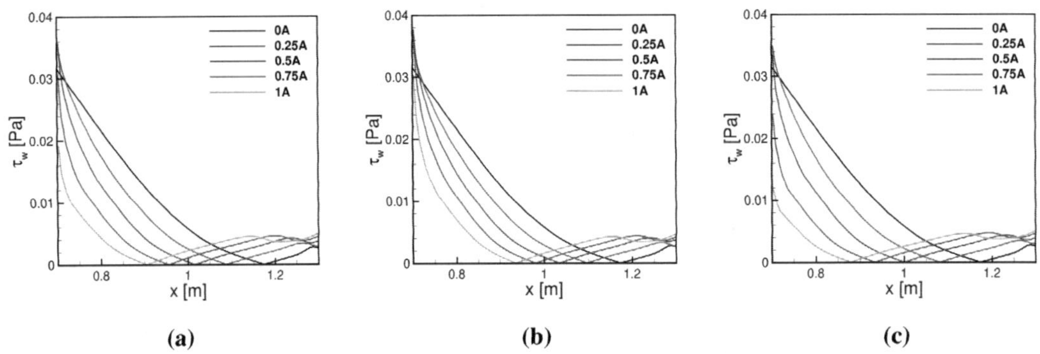

Figure 9: RANS solution - wall shear stresses distribution for Re_θ=4900: a) 5T, b) 4T, c) 6T.

Figure 9 presents the influence of wavy wall amplitude A and number of periods T on the wall shear stresses distribution. It can be observed that increasing the amplitude and number of periods significantly lowers the wall shear stresses.

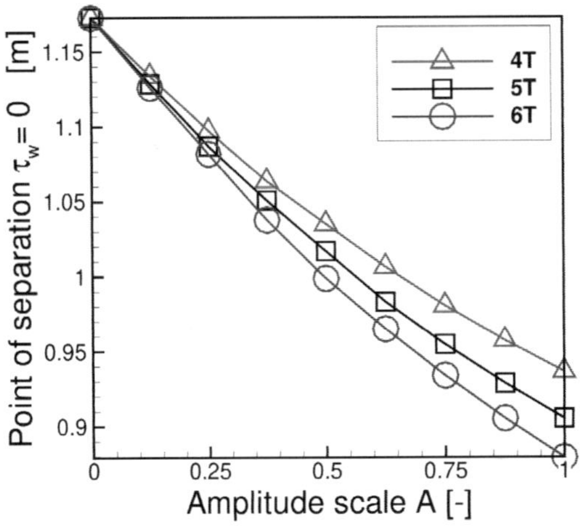

Figure 10: Point of separation as a function of amplitude for Re_θ=4900 obtained with RANS.

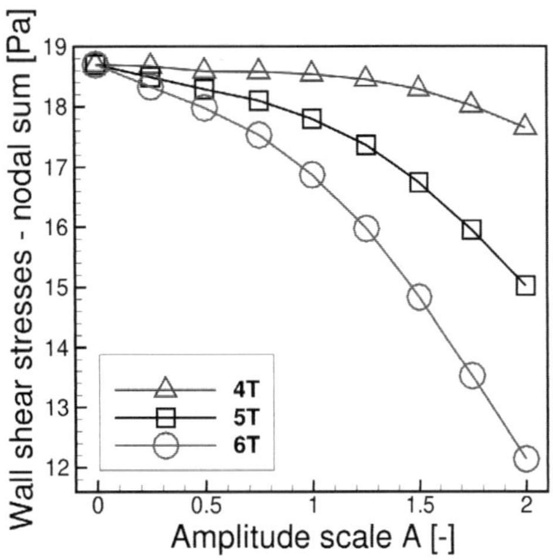

Figure 11: Nodal sum of the wall shear stresses for Re_θ=14600 obtained with RANS.

Figure 10 presents the point of separation as a function of the amplitude for various numbers of periods. It can be observed that, for a given number of periods T, the point of separation moves upstream with increasing the amplitude A. The bigger the number of periods, the more pronounced this negative effect is. The obtained solution indicates that the RANS model might not be the proper tool to use for optimisation of the wall roughness as it is not able to simulate the effect shifting the separation point downstream, as it has been observed in [6]. While for some amplitudes the wall shear stresses correctly decrease, in a range between 0A and 0.25A the separation should at least stay constant. Further tests for Re_θ=14600 have confirmed that RANS model should probably not be used for such studies.

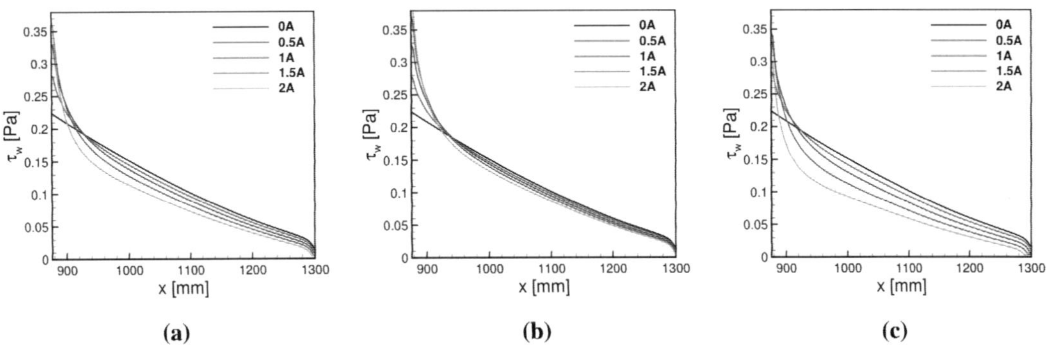

Figure 12: RANS solution - wall shear stresses distribution for Re_θ=14600: a) 5T, b) 4T, c) 6T.

Figure 12 presents the influence of the wavy wall amplitude A and number of periods T on the wall shear stresses for the Re_θ=14600. While for the Re_θ=4900 case the point of separation (figure 10) was shown, in this case no separation was observed due to the insufficient length of the channel. Therefore, a sum of the wall shear stresses from all mesh nodes (nodal sum) of the wall behind waviness was presented in the figure 11. It can be observed that with the increase of the amplitude A, for each of the period numbers, the wall shear stresses decrease. Despite the fact that some increase can be seen at x=900 mm, the total shear stresses on the wall behind the waviness are low. This effect observed is similar to the case with Re_θ=4900.

3.2. LES solution

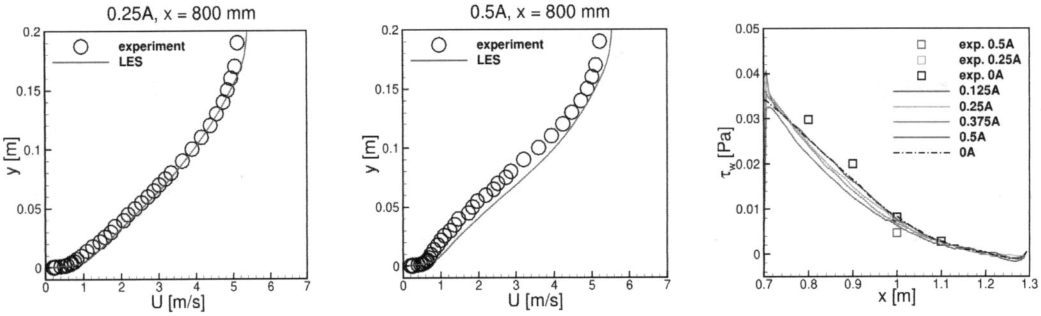

(a) (b) (c)

Figure 13: LES solution for the wavy wall and Re_θ=4900: a) 0.25A, b) 0.5A, c) wall shear stresses.

Figure 13 presents the LES results obtained for the wavy wall cases. It can be observed that for the amplitude 0.25A the magnitude velocity is well predicted. For the 0.5A case, the obtained velocity is larger, this is probably due to not increasing the suction velocity on the *top* boundary condition. In reality, the presence of the wavy wall changes the velocity field even as high as y=0.2m. In order to keep the boundary conditions constant for all cases, the *top* boundary was also kept constant for all of the wavy wall simulations.

4. Conclusions

The two-dimensional solution obtained with the standard $k - \epsilon$ model show that with the increase of the A and T, the point of separation is moved upstream, which does not agree with the results obtained in [6]. The $k - \epsilon$ RANS model, yet providing the best solutions compared to all other tested RANS models, was also unable to provide appropriate resolution of the near-wall velocity magnitude profiles for the wavy wall (figure 8.c). Due to that the LES were performed. The preliminary tests have shown that the Large Eddy Simulations are much better at predicting the flow changes induced by the presence of the wavy wall. The future research will be aiming at determination of the A and T parameters and their mutual relation allowing a smooth control of the separation point for various Reynolds numbers and different wavy wall geometries.

Acknowledgments

This work was supported by the National Science Center in Poland (Grant 2020/39/B/ST8/01449). The LES simulations were carried out using the PL-Grid infrastructure.

References

[1] Wenzel C, Gibis T, Kloker M and Rist U 2019 *Self-similar compressible turbulent boundary layers with pressure gradients. Part 1. Direct numerical simulation and assessment of Morkovin's hypothesis* J. Fluid Mech, **880**, pp 239-283.

[2] Kermode A.C 1972 *Mechanics of flight - 8th edition* (London: Pitman Publishing), pp 350.

[3] Lorenz RD 2006 *Spinning Flight* (New York: Springer), pp 33.

[4] Mathis R, Hutchins N and Marusic I 2009 *Large-scale amplitude modulation of the small scale structure of turbulent boundary layers* J. Fluid Mech, **628**, pp 311-337.

[5] Baars W, Hutchins N and Marusic I 2017 *Reynolds number trend of hierarchies and scale interactions in turbulent boundary layers* Phil. Trans. R. Soc. A, **375**.

[6] Dróżdż A, Niegodajew P, Romańczyk M, Sokolenko V and Elsner W 2021 *Effective use of the streamwise waviness in the control of turbulent separation* EExp. Therm Fluid Sci., **121**.

[7] Koyama S, Takashima K and Hagiwara Y 2007 *Turbulence modification in flow around a periodically deforming film* J. Turbul., **8**.

[8] Patel VC, Chon JT and Yoon JY 1991 *Turbulent Flow in a Channel with a Wavy Wall* J. Fluids Eng, **113**, pp 579-586.

Numerical analysis of heat and mass transfer through beds of spherical and non-spherical elements

Ewa Szymanek [1], **Artur Tyliszczak** [1] **and Maciej Marek** [1]
[1] Czestochowa University of Technology, The Faculty of Mechanical Engineering and Computer Science, al. Armii Krajowej 21, 42-201 Czestochowa, Poland
E-mail: ewa.szymanek@pcz.pl

Abstract. Many issues related to mass and heat transfer through beds of granular materials are still not fully understood. In this work, non-isothermal turbulent flow is analysed within granular layers of spherical and non-spherical elements. We apply a volume penalization (VP) approach formulated in the framework of an immersed boundary technique (IB) on Cartesian computational meshes. It allows modelling flows around solid objects with almost arbitrarily complex shapes and in any form of contact. The validation of the solution accuracy is performed against ANSYS Fluent simulations using body-fitted meshes and experimental literature data. It shows the capability of the IB-VP approach for the simulations of flows in complex geometries. The main research focuses on the comparison of the influence of various types of particles and their temperature on vorticity, turbulence level and pressure drop inside and behind the granular bed. In particular, we analyse how the shape of the solid particles affects the efficiency of heat transfer in different flow conditions. The obtained results reveal the occurrence of very complex flow structures (recirculation and stagnation regions) inside beds. Comparison of results also point out preferred configurations of the beds.

1. Introduction

In recent years, there has been a growing interest in the applications of granular materials and granular beds in various industries. They are used for instance in the construction industry (insulators), energy systems (heat storage, heat carriers) or catalytic reactors (intensification of chemical reactions). Depending on the application of granular layers, their building material may be one type of material (e.g. carbon in the filter column, glass spheres in absorption columns) or they may consist of various combinations of them (e.g. sand-gravel-carbon in the filter beds, gravel-stones in insulation and road underlay). Various applications of granular layers and methods of their construction necessitate a better understanding of the basics of the phenomena of mass and heat transport inside them. The structures of granular layers are complicated, and it is practically impossible to determine the configuration of material arrangement during their construction in real conditions. Regarding a fluid flow through the granular beds, many papers focus on flows around simple shape single elements, e.g., spherical [1]. However, more and more authors analyse the use of more complex elements such as Raschig rings, tetrahedrons, cylinders [2]. Elements of various shapes (cubes, ellipsoids, cylinders, hollow cylinders) with different orientation in relation to the flow direction have been analyzed by many authors [3, 4, 5]. The description of the heat flow using the discrete element method is particularly complicated due to the fact that the microstructure of the bed, the description of which is quite difficult even in the simplest case, may change due to the influence of external forces or the flow. Most of the verification tests presented in the past have been limited to simple shapes of granular elements. On the other hand, more complex shapes of granules can, among others, increase the reactor capacity. It was shown in [6] that the pressure drop in the bed of elements in the form of metal foam is lower compared to the bed of Raschig rings. The metal foam granules are highly porous, allowing a significant internal flow followed by a low pressure drop. CFD simulations can provide detailed insight into heterogeneous bed structure and fluid flow behavior. Whenever possible, numbers of tetrahedral cells are used that allow for the best reproduction of the original surfaces of granular elements and walls. By analyzing the distribution of granules in the beds, it can be concluded that numerical simulations of the flow in granular layers are carried out for configurations with both deterministic and chaotic, random arrangement of elements. An example can be the work [7], where pressure drops are investigated in a fixed bed of spheres in various configurations. In paper [8], the authors applied the FV method on unstructured meshes and analyzed the flow in a catalytic reactor with different types of granular elements arrangement. The authors considered layers with spherical elements packed in an orderly (one after another), shifted and random manner. Randomly packed layers were analyzed by Yang [9] and Bale [10] using Direct Numerical Simulation (DNS) and the FV method on solid tetrahedral meshes fitted to solid elements and by Wood [11] using DNS in conjunction with the immersed boundary (IB) method. Granular material is a complex structure in which a mesh generation process can be very arduous. Therefore, a good alternative is an IB method. The computational meshes have a simple Cartesian structure, whereas the impact of solid objects (granular material) on the fluid is mimicked by source terms added to the energy equation and the Navier-Stokes equations. In this paper, we focus on the analysis of the flow and heat transfers in beds filled with granular layers composed of objects of various shapes, i.e., spheres, cylinders and Raschig rings (elements are commonly used in industry [12]. We use the large eddy simulation method (LES) in combination with the immersed boundary-volume penalization (IB-VP) technique [13, 14, 15]. We analyse the flow and temperature distributions around separate elements and also check how the structure of the layer impacts the flow field, e.g. turbulence or vorticity intensity inside and behind the granular layers.

2. Mathematical model

We consider variable temperature, variable density, low Mach number flows described by the continuity equation, the Navier-Stokes equations and the energy equation defined as:

$$\partial_t \bar{\rho} + \nabla \cdot (\rho \tilde{\mathbf{u}}) = 0 \tag{1}$$

$$\bar{\rho}\left(\partial_t \tilde{u} + (\tilde{u} \cdot \nabla)\tilde{u}\right) + \nabla \bar{p} I = \nabla \cdot \tau_{ef} + f^{IB} \tag{2}$$

$$\bar{\rho} C_p \left(\partial_t \widetilde{T} + (u \cdot \nabla)\widetilde{T}\right) = \nabla \cdot \left(\kappa_{ef} \nabla \widetilde{T}\right) + \mathrm{f}_T^{IB} \tag{3}$$

where bar and tilde operators denote the LES and Favre filtered quantities. The symbol ρ stands for the density, p - hydrodynamic pressure, T - temperature, $\mathbf{u}$ - velocity vector, C_p - specific heat capacity and κ_{ef} - effective heat conductivity $(\kappa + \kappa_T)$, where $\kappa = \mu / C_p \mathrm{Pr}$ and $\kappa_T = \mu_T / C_p \mathrm{Pr}_T$, where the molecular viscosity (μ) is computed from the Sutherland law, the sub-grid viscosity is calculated using the Vreman model [17], the Prandtl number and its turbulent counterpart are equal to $\mathrm{Pr} = 0.71$ and $\mathrm{Pr}_T = 0.9$. The stress tensor τ_{ef} includes the effective viscosity $\mu + \mu_T$. The set of equations (1)-(3) is complemented with the equation of state $p_0 = \rho R T$, where p_0 denotes the thermodynamic pressure and R - is the specific gas constant. In open flows with inlet/outlet boundaries p_0 is constant in space and time and in this work it is assumed to be 101325Pa.

IB-VP source term. The source terms $\mathbf{f}^{\mathrm{IB}}$ and f_T^{IB} originate from the immersed boundary method. In IB-VP variant the role of these terms is to penalize a difference between the actual and assumed velocity and temperature of the solid body. They act on a fluid in such a way as if there were solid objects immersed in the flow domain and they are defined as:

$$\mathbf{f}^{\mathrm{IB}} = -\frac{\rho}{\eta}\Gamma(\mathbf{x})(\mathbf{u}-\mathbf{u}_s), \quad f_T^{\mathrm{IB}} = -\frac{\rho C_p}{\eta}\Gamma(\mathbf{x})(T-T_s) \tag{4}$$

where $\mathbf{u}_s$ and T_s are the velocity and temperature of the solid body, the parameter $\eta \ll 1$ is the so-called penalization parameter with dimension of time unit and Γ - the phase indicator defined as:

$$\Gamma(\mathbf{x}) = \begin{cases} 0, & \text{for } \mathbf{x} \in \Omega_f \\ 1, & \text{for } \mathbf{x} \in \Omega_s \end{cases} \tag{5}$$

where Ω_f and Ω_s are the regions of fluid and solid part of the computational domain. For $\Gamma(\mathbf{x}) = 1$ discrete forms of equation (2) and (3) with the time step Δt reduce to $\mathbf{u}^* \approx \Delta t \mathbf{u}_s/(\eta + \Delta t)$ and $T^* \approx \Delta t T_s/(\eta + \Delta t^n)$ which for $\eta \ll \Delta t$ leads to $\mathbf{u}^* \approx \mathbf{u}_s$ and $T^* \approx T_s$. The consequence of the simplicity of implementation of the IB-VP method is a lowering of the solution accuracy. Similarly, as in the classical IB method with a stepwise approach, i.e., without the interpolation [16] near solid boundaries, the order of the IB-VP method is at most equal to one [14, 15, 18].

The solution algorithm for equations (1-3) is formulated in the framework of a projection method for pressure-velocity coupling. The time integration is based on a predictor-corrector approach (Adams-Bashforth/Adams Moulton) and the spatial discretisation is performed using 6th/5th order compact difference and WENO (Weighted Essentially Non-Oscillatory) schemes on half-staggered meshes [19, 20].

3. Assessment of the IB-VP accuracy

This part of article shows how the theoretical limitation of the solution accuracy of Immersed Boundary - Volume Penalization method translates on the results obtained for experiment [21]. The considered computational domain is a rectangular channel in which the layers of spheres are placed in the corners of the channel (see figure 1) and a fluid flows in its central part. The dimensions of the domain was $1D \times 1D \times 15D$, where D - the diameter of the sphere was 28mm. In figure 1(a) a fragment of the channel with spheres and velocity distributions in several planes and the vorticity iso-surface is presented. Reynolds numbers for the simulations were equal $Re = UD/\nu = 59.78$ (U - inlet velocity, ν - kinematic viscosity) and $Re = 204.74$. Figure 1(c) shows the numerical results for both methods and literature data [21]. The computations were performed on uniform computational meshes consisting of $40 \times 160 \times 40$ (mesh M1), $80 \times 160 \times 80$ (mesh M2) and $80 \times 320 \times 80$ (mesh M3) nodes in the IB-VP simulations and the body-fitted meshes counting of approximately $2.5 \cdot 10^6$ (mesh M4) and $7.0 \cdot 10^6$ (mesh M5) tetrahedral cells in the simulations performed using the ANSYS Fluent code. We verify the solution based on the vertical (streamwise) velocity profiles along the flow direction from the locations in between 4th and 5th sphere (see figure 1b). The numerical results agree with the experimental data well. Moreover, the accuracy is better for the larger Reynolds number. It is important, the results using the Ansys Fluent code and proposed the IB-VP approach practically overlap and are virtually independent of the grid density. So it can be said, that the IB-VP method used allows for obtaining accurate results. Some additional test, for example for laminar flow in a simplified model of granular layer $10 \times 10 \times 10$ spheres also confirmed good characteristics of the applied IB-VP method [22].

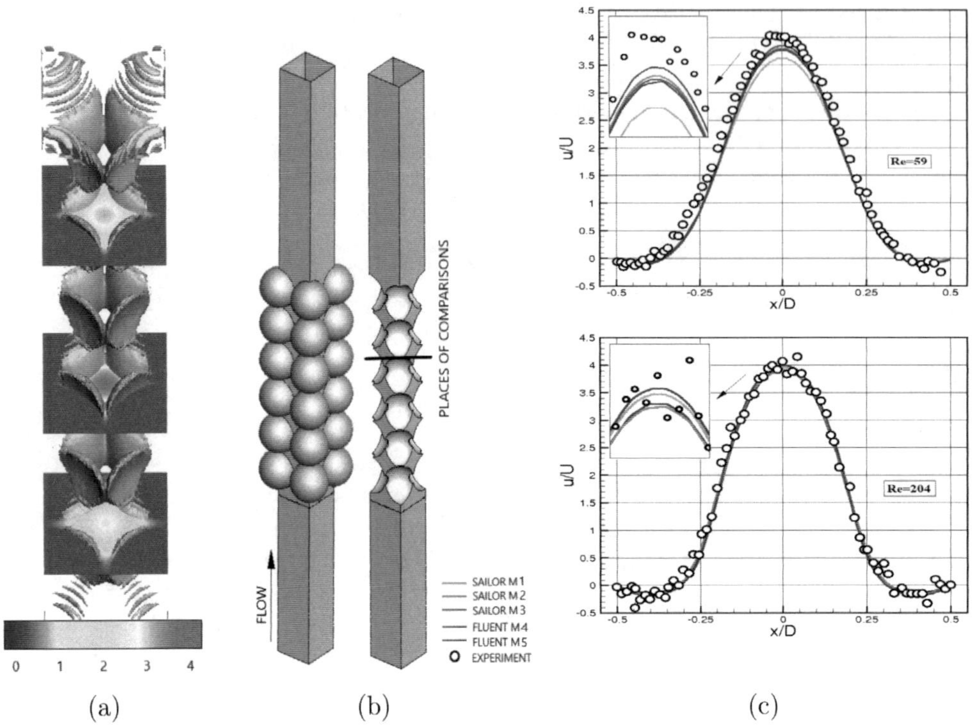

Figure 1. Velocity distribution presented at several heights and an isosurface of vorticity equal to 30 s^{-2} (Re=59.78) (a), the computational domain (b) and the velocity profiles at $y = 0.265$m for Re = 59.78 and Re = 204.74 (c)

4. Flow in granular layers

The structure of the beds analysed in this paper is random and is obtained in a separate simulation of a container filling process. The cylinders, rings or spheres are dropped sequentially into the container. They move due to gravity and reaction forces when they come in contact with the remaining particles or the container walls. After the mechanical equilibrium of the particle is attained it remains at rest for the rest of the process and the following elements do not disturb the established equilibrium (a "frozen bed" assumption). This allows for tracking of one particle at a time and greatly reduces the computational costs [12]. It has been shown in [23, 24] that the generated beds have both realistic porosity profiles and particle orientation distribution. The computational domain and the granular beds are presented in figure 2. The beds analysed in this paper consist of three type of elements: spheres, cylinders and Raschig rings. The reference configuration is the layer of 2000 spheres ($D = 0.008$ m). The elements are packed in a pipe of diameter 0.1m, which is the central part of the computational domain (rectangular box $0.12 \times 0.35 \times 0.12\text{m}^3$), where the grey colour represents the boundary of the pipe. The solid regions between the domain boundaries (blue edges) and the pipe wall were modelled in the same way as the granular beds (applying the IB-VP method). The fluid was injected from the bottom of the domain (with a uniform or slightly perturbed velocity), it moves in between and around elements and leaves the channel.

It should be noted that the process of preparing the mesh in case of complex geometries or objects attached one to another would be tremendously difficult and would require special treatment of all contact points. Otherwise, the grid would be very distorted near the such critical points. There are no such problems when using the IB-VP method. For the simulationsthe used

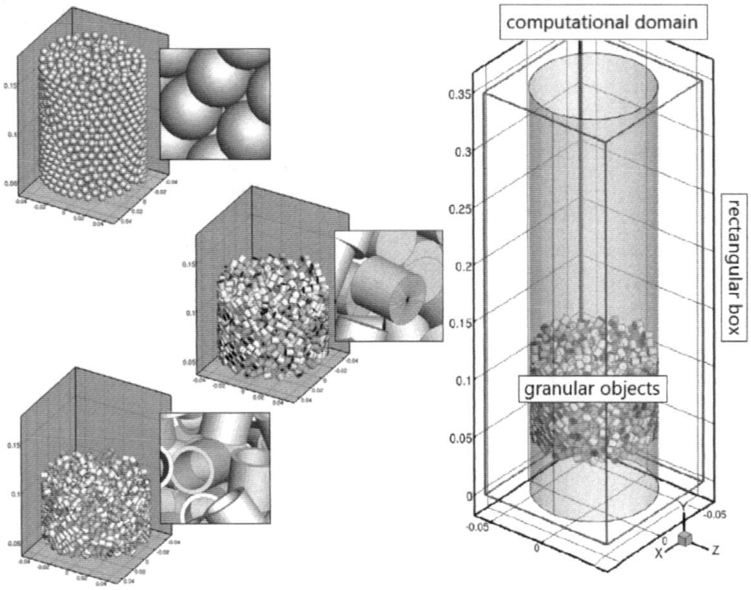

Figure 2. Computational domain and three types of analyzed elements: spheres, cylinders and rings

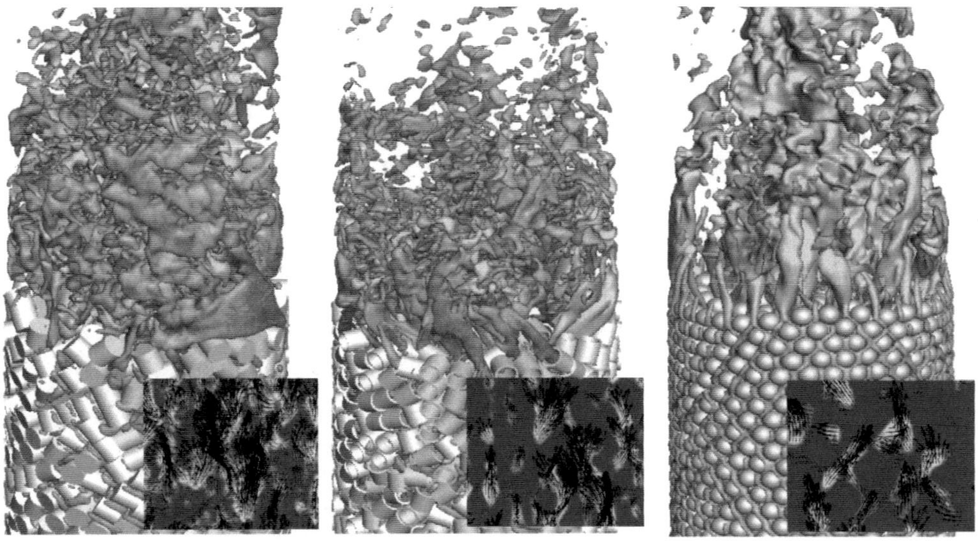

(a) (b) (c)

Figure 3. View of the granular layers, isosurfaces of velocity: orange -3 m/s, green 5 m/s. Velocity and vorticity distribution along the axial direction

mesh was regular and uniform and consisted of $160 \times 320 \times 160$ nodes. The granular objects can be in any form of contact and can even be deformed. The granular layers extend from $y = 0.05$ m towards the upper part of the pipe up to the distance $y = 0.12 - 0.18$ m, depending on the type of element. We consider cylinders and rings with equal heights and external diameters. It was assumed that their total area is the same and equal to the area of the spheres. The Reynolds number was equal to 1000 and was defined based on the inflow velocity (1.959 m/s) and the sphere diameter. The inlet temperature was equal to 300 K. We consider two cases where temperature of the elements in the layers and the temperature of the pipe wall were equal 350 K and 673 K appropriately. We kept them constant assuming that their heat capacity is much larger compared to the heat capacity of the flowing medium. The flow inside the layers is characterised by very complex structures and we analyse how it changes depending on the type of the element. On figure 3 showing iso-surfaces of the positive and negative velocity along the y-direction over the layers. We can observe that in case of bed with the Raschig rings the flow moves around and also inside the objects. It can be also seen that the regions of positive (green colour) and negative (orange colour) velocity interpenetrate each other for all elements. The orange iso-surfaces visualise recirculation regions, which in case of the layer with the cylinders seem to be larger and stronger. Small embedded subfigures present the velocity vectors in the central slice of the domain.

Figure 4. Velocity distribution in the cross-sections (1-4) for three granular bed configurations (cylinders (a), Raschig rings (b) and spheres (c))

The velocity distributions at four heights for all analyzed beds can be see figure 4. The planes 1-3 cross the granular beds at different heights, while the plane 4 lies above the bed. This figure illustrates the velocity distribution of the flow passing through the bed. One can see the existence of high velocity peaks especially for the rings (figure4(2b)). The fluid flows through empty intergranular spaces and increases velocity while flowing in these narrow gaps.

Figures 5-6 show the profiles of temperature, the magnitude of velocity components and vorticity magnitude along the streamwise direction for two different temperature of elements of beds: 350 K and 673 K. In figure 5 we can see that the temperature rises quickly and reaches 350 K near from the entrance to the granular beds independently of the type of elements. Taking into consideration to the intensity of heating of the flowing medium, seems to that all layers have similar properties. We observe the same behaviour for the higher temperature of the elements (673 K). Perhaps, significant differences would occur for less packed or colder layers. In contrast to the small differences in temperature profiles, all remaining quantities differ significantly. The vorticity changes in a wavy manner and shows very complex behaviour. It can be seen that its bigger values are obtained for the bed composed of the cylinders.

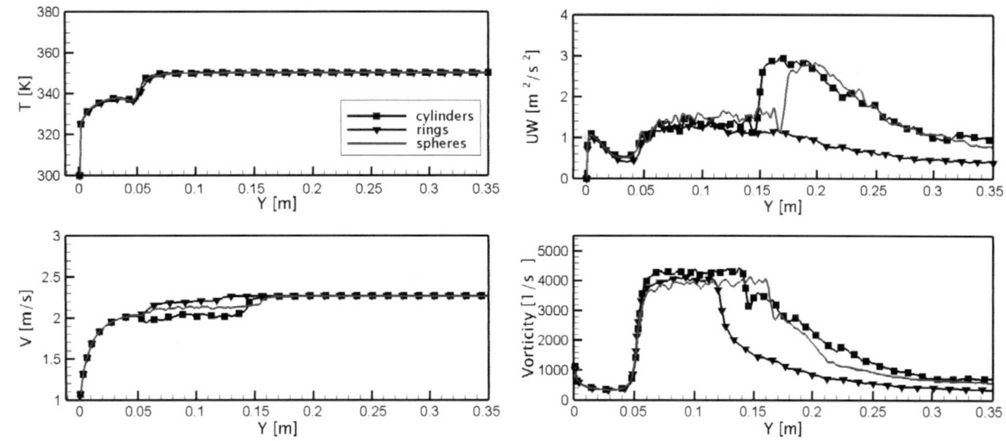

Figure 5. Profiles of mean temperature values, UW velocity components, axial velocity, vorticity modulus for $T = 350$ K and $Re = 1000$.

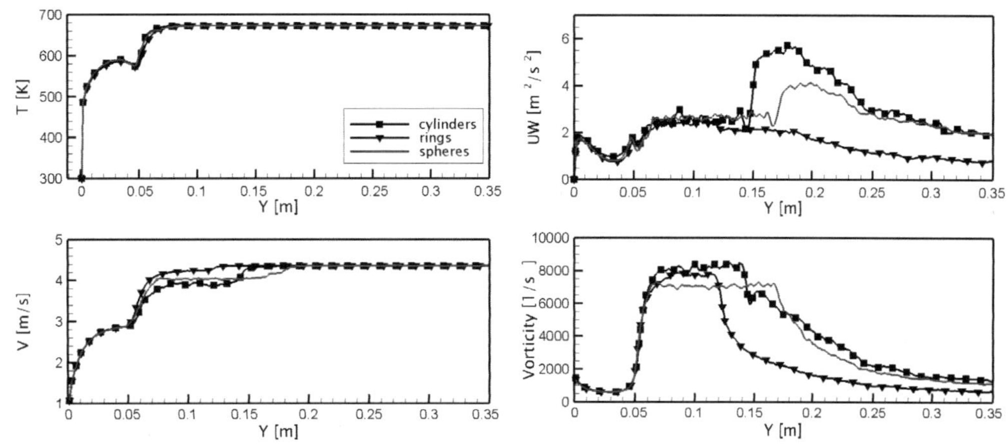

Figure 6. Profiles of mean temperature values, UW velocity components, axial velocity, vorticity modulus for $T = 673$ K and $Re = 1000$

By analyzing the results, we can conclude that the higher vorticity values denote regions of increased mixing. Hence, the heat transfer occurring in such beds configurations is the most effective. It has been observed, that increasing the temperature causes a decrease in the vorticity for the granular beds with spheres compared to the other two cases. The increase of the flow velocity is the consequence of both limited space for flow (porosity$<$1) and density drop due to heating. Similar and higher velocity magnitude values were observed for the rings, while the lowest for cylinders. In this case, densely packed elements block the flow the most, inducing the highest vorticity and definitely the highest values of radial UW velocity components. Significant differences are visible mainly near the outlet from the layer with the maximum value in the case of cylinders. Relatively small values of UW velocity components and vorticity in the case of the bed composed of the Raschig rings are due to the fact that this is the only case in which the fluid can flow through the central parts of these elements. Thus, this bed does not induce large local velocity values.

5. Conclusions

The paper focused on the mass and heat transfer in granular layers. The verification tests have shown that the applied IB-VP method can be used to simulate flows in such configurations. The applied numerical method based on high-order discretisation combined with the immersed boundary method allowed for a deep analysis of the flow and temperature distributions in beds (between the granular material) and above them. It was shown that beds with granular materials characterise very complex structures. Additional, the configuration of the layer significantly affects the effectiveness of the mixing and heat transfer. It could be seen that in the case of the Raschig rings the flow moves around and also inside the objects. The presented iso-surfaces visualised existence of recirculation regions, which in case of the layer with the cylinders in the layers seem to be larger and stronger. This translates to higher values of the vorticity at the end of the computational domain. In the considered cases of flow in granular layers, cylindrical particles introduced the largest values of vorticity and magnitude of velocity components behind the layer, but also the smallest values of velocity in the flow in the layer regardless of the bed temperature. Due to the complexity of the issues considered in the paper and the use of precise computational methods allowing for their in-depth analysis, the problems raised in it require further research and analysis.

Acknowledgments

This work was supported by NAWA ANIMATE Project (PPI/APM/2019/1/00062/U/00001) and National Science Center, Poland (grant no. 2017/27/N/ST8/02318). PL-Grid infrastructure was used to carry out the computations.

References

[1] Wongkham J et al. 2020 *Fusion Engineering and Design* **161** 111953
[2] Dorai F et al. 2015 *Chemical Engineering Science* **129** 180-192
[3] Richter A, Nikrityuk P A 2013 *Powder Technology* **249** 463-474.
[4] Nijemeisland M, Dixon A G 2004 *American Institute of Chemical Engineers Journal* **50(5)** 906-921.
[5] Minhua Z et al. 2018 *Chemical Engineering Research and Design* **132** 149-161.
[6] George G R et al. 2021 *American Institute of Chemical Engineers Journal* 17284.
[7] Ahmadi S, Sefidvash F 2018 *Chemical Engineering Journal* **2** 14.
[8] Calis H P A et al. 2011 *Chemical Engineering Science - Journals* **56(4)** 1713-1720.
[9] Yang J et al. 2016 *Nuclear Engineering and Design - Journals* **300** 85-96.
[10] Bale S et al. 2017 *International Journal of Heat and Mass Transfer* **110** 406-415.
[11] Wood B D et al. 2015 *Advances in Water Resources - Journals* **85** 45-63.
[12] Marek M 2013 *Chem. Process Eng.* **34(3)** 347-359
[13] Tyliszczak A, Szymanek E 2019 *Numer. Heat Transf. Part A Appl.* **76** 737-759
[14] Khadra K, Angot P, Parneix S, Caltagirone J P 2000 *Inter. J. for Num. Methods in Fluids* **34(8)** 651-684
[15] Kadoch B, Kolomenskiy D, Angot P,Schneider K A 2012 *J. of Comp. Physics* **231(12)** 4365-4383
[16] Fadlun E A, Verzicco R, Orlandi P, Mohd-Yusof J 2000 *Journal of Computational Physics* **161(1)** 35-60
[17] Vreman A W 2004 *Physics of Fluids* **16** 3670–3681
[18] Tyliszczak A, Ksiezyk M *International Journal for Numerical Methoda in Fluids* **87(7)** 358-381
[19] Tyliszczak A 2014 *Journal of Computational Physics* **276** 438-467
[20] Tyliszczak A 2016 *Computers and Fluids* **127** 131-145
[21] Suekane T, Yokouchi Y, Hirai S 2003 *AIChE Journal* **49(1)** 10-17
[22] Szymanek E, Tyliszczak A 2018 *Journal of Physics: Conference Series* **1101** 012043
[23] Marek M 2017 *Chemical Engineering Science* **161** 382-393
[24] Niegodajew P, Marek M 2016 *Powder Technology* **297** 193-201

Liquid Phase Identification in Wet Steam Transonic Flows by Means of Measurement Methods

S Shabani[1], M Majkut[1], S Dykas[1], K Smołka[1] and X Cai[2]

[1] Silesian University of Technology, Department of Power Engineering and Turbomachinery, Konarskiego 18, 44-100 Gliwice, Poland

[2] University of Shanghai for Science and Technology, Shanghai, 200093, China

E-mail: Sima.Shabani@polsl.pl

Abstract. The major goal of this study is comparing the results of heterogeneous and homogeneous condensation through a CD nozzle. The experimental technique for liquid phase assessment in wet steam transonic flow is tested under real conditions in an in-house steam tunnel test rig. The flow was modelled numerically using an in-house CFD code for performing the heterogeneous and homogeneous condensation models. For the comparison between CFD results and experimental data, the static pressure distribution throughout the nozzle bottom wall were chosen. One may observe a good agreement of the CFD results with experimental data. The presented approach can be used in the future for searching the reasons for the lack of accuracy between experimental and numerical static pressure distribution in the condensing steam flows.

1. Introduction

The present standards for pollutant emissions reduction and steam power unit flexibility necessitate the constant improvement of power equipment and devices. The wetness mass fraction y (or dryness x) at the turbine outlet is one of the most important quantities in condensing steam turbines. The liquid mass fraction at the turbine outlet is determined using a variety of experimental procedures. Since the early 1960s, these experimental methods have been developed and they still not are ready for industrial applications. However the experimental methods for liquid phase measurement in the wet steam flow are commonly used in the laboratory measurements. There are many techniques, each technique has the pros and cons, and it is beneficial in a certain application. The methods used mainly in research are definitely large in size, which generally require the use of specially designed and adapted research tunnels with wide access to the tested measuring areas for the optics of measuring systems. In recent years, among the tested methods, particularly interesting techniques based on the phenomenon of light extinction are also used, attractive due to the relative ease of implementation and interpretation of data, as well as the possibility of continuous measurement with a fast time response and low optical requirements for access to the measurement area.

The light extinction method for estimating the liquid mass fraction in the wet steam area, is utilized in experimental studies on the steam tunnel test rig at the Silesian University of Technology's Department of Power Engineering and Turbomachinery [1]. Measuring systems based on light extinction principle were tested with the results of similar solutions from several large research centers [2, 3, 4, 5]. However, in the conducted research, there was a discrepancy in static pressure distribution on the nozzle wall for fairly similar sets of measured total parameters ahead of the nozzle, total temperature and total pressure, over several measuring campaigns for the nozzles. We found that the presence of a liquid phase in the incoming steam to the test section could be the cause. As the in-house wet steam probe [6] cannot be used to measure the liquid phase ahead of the nozzle in the steam tunnel's test section, the notion of using CFD modelling to calculate the data of the liquid phase present at the nozzle arises. In LP stages, high-velocity steam becomes supercooled, and the subcooled vapor loses its latent heat, forming liquid droplets with small diameters. The small droplets then develop or vanish depending on the thermodynamic conditions. As a result, the numerical approach for modelling such flow must solve the governing compressible flow equations, as well as the real gas equation of state, and include additional phase transition relations.

In the models presented here, the two phases are considered to be controlled by the same pressure nozzle inlet. The condensation model was employed with the in-house CFD code [7, 8], where the liquid phase content is set at the inlet boundary conditions. The good agreement of the static pressure distribution between the CFD simulation and experiment will return the information about the liquid phase content at nozzle inlet.

2. Physical and numerical model

The following equations link the liquid and vapor properties:

$$p = p_v = p_l \tag{1}$$

$$\alpha = \frac{V_l}{V_m},\ y = \alpha\frac{\rho_l}{\rho_m} \tag{2}$$

Where *p, V,* α*, y,* ρ are pressure, volume of the phases, the volume fraction, the mass fraction and the density, respectively. Subtitles *v, l, m* represents vapor phase, liquid phase and two-phase mixture, respectively. For calculating the mixture properties, the following equations can be used. In these equations, the density and enthalpy of the mixture (ρ_m, h_m) are dependent on the liquid and vapor density and enthalpy, respectively:

$$\begin{aligned}\rho_m &= (1-\alpha)\rho_v + \alpha\rho_l \\ h_m &= (1-y)h_v + yh_l\end{aligned} \tag{3}$$

2.1. Gaseous phase equation of state

Nowadays, the "IAPWS Industrial Formulation 1997" is the most accurate equation of state for steam which is used extensively for industrial application (IAPWS-IF97). Because of the sophisticated form of this equation, it is very difficult to use it directly in CFD codes. For that reason, the following 'local' real gas state equation, was employed [9]:

$$\frac{p}{RT\rho_v} = z(T,\rho_v) = A(T_v) + B(T_v)\rho_v \tag{4}$$

Where *p,* ρ*, T and z* are pressure, density, temperature and compressibility coefficient, respectively. Individual gas constant *R* amounts to 461.5 J/(kg K). The polynomials *A(T), B(T)* are specified as:

$$\begin{aligned}A(T) &= a_0 + a_1T + a_2T^2 \\ B(T) &= b_0 + b_1T + b_2T^2\end{aligned} \tag{5}$$

Where the coefficients a_i, b_i ($i = 0...2$) are temperature-dependent and are derived using an approximation of steam thermodynamic properties based on the IAPWS-IF97.

2.2. Condensation model

The applied homogeneous condensation model consists of the nucleation and droplet growth processes. The nucleation rate is determined based on the equation of the classical nucleation theory [10]:

$$J_{hom} = C(\frac{2\sigma}{\pi})^{1/2} m_v^{-3/2}\frac{\rho_v}{\rho_l}\exp\left(-\beta\frac{4\pi r^{*2}\sigma}{3kT_v}\right) \tag{6}$$

Where *J* is the nucleation rate presenting the number of nuclei with critical radius *r** formed in 1 kg of moist air in one second; *r* is the droplet radius. σ, m_v, β, ρ_l, *k* are the surface tension, the mass of a water molecule, the correction factor, the liquid density at temperature T_v , and Boltzmann constant, respectively. In equation (6) C is called as a non-isothermal correction factor and was suggested by Kantrowitz [11]. This coefficient is determined using the following equation since the isothermal model does not apply to vapor:

$$C = \left[1 + 2\frac{\gamma - 1}{\gamma + 1}\frac{h_v - h_l}{RT_v}\left(\frac{h_v - h_l}{RT_v} - \frac{1}{2}\right)\right]^{-1} \quad (7)$$

Where γ stands for the ratio of specific heat capacities.

The continuous model of droplet growth proposed by Gyarmathy is employed in proposed condensation model[12]:

$$\frac{dr_{hom}}{dt} = \frac{1}{\rho_l}\frac{\lambda_v}{(1 + 3.18Kn)}\frac{r - r^*}{r^2}\frac{T_s - T_v}{h_v - h_l} \quad (8)$$

Where Kn is the Knudsen number, which is the ratio of the mean free path of vapor molecules to the droplet diameter. λ and T_s are the heat transfer coefficient and saturation temperature, respectively.

In the heterogeneous condensation model, the nucleation process is ignored. Foreign solid particles promote nucleation, owing to a lower thermodynamic barrier (compared to the homogeneous nucleation). The equation (8) is also used to describe the growth of droplets on particle impurities, which are assumed to be spherical with the given initial mean radius and concentration in mass unit.

2.3. liquid phase equations:

In this research, a single-fluid model is applied which consists of the mass, momentum and energy equations for a vapor/liquid mixture, as well as two equations for the liquid phase explaining the creation and growth of liquid droplets caused by homogeneous and heterogeneous condensation:

$$\begin{gathered}\frac{\partial(\rho_m y_{hom})}{\partial t} + \frac{\partial(\rho_m u_{mj} y_{hom})}{\partial x_j} = \Gamma_1 + \Gamma_2 \\ \frac{\partial(\rho_m n_{hom})}{\partial t} + \frac{\partial(\rho_m u_{mj} n_{hom})}{\partial x_j} = J \\ \frac{\partial(\rho_m y_{het})}{\partial t} + \frac{\partial(\rho_m u_{mj} y_{het})}{\partial x_j} = \Gamma_3\end{gathered} \quad (9)$$

Where u_m is the velocity of the vapor/liquid mixture. It should be noted that Γ_1 is the mass source of critical droplets formed by the nucleation process, and Γ_2, Γ_3 is the mass condensation rate of all droplets per volume unit of the two-phase mixture for homogeneous and heterogeneous condensation respectively, and n is the concentration of liquid droplets per unit mass. Γ_1, Γ_2, Γ_3 is calculated using following equations.

$$\begin{gathered}\Gamma_1 = \frac{4}{3}\pi\rho_l r_{hom}^{*3} J \\ \Gamma_2 = 4\pi\rho_l\rho_m n_{hom} r_{hom}^2 \frac{dr_{hom}}{dt} \\ \Gamma_3 = 4\pi\rho_l\rho_m n_{het} r_{het}^2 \frac{dr_{het}}{dt}\end{gathered} \quad (10)$$

3. Test rig and instrumentation

From the point of view of the practical application of the presented research, tests carried out in real conditions of wet steam flow in a steam tunnel, which gives the greatest overview of its measurement capabilities. It is also valuable to compare the experimental measurement with advanced methods of numerical fluid dynamics carried out in parallel. For this purpose, the measurements also performed in the steam tunnel located in the Thermal Machines Hall, used for a wide range of water vapor flow tests, with the use of various types of nozzles and steam turbine blade channels. The geometry tested in this research, is a converging-diverging nozzle (CD nozzle) with the geometrical parameters [13]:

- Expansion rate of the diverging section 3000 s−1
- Total throat height of 40 mm.

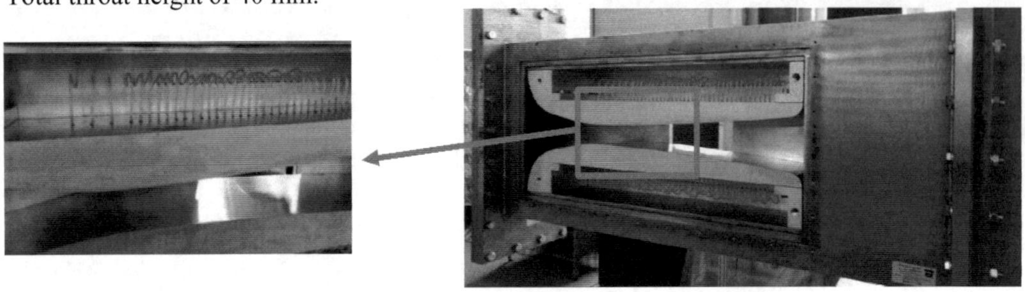

Figure 1. The view of the geometry of the investigated nozzle.

On the walls of the tested nozzle, there were 40 impulse holes connected with pressure sensors, used to measure the static pressure distribution on both the lower and upper tunnel walls, 20 pieces on each side. The holes are evenly spaced every 10 mm, starting from the position x = 0m of the nozzle (figure 1). The test section has a possibility to measure the liquid phase downstream from the nozzle throat only. To this end, a speculum in the form of windows on both lateral walls was manufactured. At the nozzle inlet, the total pressure and temperature are measured, there is no wetness mass fraction measured there.

The presented results of the static pressure values for each of the measured points were averaged from the specified number of samples achieved for the three seconds duration of the measurement, and then the pressure pipes were purged in order to eliminate the influence of water drops on the measurement results, and next series of tests was performed. The uncertainty of the pressure results, after carrying out and applying additional calibration corrections before the measurement, was estimated at the level of ± 250 Pa. The static pressure difference between the bottom and top walls was within the estimated measurement uncertainty for the entire presented range of measurement data. Therefore, in order to simplify the conducted analyses, only the experimental results for the tunnel bottom wall are presented here. Table 1 shows the results of an exemplary measurement of the steam condensing flow through the CD nozzle. The steam at the test section's inlet was superheated in this situation, which signifies the degree of superheat $\Delta t = t_0 - t_{sat}(p_0) > 0$. For this set of inlet total parameters, there were obtained static pressure distributions that differ in the location of the condensation shock. Some of them agree with the static pressure distribution determined from CFD modeling for the homogeneous condensation model. Presented herein pressure experimental data underpredict the location of homogeneous condensation onset.

Table 1. Experimental data (p_0=130kPa$_{(a)}$, t_0= 111°C, Δt=13.8°C).

X (cm)	P (kPa$_{(a)}$)	X (cm)	P (kPa$_{(a)}$)
-0.06	102.05	0.17	37.93
-0.04	92.53	0.18	36.53
-0.02	82.02	0.19	35.17
0	72.18	0.20	34.61
0.01	65.87	0.21	33.39
0.02	64.03	0.22	32.52
0.03	62.95	0.23	31.74
0.04	63.00	0.24	30.49

0.05	65.70	0.25	28.88
0.06	64.87	0.26	27.34
0.07	61.09	0.27	26.65
0.08	58.40	0.28	25.39
0.09	55.72	0.29	25.01
0.10	51.62	0.30	24.26
0.11	49.48	0.31	23.87
0.12	46.34	0.32	23.11
0.13	44.24	0.33	22.54
0.14	42.26	0.34	21.56
0.15	40.82	0.35	20.56
0.16	39.21	0.36	19.03

4. Results

For experimental results the CFD calculations were performed. To this end the CFD code available at the Power Engineering and Turbomachinery department of Silesian University of Technology was employed. Many technical applications of condensing flows have been successfully implemented using this code.

Figure 2 shows the static pressure distribution on the bottom wall of the investigated nozzle. The experimental results were compared with CFD simulation for homogeneous and heterogeneous models. For the heterogeneous condensation model it was assumed at the inlet 7.5×10^{13} number of droplets of the radius $10^{-8}m$ in kilogram of steam ($n_{het}=7.5\times10^{13}kg^{-1}$ $r_{p,het}=10^{-8}m$). It corresponds to the wetness mass fraction at the inlet a little more than 0.3×10^{-6}, which seems to be very low value. However, it makes the CFD results for heterogeneous condensation model very close to the experimental data of static pressure distribution on the nozzle wall (figure 2).

Figure 3 shows the mean droplet diameter distribution along the nozzle centreline. It can be seen that before the onset of condensation, there is a significant difference between homogeneous and heterogeneous models, however after the onset of condensation, it is not very noticeable. Figure 4 shows the wetness mass fraction along the nozzle centreline. A significant difference also can be seen between the homogeneous and heterogeneous models. Similarly, for the enthalpy-entropy diagram of the expansion line (figure 5), all results are very similar until the condensation wave. After the condensation process especially in condensation shock region, significant differences are observed between the homogeneous and heterogeneous models. The Mach number distributions showed in figure 6 differ only in the region of the condensation onset. The isentropic efficiency for both model amounts to the same value:

$$\eta = \frac{h_1 - h_2}{h_1 - h_2 + T_2 ds_{1-2}} = 0.944 \quad (11)$$

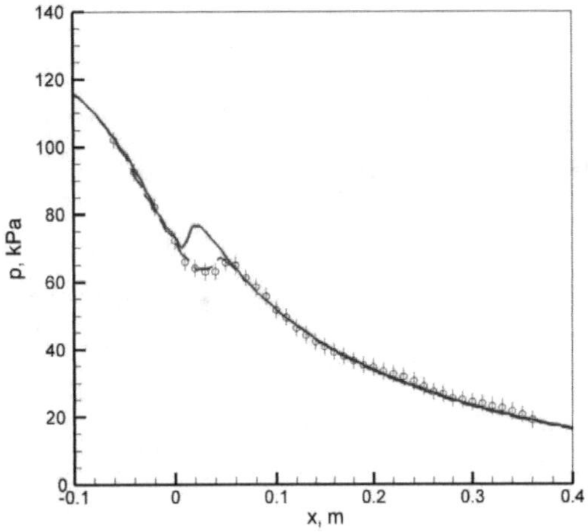

Figure 2. Pressure distributions on the bottom wall.
(— homogeneous model; - - - heterogeneous model; ° ° ° experiment)

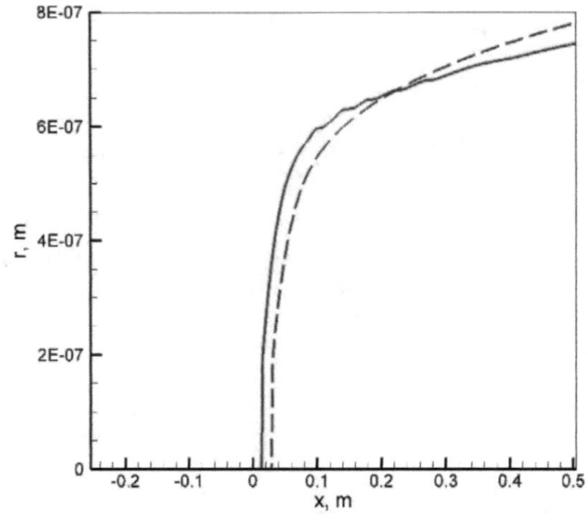

Figure 3. Droplet diameter distributions along the midline.
(— homogeneous model; - - - heterogeneous model)

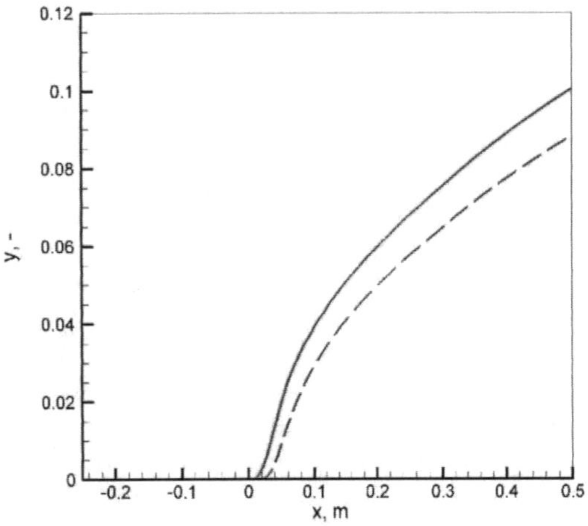

Figure 4. Liquid mass fraction distributions along the midline.
(— homogeneous model; - - - heterogeneous model)

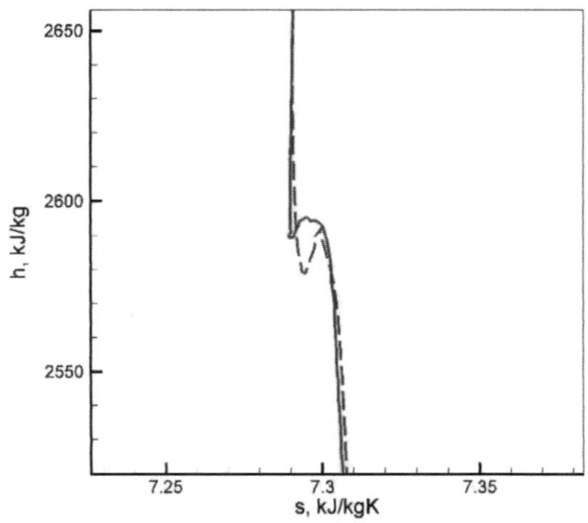

Figure 5. Expansion lines along the midline.
(— homogeneous model; - - - heterogeneous model)

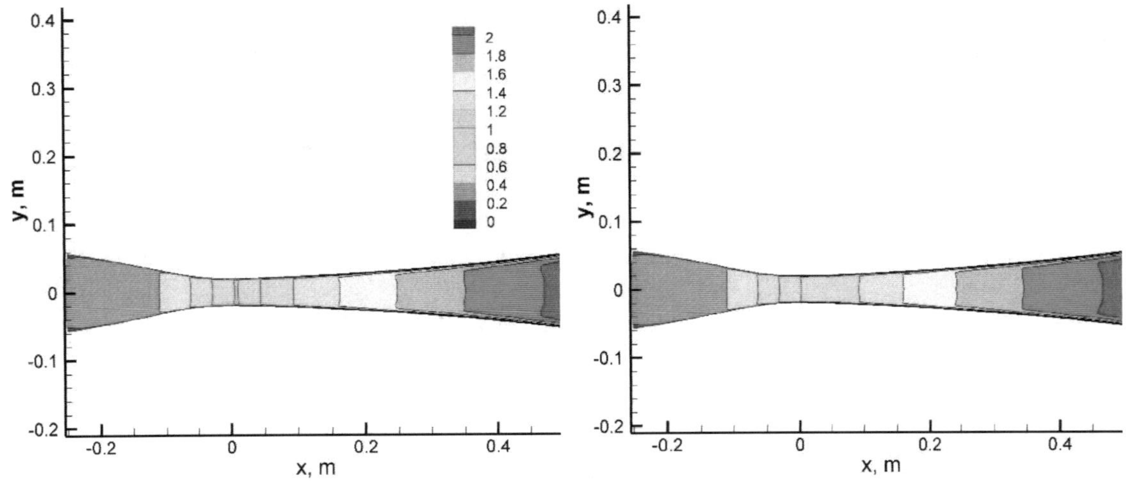

(a) (b)

Figure 6. Mach number distributions ((a) homogeneous model, (b) heterogeneous model).

5. Conclusions

The major goal of this study is comparing the results of heterogeneous and homogeneous condensation through a CD nozzle. The experimental technique for liquid phase assessment in wet steam transonic flow is tested under real conditions in an in-house steam tunnel test rig. The investigated geometry is a new converging-diverging nozzle (CD nozzle) with the expansion rate of the diverging section 3000 s−1 and the total throat height of 40 mm. The flow was modelled numerically using an in-house CFD code for performing the heterogeneous and homogeneous condensation models. The nucleation mechanism is ignored in the heterogeneous condensation. Because of the lower thermodynamic barrier, the presence of foreign solid particles encourages nucleation (compared to the homogeneous nucleation). For the comparison between CFD results and experimental data, the static pressure distribution throughout the nozzle bottom wall were chosen. One may observe a good agreement of the CFD results with experimental data. It points out both good robustness and accuracy of the CFD code and high quality of the performed experiment. Furthermore, the numerical results of liquid mass fraction distribution, mean droplet diameter distribution, expansion line along the nozzle centerline and Mach number distribution is presented for both heterogeneous and homogeneous condensation, and also compared. The presented approach can be used in the future for searching the reasons for the lack of accuracy between experimental and numerical static pressure distribution in the condensing steam flows.

Acknowledgments
Research reported in this publication was supported by the National Science Centre under number 2020/37/B/ST8/02369, and by statutory research funds for young scientists.

References

[1] Dykas S, Majkut M, Strozik M and Smołka K 2018 An attempt to make a reliable assessment of the wet steam flow field in the de Laval nozzle *Heat and Mass Transfer* **54** pp 2675-2681

[2] Eberle T and Schatz M 2011 Experimental study of droplet size and wetness in a model steam turbine using light extinction measurements *25th Turbomachinery Workshop* (Gdańsk/Krzeszna, Poland)

[3] Cai X.S, Wang N.N, Wei J.M and Zheng G 1992 Experimental investigation of the light extinction method for measuring aerosol size distributions *Journal of Aerosol Science* **23(7)** pp 749-757

[4] Cai X, Wanc L, Ouyang X and Pan Y 1992 A novel integrated probe system for measuring the two phase wet steam flow in steam turbine *Journal of Engineering Thermophisics* **22(6)** pp 52-57

[5] Kolovratník M and Bartoš O 2015 *EPJ Web of Conferences* **92** 02035

[6] Majkut M 2019 Zaawansowane techniki badań eksperymentalnych i wizualizacji przepływu gazów ściśliwych w maszynach energetycznych (Advanced techniques of experimental investigations and visualization of the flow of compressible gases in energy machines) Wydawnictwo Politechniki Śląskiej (Gliwice, Poland) ISBN 978-83-7880-623-3

[7] Dykas S, Majkut M, Smołka K and Strozik M 2015 Experimental research on wet steam flow with shock wave *Experimental Heat Transfer* **28(5)** pp 417-429

[8] Dykas S, Majkut M, Strozik M and Smołka K 2015 Experimental study of condensing steam flow in nozzles and linear blade cascade *International Journal of Heat and Mass Transfer* **80** pp 50-57

[9] Dykas S 2001 Numerical calculation of the condensing steam flow *Task Quarterly* **5(4)** pp 519–535

[10] Frenkel J 1955 Kinetic Theory of Liquids (Dover, New York)

[11] Kantrowitz A 1951 Nucleation in very rapid vapor expansions *Journal of Chemical Physics* **19(9)** pp 1097–1100

[12] Gyarmathy G 1960 Grundlagen einer Theorie der Nassdampfturbine *Juris Verlag* (Zürich)

[13] Zhang G, Dykas S, Yang S, Zhang X, Li H and Wang J 2020 Optimization of the primary nozzle based on a modified condensation model in a steam ejector *Applied Thermal Engineering* **171** 115090

Development of the LES-ADM approach for combustion modelling using high-order filters

Lena Caban[1] **and Artur Tyliszczak**[1]

[1] Czestochowa University of Technology, al. Armii Krajowej 21, 42-201 Czestochowa, Poland

E-mail: lena.caban@pcz.pl

Abstract. The paper focuses on the development of an approximate deconvolution method (ADM) for large eddy simulation (LES) of turbulent reactive flow. We derive the effective filter as a product of the LES filter and the filter induced by spatial discretization performed using high-order compact difference schemes. We analyze the accuracy of the ADM depending on the effective filter definition in the case of modelling unsteady combustion phenomena (ignition and flame extinction) in a forced homogeneous isotropic turbulent field.

1. Introduction

The Large Eddy Simulation (LES) method is an intensively developed mathematical tool for modelling turbulent reactive flows [1]. It is based on spatial filtering of the governing equations, which eliminates flow scales smaller than a specified filter width (Δ). These scales, called the sub -grid scales, are not present in the solution, however, their influence on large resolved scales must be taken into account. In the present paper, we concentrate on modelling nonlinear reaction terms in the species transport equations using the Approximate Deconvolution Method (ADM) formulated by Stolz et al. [2] for incompressible constant density flows. The unknown terms resulting from the filtration of the convective terms of the momentum, species and energy transport equations are modelled by applying the classical approach using the sub-grid viscosity model.

The deconvolution method is based on an inverse filtering operation, which allows the estimation of the unfiltered variables from the filtered ones. We consider the combined effect of the LES filtering and the filtering induced by a numerical approximation. For the mesh spacing (h) of the order of the filter width ($\Delta/h = \mathcal{O}(1)$) both these operations can have significant impact on the solution [3].

2. LES and ADM formulation

We consider a three-dimensional low Mach number reactive flow for which the governing equations in the framework of the LES method are as follows:

$$\partial_t \overline{\rho} + \nabla \cdot (\overline{\rho} \widetilde{\mathbf{u}}) = 0 \tag{1}$$

$$\overline{\rho} \partial_t \widetilde{\mathbf{u}} + (\overline{\rho} \widetilde{\mathbf{u}} \cdot \nabla) \widetilde{\mathbf{u}} + \nabla \cdot \overline{p} = \nabla \cdot (\tau + \tau^{\mathrm{SGS}}) + \mathbf{g} \tag{2}$$

$$\overline{\rho} \partial_t \widetilde{Y}_\alpha + (\nabla \cdot \overline{\rho} \widetilde{\mathbf{u}}) \widetilde{Y}_\alpha = \nabla \cdot (\overline{\rho}(D_\alpha + D_\alpha^{\mathrm{SGS}}) \nabla \widetilde{Y}_\alpha) + \overline{\dot{\omega}}_\alpha \tag{3}$$

$$\overline{\rho} \partial_t \widetilde{h} + (\nabla \cdot \overline{\rho} \widetilde{\mathbf{u}}) \widetilde{h} = \nabla \cdot (\overline{\rho}(D + D^{\mathrm{SGS}}) \nabla \widetilde{h}) \tag{4}$$

where the bar and tilde symbols denote LES and Favre filtered quantities [4] defined through the convolutions: $\bar{f}(\mathbf{x},t) = \mathcal{G}(\mathbf{x};\Delta) * f(\mathbf{x},t) = \int_\Omega \mathcal{G}(\mathbf{x}-\mathbf{x}';\Delta)f(\mathbf{x}',t)\mathrm{d}\mathbf{x}'$, $\widetilde{f}(\mathbf{x},t) = \overline{\rho(\mathbf{x},t)f(\mathbf{x},t)}/\bar{\rho} = 1/\bar{\rho}(\mathbf{x},t)\int_\Omega \mathcal{G}(\mathbf{x}-\mathbf{x}';\Delta)\rho(\mathbf{x}',t)f(\mathbf{x}',t)\mathrm{d}\mathbf{x}'$, where f stands for a general variable and $\mathcal{G}$ is the filter kernel such that $\int_\Omega \mathcal{G}(\mathbf{x};\Delta)\mathrm{d}\mathbf{x} = 1$. The filter width is equal to $\Delta^{-1} = \int_\Omega \mathcal{G}^2(\mathbf{x};\Delta)\mathrm{d}\mathbf{x}$. Most often $\mathcal{G}$ is assumed as a top-hat filter defined as:

$$\mathcal{G}(\mathbf{x};\Delta) = \begin{cases} 1/\Delta & \text{if} \quad |\mathbf{x}-\mathbf{x}'| < \Delta/2 \\ 0 & \text{otherwise} \end{cases} \tag{5}$$

Its width is related to the mesh size $\Delta = Vol^{1/3}$ where Vol is the volume of a numerical cell. The symbol ρ in Equations. (1-4) is the density, $\mathbf{u}$ is the velocity vector, p denotes the hydrodynamic pressure, τ is the viscous stress tensor, and h stands for the total enthalpy. The variables Y_α represent the species mass fractions and α is the index of the species, $\alpha = 1,\ldots,N$-species. The unresolved sub-grid stress tensor, resulting from the filtering of the non-linear advection terms, is defined as $\tau^{\mathrm{SGS}} = 2\mu_t S$, where S is the rate of strain tensor of the resolved velocity field and μ_t is the sub-grid viscosity computed with the help of the model proposed by Vreman [5]. The term $\overline{\dot{w}_\alpha(Y_{\alpha=1,\ldots,N_s},h)}$ represents the chemical source terms, which in the present work are modelled using the ADM method and for comparison, also applying the Eulerian Stochastic Field method [6]. The sub-grid diffusivities in Equations. (3-4) are computed as $D^{\mathrm{SGS}} = D^{\mathrm{SGS}}_\alpha = \mu_t/\bar{\rho}\sigma$ where σ is the turbulent Schmidt/Prandtl number assumed to be equal to 0.7 [10]. The set of Equations (1-4) is complemented by the equation of state $p_0 = \bar{\rho}R\widetilde{T}$, where p_0 is the thermodynamic pressure constant in space and time, R - gas constant, T - temperature.

The computational configuration is a forced homogeneous isotropic turbulence (HIT) in which the energy level is maintained using the method developed by Lundgren T. [11] in spectral space and then formulated in physical space and thoroughly analyzed by Rosales and Meneveau [12]. It consists in adding a source term $\mathbf{g}$ to the Navier-Stokes equations (2). The i-th component of $\mathbf{g}$ is equal to $g_i = A\bar{\rho}\widetilde{u}_i$ where the parameter A can be assumed constant or it can be determined in a more sophisticated way based on a balance of kinetic energy for a statistically stationary state as follows $A = \theta\varepsilon/3u^2_{\mathrm{rms}}$, where $\varepsilon = -\nu\langle\widetilde{\mathbf{u}}\cdot\nabla\widetilde{\mathbf{u}}\rangle$ is the mean energy dissipation rate per unit mass, $u^2_{\mathrm{rms}} = \langle\widetilde{\mathbf{u}}\cdot\widetilde{\mathbf{u}}\rangle/3$ and θ is the strength of the forcing [12]. For the purpose of the present comparative studies the first approach is sufficient.

In LES, the governing equations are discretized on a numerical mesh, which means that the partial derivatives (e.g., ∂_x, ∂^2_x) at the PDE level are not equal to their discrete counterparts ($\delta_x \neq \partial_x$, $\delta^2_x \neq \partial^2_x$). The approximation of derivatives induces their filtering with an additional numerical filter operator ($\mathcal{L}_h$) with an associated filter kernel $\mathcal{G}_h$. Its shape and properties are largely dependent on the numerical scheme and can be easily assessed in the Fourier space based on the modified wavenumber analysis. Assuming that the induced filter acts on the flow variables much like LES filters, one can write $\delta_i\bar{f} = \partial_i[\mathcal{L}_h(f)]$, which means that solving the discretized LES equations one obtain the variables $\widehat{\bar{f}}$ defined as:

$$\begin{aligned}\widehat{\bar{f}}(\mathbf{x},t) &= \int_\Omega \mathcal{G}_h(\mathbf{x},\mathbf{z}-\mathbf{z}';\widehat{\Delta})\left[\int_\Omega \mathcal{G}(\mathbf{x}-\mathbf{x}';\Delta)f(\mathbf{x}')d\mathbf{x}'\right]d\mathbf{z}' \\ &= \int_\Omega \underbrace{\left[\int_\Omega \mathcal{G}_h(\mathbf{x},\mathbf{z}-\mathbf{z}';\widehat{\Delta})\mathcal{G}(\mathbf{x}-\mathbf{x}';\Delta)d\mathbf{z}'\right]}_{\mathcal{G}_e(\mathbf{x},\mathbf{z}';\widehat{\Delta},\Delta)} f(\mathbf{x}')d\mathbf{x}'\end{aligned} \tag{6}$$

where $\mathcal{G}_e(\mathbf{x},\mathbf{z}';\widehat{\Delta},\Delta)$ is an effective filter kernel whose width $\widehat{\Delta}$ depends on the numerical scheme. In this paper, we consider the impact of induced filtering associated with the first derivative approximation, because it can be assumed that it introduces the strongest induced filtering. This is certainly the case with large Reynolds number flows and in numerical algorithms in which the convective terms (the first order derivatives) are discretized using lower-order upwind-like methods than any other discretization method applied for the remaining terms in the governing equations. In the next section, we will explain how the induced filters formulated on a numerical mesh are used in the ADM.

2.1. Induced filters and ADM

A finite and compact difference discretization method (FD, CD) for the 1st derivative approximation (f') can be written for the i-th mesh node in a general form as:

$$f'_i + \sum_{k=1}^{M} a_k \left(f'_{i-k} + f'_{i+k}\right) = \frac{1}{h}\sum_{j=1}^{N} b_j \left(f_{i+j} - f_{i-j}\right) \tag{7}$$

where M, N defines the computational stencils, a_k and b_j are the coefficients calculated based on the Taylor series expansions of f' and f [8, 9]. Writing (7) for each mesh node along a node line leads to the system of equations $\mathbf{Af}' = \mathbf{Bf}$ of which the solution gives a matrix operator $\mathbf{D}$ of the 1st derivative approximation such that $\mathbf{D} = \mathbf{A}^{-1}\mathbf{B}$. The matrices $\mathbf{A}$, $\mathbf{B}$ are composed of the coefficients a_k and b_j (see [8, 9]). For the classical finite difference approximation $a_k = 0$ and $\mathbf{A} = \mathbf{I}$ ($\mathbf{I}$ - the identity matrix). The induced filter associated with $\mathbf{D}$, denoted as $\mathbf{G}_h$, can be formulated in terms of a superposition of the top-hat filters of varying filter widths (h, $2h$,..., etc.) that define $\widehat{\Delta}$ [3]. Thus, assuming the exact differentiation operator $\boldsymbol{\mathcal{D}}$ one can write $\mathbf{D} = \boldsymbol{\mathcal{D}}\mathbf{G}_h$. Applying this operator to the filtered variables ($\mathcal{G}(\mathbf{x};\Delta) * f(\mathbf{x},t) \longrightarrow \mathbf{Gf}$) we finally get $\widehat{\overline{\mathbf{f}'}} = \mathbf{D}\overline{\mathbf{f}} = \boldsymbol{\mathcal{D}}\mathbf{G}_h\mathbf{Gf} = \boldsymbol{\mathcal{D}}\mathbf{G}_e\mathbf{f}$ where $\widehat{\overline{\mathbf{f}}} = \mathbf{G}_e\mathbf{f}$ and $\mathbf{G}_e = \mathbf{G}_h\mathbf{G}$ are the discrete analogies of the double-filtered variables and the effective filter defined in (6).

In the ADM, the inverse of $\mathbf{G}_e$ must be computed ($\mathbf{G}_e \rightarrow \mathbf{G}_e^{-1}$), which allows the calculation of the unfiltered variables based on the filtered ones from the following relations:

$$\mathbf{f} = \mathbf{G}_e^{-1}\widehat{\overline{\mathbf{f}}}, \quad \mathbf{f} = \frac{\mathbf{G}_e^{-1}(\widehat{\overline{\rho\mathbf{f}}})}{\mathbf{G}_e^{-1}\widehat{\overline{\rho}}}. \tag{8}$$

Knowing $\mathbf{f}$ one can calculate the unknown sub-grid tensor and the reaction terms in Equations (2-4). In this work, the unfiltered species mass fraction and enthalpy are needed to calculate $\overline{\dot{w}_\alpha(Y_{\alpha=1,\dots,N}, h)}$. The inversion procedure relies on the van Cittert series $\mathbf{G}_e^{-1} \approx \mathbf{Q} = \sum_{n=0}^{N_{\mathrm{ADM}}}(\mathbf{I} - \mathbf{G}_e)^n$. As shown in [2], the number of terms $N_{\mathrm{ADM}} = 5$ is sufficient to approximate $\mathbf{G}_e^{-1}$ with a good accuracy. Nevertheless, as a consequence we have $\mathbf{f} \approx \mathbf{f}^* = \mathbf{Q}\widehat{\overline{\mathbf{f}}}$ and thus the reaction terms are computed based on the $\mathbf{f}^*$ approximations of $\mathbf{f}$, i.e., $\overline{\dot{w}_\alpha(Y^*_{\alpha=1,\dots,N}, h^*)}$. In the performed simulations we assume that the basic LES filter corresponds to the top-hat filter approximated by the three point trapezoidal filter, while the induced filter is associated with the 6th order CD approximation of the first derivative [8, 9] and is computed using the method proposed in [3]. We compare two approaches with: (i) the effective filter determined according to the above procedure; (ii) the induced filter neglected in $\mathbf{G}_e$, and thus, $\mathbf{G}_e \equiv \mathbf{G}$.

3. Computational configurations and numerical code

The correctness and accuracy of the ADM formulated above are verified based on the example of simulation of an ignition in a field of forced homogeneous isotropic turbulence (HIT) in which a hot oxidizer is partially mixed with fuel. Maintaining a constant energy level is necessary to maintaining the intense mixing process before and during the ignition phase. To some extent, the forced HIT configuration mimics a jet ignition system schematically presented in Figure 1. Hot combustion products from a pre-ignition chamber are injected into the main combustion chamber where they intensively mix with the fuel and the mixture ignites. In the performed computations we consider a small 3D cubic volume L^3 ($L = 0.01$ m) in which the combustion products at $T_P = 1100$ K, with the composition defined in section 4.2, are mixed with the fuel being a nitrogen diluted hydrogen with $Y_{\mathrm{H2}} = 0.1$, $Y_{\mathrm{N2}} = 0.9$ ($T_F = 300$ K). The simulations were performed with the help of an in-house high-order numerical code SAILOR in which all but one spatial term are discretized using the 6th order CD method. The exception is the convective term in the species and enthalpy transport equations discretized using a 5th order WENO scheme. The time integration is performed by applying a predictor-corrector approach (Adams-Bashforth/Adams-Moulton methods) coupled with the projection method for pressure-velocity coupling. The chemical mechanism of the hydrogen combustion consists of 9 species reacting through 19 reactions [7]. The SAILOR code has been used before and thoroughly verified in numerous studies and has always shown excellent accuracy. Although two different discretization methods are used, the induced filter $\mathbf{G}_h$ is defined based on the CD approximation, as mentioned in the previous section. In this work, instead of trying to derive its replacement form of $\mathbf{G}_h$ for CD and WENO schemes we examine to what extent the inclusion of $\mathbf{G}_h$ into the $\mathbf{G}_e$ definition affects the ADM results.

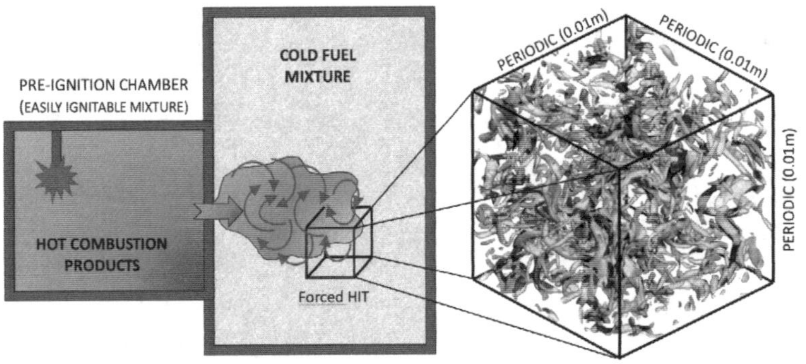

Figure 1. Schematic illustration of a jet ignition system and flow configuration.

4. Results

4.1. HIT initialization

The linear forcing method with a given constant forcing parameter A has such feature that starting from arbitrary initial conditions it always leads to a developed flow field with the same average energy spectrum [12]. The only difference is the time period from the beginning of the simulation to the time moment when this pseudo-steady state is reached. Figure 2a shows contours of the initial vorticity field (Ω) in the main 'x-y' cross-section of the computational domain. This flow field was generated based on the von Karman spectrum of the turbulent kinetic energy (TKE) represented by the red line in Figure 2b. When the simulations begin, the flow starts to evolve, the viscous forces suppress the small scales, the flow becomes smoother and the spectrum slightly oscillates around the blue line. This spectrum corresponds to the flow field with the vorticity distribution shown in Figure 2c obtained with the parameter $A = 300\ \mathrm{s}^{-1}$. As an interesting finding, we report that even starting from a randomly generated velocity field, the solution converges to a flow field similar to that shown in Figure 2c with the energy spectrum presented in Figure 2b.

The main computations were performed on two numerical meshes consisting of 256^3 nodes for DNS and 64^3 nodes for LES. Figure 3a shows time evolution of the total TKE as a function of

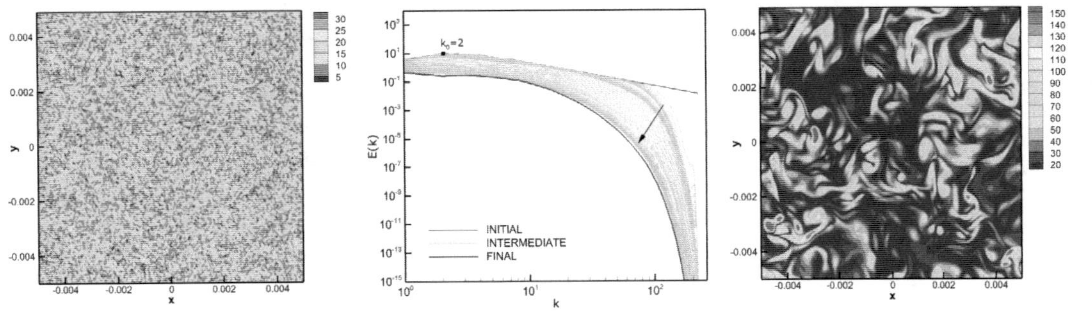

(a) (b) (c)

Figure 2. Vorticity contours in HIT configuration and time evolution of the energy spectrum.

the number of nodes and the A parameter ($A = 100$, 300, 500 s^{-1}). It can be seen that initially TKE is large, but after a transient phase it stabilizes on constant levels depending on the forcing strength A. The larger it is, the higher and more diversified the level of TKE. For all three A values the impact of the mesh density is small, yet, there are significant differences regarding the ratio of the mesh size (h) to the Kolmogorov length scale (η). On the dense mesh, for the cases with $A = 100$, 300, 500 s^{-1} it is equal to $h/\eta = 0.6$, 1.4, 2, and it is four times larger on the coarse mesh. Figures 3b,c show the TKE spectra calculated at time $t = 0.05$ s. In a range $k \leq 10$, where k stands for an index of the wave number $\omega_k = k2\pi/L$, the spectra are very similar on both meshes. Significant differences are visible for $k > 30$, but in this range the values of $E(k)$ are small. Hence, their contribution to the total TKE do not change it significantly.

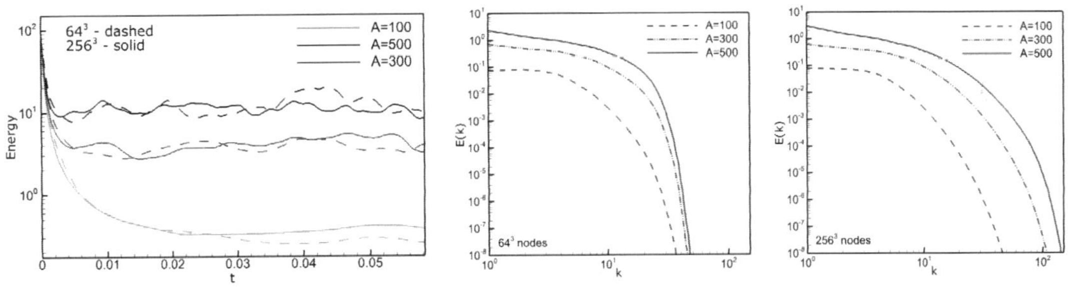

(a) (b) (c)

Figure 3. Time evolution of TKE and its spectra calculated on two computational meshes.

4.2. Species and temperature initialization

Simulations for the cold flow conditions on the mesh consisting of 256^3 nodes showed that only with the parameter $A = 100$ s^{-1} the obtained solution can be treated as a well resolved from the point of view of DNS requirements ($h < \eta$). Hence, in the rest part of the paper we analyze and compare the LES and DNS results for this case. Figure 4 shows the velocity vectors and the corresponding contours of the vorticity field, which indirectly serve to initialize the mixture composition and enthalpy. We connect $|\Omega|$ to the mixture fraction distribution assumed as $Z = (|\Omega|_{\mathrm{INI}}/|\Omega|_{\mathrm{MAX}})^n$, which is in the range $Z = 0 - 1$. The parameter n sets the sharpness of the initial Z distribution and is taken as $n = 2$. Figure 5 on the left side shows the initial H_2

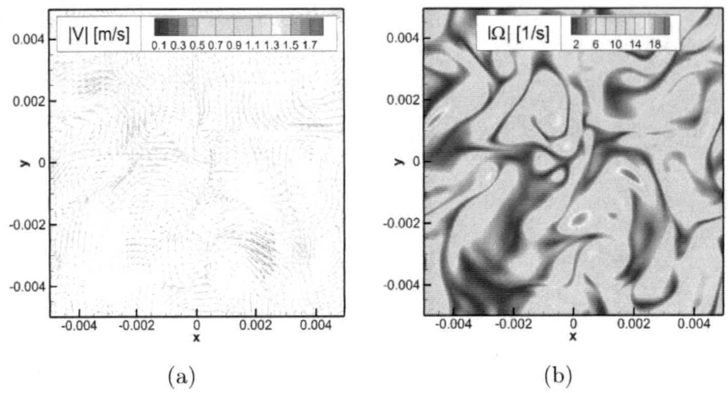

Figure 4. Velocity vectors (a) and vorticity module contours (b).

and O_2 fractions calculated based on Z as

$$\begin{aligned} Y_k &= (1 - Z) \cdot Y_{k,P} + Z \cdot Y_{k,F} \\ h &= (1 - Z) \cdot h_P + Z \cdot h_F \end{aligned} \tag{9}$$

where the subscripts $\cdot_P$ and $\cdot_F$ denote the combustion products and fuel composition. These formulas assign the fuel to high vorticity regions and the products, which play the role of ignition kernels, to the low vorticity regions. The values of Y_K are extracted from a steady state flamelet solution presented in Figure 5 from the place marked by the blue line. The temperature at this point is $T_P = 1100$ K and the species composition is given in the figure. The relatively high content of the oxygen ($Y_{O2} = 0.168$) causes the products to act as the oxidizer in this case, as there is no oxygen in the fuel mixture. Such an initialization method leads to a relatively quick ignition followed by extinction and finally the mixing of the hot combustion products.

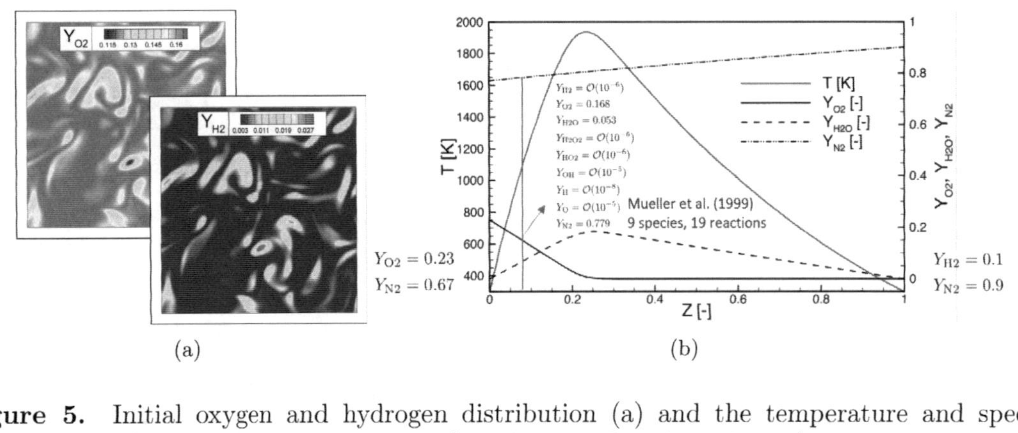

Figure 5. Initial oxygen and hydrogen distribution (a) and the temperature and species composition (b) in mixture fraction space (steady state flamelet solution).

4.3. Ignition and extinction scenario

Figure 6 shows the contours of Y_{H2}, Y_{OH} and temperature at the time moments before and during the ignition phase and when the flame vanishes. Note that the high oxidizer temperature causes a significant molecular velocity increase that dampens the turbulence intensity, and hence, compared to the initial conditions the flow field becomes 'smoother'. The blue regions in the temperature field correspond to the location of the fuel. The ignition is preceded by the appearance of the OH species, high concentrations of which are found in regions of the lean mixture. Shortly after the temperature rises, maxima of OH are found in places with initially high values of H_2. Then the OH disappears, the hydrogen is fully burned, and the flame vanishes.

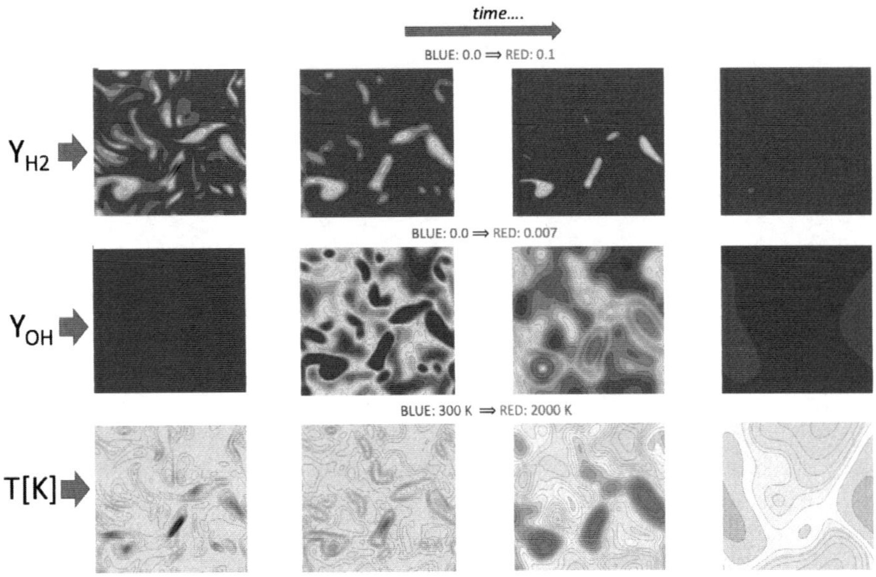

Figure 6. Countours of the species mass fractions and temperature.

The temporal evolution of the maximum temperature (T_{MAX}) in the domain is shown in Figure 7a. The presented LES results have been obtained using a no-model approach with $\dot{w}_\alpha(Y_{\alpha=1,\dots,N}, h) \approx \dot{w}_\alpha(\widetilde{Y}_{\alpha=1,\dots,N}, \widetilde{h})$, the ESF model on 4 fields and two variants of the ADM. Initially, in each case, a slight drop in the temperature occurs due to convective and diffusive mixing of the hot oxidizer and cold fuel. Then, around 7×10^{-5} s, the temperature suddenly rises and this is the ignition moment. In all cases it happens at almost the same time. The temperature rises rapidly, reaches its maximum value, and then drops immediately after the flame is extinguished. The differences between particular LES results may seem surprisingly small and even the no-model approach leads to a good solution. A closer analysis of the solutions shows that the reason for this is a low turbulence intensity resulting from the above-mentioned viscous damping. We note that without the forcing the flow field would be even smoother and would appear frozen compared to a very fast ignition process. The discrepancies between DNS and LES results during the flame vanishing phase are due to the slightly higher initial fuel content generated on the denser mesh. However, this has no effect on the ignition process and it can be seen that generally before and during the ignition phase, all LES results predict the temperature evolution and its maximum value relatively well. However, when the ADM model is formulated based on $\mathbf{G}_e = \mathbf{G}_h\mathbf{G}$ the temperature rises fastest and its level is the highest. Figure 7b shows the spectra of Y_{OH} in time instances close to the ignition moment. The solid lines represent the solution before deconvolution, and the dashed lines after its execution. These values are used to calculate the reaction terms. The thick and thin lines correspond to the two ADM variants. It can be seen that in the case with $\mathbf{G}_e = \mathbf{G}_h\mathbf{G}$ the spectral amplitudes in the range of higher wavenumbers are larger. This intensifies the mixing process within the small scales range, which accelerates the flame development and causes a slightly higher level of its maximum.

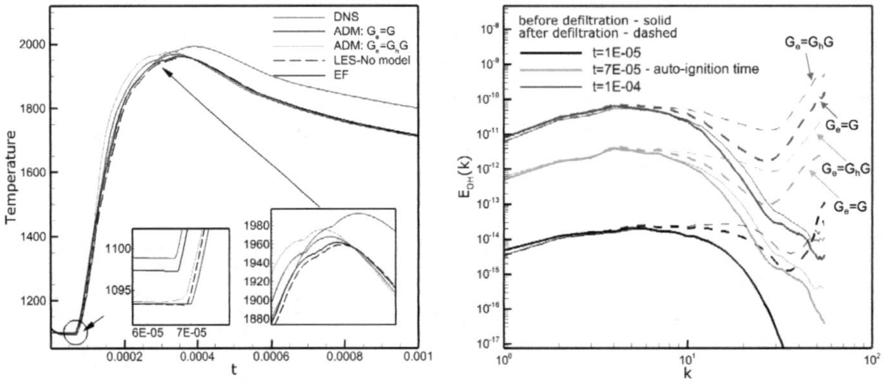

(a) (b)

Figure 7. Maximum temperature (a) and Y_{OH} spectra at the selected time instances (b).

5. Conclusions

The paper focused on the analyzis of the ADM applied for modelling the reaction terms in LES of the ignition phenomenon in the forced homogeneous isotropic turbulent field. It has been shown that numerical discretization induces the filtering effectthe inclusion of which in the formulation of the effective filter affects the values of the deconvolved variables. Their spectra in the range of higher wavenumbers are characterized by higher amplitudes. This has an impact on the mixing of the small scales which becomes more intense. The comparison with the DNS and LES results obtained using different approaches was made only on the basis of the global parameter, i.e., the maximum temperature in the domain, and it did not allow for unambiguous conclusions. We found that it increases faster in the ignition phase, but its maximum value is closer to that predicted by DNS. In the simulations performed, the induced filter $\mathbf{G}_h$ was defined on the basis of the 6th order compact difference approximation and did not exactly reflect the discretization schemes used in the scalar transport equations. It can be assumed that with careful formulation of $\mathbf{G}_h$ the accuracy of the LES-ADM approach can be improved.

6. Acknowledgements

This work was supported by the National Science Centre, Poland (Grant no. 2018/29/B/ST8/00262) and statutory funds of Czestochowa University of Technology BS/PB 1-100-3011/2021/P. The authors acknowledge Prof. Bernard J. Geurts for fruitful discussions stimulating this work within the project No. PPI/APM/2019/1/00062 sponsored by NAWA.

References

[1] Pitsch H 2006 *Annu. Rev. Fluid Mech.* **38** 453
[2] Stolz S and Adams N A 1999 *Phys. Fluids* **11** 1699.
[3] Geurts B J and van der Bos F 2005 *Phys. Fluids* **17** 125103.
[4] Geurts B J 2004 *Elements of Direct and Large-Eddy Simulation* (Edwards Publishing)
[5] Vreman A W 2004 *Phys. Fluids* **16** 3670
[6] Valiño L 1998 *Flow Turbul. Combust.* **60** 157
[7] Mueller M A, Kim TJ, Yetter, RA and Dryer FL 1999 *Int. J. Chem. Kinet.* **31** 113
[8] Lele SK 1992 *J. Comput. Phys.* **103** 16
[9] Caban L and Tyliszczak A 2022 *Comput. Math. Appl.* **108** 123
[10] Triantafyllidis A and Mastorakos E 2010 *Flow Turbul. Combust.* **84** 481
[11] Lundgren T S 2003 *CTR Report* 461
[12] Rosales C and Meneveau C 2005 *Phys. Fluids* **17** 095106

Journal of Physics: Conference Series **2367** (2022) 012015 doi:10.1088/1742-6596/2367/1/012015

CO_2 methanation: on the modeling of reacting laminar flows in structured $Ni/MgAl_2O_4$ catalysts

J A Medina Méndez[1], M González-Castaño[2], F M Baena-Moreno[2,3] and H Arellano-García[2]

[1] Numerical Fluid and Gas Dynamics, BTU Cottbus-Senftenberg, Siemens-Halske-Ring 15a, 03046 Cottbus, Germany

[2] Process and Plant Technology, BTU Cottbus-Senftenberg, Burger Chaussee 2 LG 4/1, 03044 Cottbus, Germany

[3] Chemical and Environmental Engineering Department, Technical School of Engineering, University of Seville, 41092 Sevilla, Spain

E-mail: medinjua@b-tu.de

Abstract. This work presents an overview of issues for the modeling of laminar flows in monolith catalysts. Both 0-D and 3-D models are evaluated for a parallel channel structured honeycomb catalyst (PC-HC), and a gyroid 3-D printed structured catalyst (G-3D). At the 0-D homogeneous reactor modeling level, the analysis is focused on the effect of the bulk porosity, as well as the model choice to represent Nusselt number effects. Results show the better suitability of a long tube Nusselt number model for the representation of the maximum temperature achievable in the 0-D homogeneous reactor, as well as a modest effect of the porosity on the catalyst CO_2 conversion. A more detailed insight on heat transfer and the core reaction zone inside the monolith can be obtained at the 3-D homogeneous reactor modeling level.

1. Introduction

Fluid dynamic treatment of structured catalysts is generally conceived within the framework of porous media flows. For homogeneous media, the most widely used closure model for the upscaling of the momentum balance using volume-averaging theory (VAT) is the Darcy-Forchheimer equation [1]. In the experimental field, the use of VAT generally leads to confusion regarding the nature of the flow. Specifically, the use of a variety of Reynolds numbers for flow characterization, as well as deviations from the classical Darcy's law in the observed regimes, have led to some confusion regarding the laminar or turbulent nature of the flow, see [1]. In this work, we focus on the macroscopic (volume-averaged) evaluation of the flow over the porous media, and its corresponding bulk flow Reynolds number Re. Within this volume-averaged point of view, structured catalysts generally operate at laminar flow regimes due to the very small cross-sectional areas of their associated monolith channels. It is clear that, on one hand, a structured catalyst is better represented in terms of a heterogeneous model, which is faithful to the dynamics of both the solid and the working fluid phases. Homogeneous reactor models, on the other hand, are still widely used for industrial applications due to its simplicity and highly efficient use of computational resources. Despite the intrinsic loss of accuracy associated to homogeneous reactor models, the use of heterogeneous reactor models may not always be justified [2].

Within the widespread field of applications for structured catalysts, one application with current strategic importance is that of emissions control. The CO or CO_2 methanation has some fundamental significance, since it is associated both to natural gas production, as well as CO and CO_2 emissions removal [3]. For the optimization of the process of natural gas production, the development of selective, reliable and efficient catalysts is required. Although structured catalysts have been received with initial skepticism due to their poor performance in comparison to pellet beds, as well as their virtual adiabatic character, specific requirements for structured catalysts can be met with dedicated designs [4]. One example of such dedicated design is the gyroid 3-D printed catalyst presented by [5]. The latter uses Ruthenium-promoted magnesium-aluminate supported Nickel as a catalyst material, 0.5%Ru-15%Ni/$MgAl_2O_4$. Preliminary evaluations of this catalyst have already demonstrated a reliable and highly efficient CO_2 conversion. However, the optimization of the design is complicated due to the multitude of factors playing a role in the catalytic efficiency: from the chemical kinetics taking place, all the way to the catalyst-enhanced flow drag, and the associated heat and mass transfer. We focus on the evaluation of the recently introduced Ru-Ni/$MgAl_2O_4$ structured catalysts [5], from the numerical fluid-dynamic point of view.

This paper is structured as follows. In section 2, we briefly detail the relevant aspects associated with a zero dimensional homogeneous model for the numerical fluid-dynamic evaluation of structured catalysts. Next, in section 3, we explain some aspects concerning the three dimensional homogeneous modeling. The results for the evaluation of a parallel channel structured honeycomb catalyst (PC-HC), and a gyroid 3-D printed structured catalyst (G-3D), as in [5], are shown in section 4. Finally, closing remarks and an outlook of potential future work is shown in section 5.

2. Zero-dimensional structured catalyst model

An intended 0-D homogeneous model for structured catalysts follows from the model developed by [6]. The velocity U_b is assumed constant and homogeneous in the 0-D reactor. The model relates streamwise advancement of plug-flow-like reactor models to time, by the spatial-to-temporal transformation $\mathrm{d}t = \mathrm{d}z/U_\mathrm{b}$, where $\mathrm{d}z$ is a differential of the reactor streamwise coordinate, and $\mathrm{d}t$ is a differential of time. The reactor time-scale t_τ is then obtained as L/U_b, where L is the monolith length. The chemical species (k-th mass fractions) Y_k and temperature T equations are formulated for one reactor volume, in the form of a balance, and are then expressed per unit volume, in order to obtain the corresponding ordinary differential equations (ODEs) for the homogeneous fluid phase,

$$\epsilon\rho\frac{\mathrm{d}Y_k}{\mathrm{d}t} = \bar{\dot{\omega}}_k W_k \tag{1}$$

$$\epsilon\rho c_p\frac{\mathrm{d}T}{\mathrm{d}t} = \sum_j \left[\rho_\mathrm{cat}\bar{\dot{\omega}}_{\mathrm{R},j}\left(-\overline{\Delta H}_{\mathrm{R},0}\right)\right] - \frac{4}{D}U_A\left(T - T_\mathrm{w}\right) \tag{2}$$

Here, ϵ is the bulk porosity of the catalyst, $\bar{\dot{\omega}}_k$ is a molar-based species reaction rate and W_k is the k-th species molecular weight. Also, ρ is the fluid density, ρ_cat is the catalyst density (catalyst mass per reactor volume), and c_p the specific heat capacity at constant pressure. $\bar{\dot{\omega}}_{\mathrm{R},j}$ is the j-th reaction net molar rate of progress per unit mass of catalyst, i.e., in kmol/(kg$_\mathrm{cat}\cdot$ s), specified by the used chemical kinetics mechanism. This is related to the species reaction rate by $\bar{\dot{\omega}}_k = \rho_\mathrm{cat}\sum_j \nu_{jk}\bar{\dot{\omega}}_{\mathrm{R},j}$, being ν_{jk} the stoichiometric coefficient of species k in reaction j. Also, $\overline{\Delta H}_{\mathrm{R},0}$ is the reaction enthalpy at reference temperature conditions, e.g., 298 K. Furthermore, D is the monolith diameter, U_A is a global convective heat transfer coefficient, and T_w is a constant reactor cooling wall temperature. Equations (1) and (2) can be integrated in time using an explicit Euler method.

2.1. Chemical kinetics

Chemical kinetics are modeled in this work by the methanation reaction mechanism developed by [7]. This is a reduced chemistry mechanism for Ni/$MgAl_2O_4$-spinel catalysts, with 3 reactions and 5 chemical species (H_2, H_2O, CH_4, CO, CO_2). N_2 is added as an inert specie. This mechanism was also used by [6] for the study of CO_2 methanation reactors. It is noted that this mechanism may not incorporate the effects of the added Ru in the intended catalyst to evaluate, i.e., the Ru-Ni/$MgAl_2O_4$ catalyst from [5]. Both the reaction mechanism and the thermodynamic data for the species contained in the mechanism are supplied in the form of a CTI file to the open source library Cantera [8], in order to handle the reaction rate calculations. The methanation mechanism developed by [7] also contains the presence of so-called DEN inhibition terms. These limit the reaction rates described by the rate coefficients supplied in [7]. The inhibition terms are calculated in this work in an explicit way, in addition to the Cantera-calculated reaction rates, and are applied as multiplicative factors to such rates afterwards.

2.2. Convective heat transfer

The global heat transfer coefficient U_A is calculated in [6] based on a Nusselt number (Nu) correlation for the heat conduction at the wall, and a radial effective transport specialized for fixed-bed reactors. More generally, U_A is related to the bulk Nusselt number as

$$U_A = \frac{\mathrm{Nu}(t)\lambda}{\Omega} \tag{3}$$

Here, λ is the thermal conductivity of the fluid, Ω is a characteristic length, and $\mathrm{Nu}(t)$ is a time-dependent Nusselt number.

We consider the analytically derived solutions to the Nusselt problem obtained by [9] and [10]. In [9], asymptotic analytical solutions are presented for short reactors with uniform inlet flow, as well as fully developed laminar parabolic inlet profiles. Short reactors are defined at negligible aspect ratio, or conversely, at very large transverse Peclet number $\mathrm{Pe_T} >> 1$, where $\mathrm{Pe_T} = U_\mathrm{b}\Omega'^2/(L\mathcal{D}_m)$, being $\Omega' = D_h/2$ the hydraulic radius in the short reactor case, and $\mathcal{D}_m$ a mass-diffusion coefficient. For $\mathrm{Pe_T} >> 1$, the Nusselt scaling considering flat uniform inlet velocity profiles in a (constant pressure) plug-flow reactor (PFR), is then given by [9], previous consideration of the space-to-time analogy discussed in this work,

$$\mathrm{Nu}(t) = 2\sqrt{\frac{\mathrm{Pe_T}}{\mathrm{Le_f}}} \tag{4}$$

Note that for this case, $\Omega = D_h$, the hydraulic diameter of the monolith, is the correct characteristic length associated to Nu or U_A, see [9]. Also, $\mathrm{Le_f}$ is the fluid Lewis number in equation (4), where $\mathrm{Le_f} = \lambda/(\rho c_p \mathcal{D}_m)$. The time-dependency for Nu in equation (4) arises from the temperature dependency of the fluid properties.

We also evaluate a long tube Nu model. In [10], a solution to the Nu problem for long reactors is obtained ($\mathrm{Pe_T} << 1$). In the case of long reactors, $\Omega' = D_h/4$, while $\Omega = \Omega'$ is the correct characteristic length to associate with Nu, see [10]. We focus again on the flat uniform inlet velocity profile, for a constant operating pressure and fixed wall temperature reactor, i.e., a developing flow consideration for which Nu has now a streamwise dependency $\mathrm{Nu} = \mathrm{Nu}(z)$, see [10],

$$\mathrm{Nu}(z) = \mathrm{Nu}_\infty + \frac{0.44\mathrm{Pr}^{-\frac{1}{6}}\frac{\mathrm{Pe_T}L}{z}}{1 + 0.314\sqrt{\frac{\mathrm{Pe_T}L}{z}}} \tag{5}$$

The streamwise-dependency can be exchanged by the time-dependency in our model. In equation (5), $\mathrm{Pr} = c_p\mu/\lambda$ is the Prandtl number, with μ being the fluid dynamic viscosity.

Also, $\mathrm{Nu}_\infty = 2.966$ is the asymptotic Nusselt number value for a honeycomb-shaped channel cavity in [10].

The hydraulic diameter, D_h, is calculated as a function of the monolith total volume, and the interfacial fluid and catalyst area, which corresponds to the volume analog of the traditional definition based on the surface area and the wet perimeter. D_h can then be estimated as

$$D_h = \frac{4\epsilon}{\frac{4}{D} + (1-\epsilon)\frac{4S}{\pi D^2 L}} \tag{6}$$

where S is the supplied catalyst surface area. It is noted that the estimation of D_h also allows the estimation of the number of channels N_{chan} in the monolith, assuming that there is one channel per unit area by D_h, that is, $N_{\mathrm{chan}} = D^2/D_h^2$.

Figure 1 shows results for a PFR evaluation case using reference data from [6]. The PFR model used by [6] is compared to the short and long 0-D reactor models explained before. Figure 1 shows that the model of [6] is more dynamically similar to a short reactor Nu model. In general, the short tube Nu model minimizes the convective heat transfer, thus allowing larger maximum process temperatures and reaction product yields for the same initial or inlet conditions. That is, the short tube Nu model imposes the dominance of fast reactions, the faster *ignition-like* behavior in figure 1.

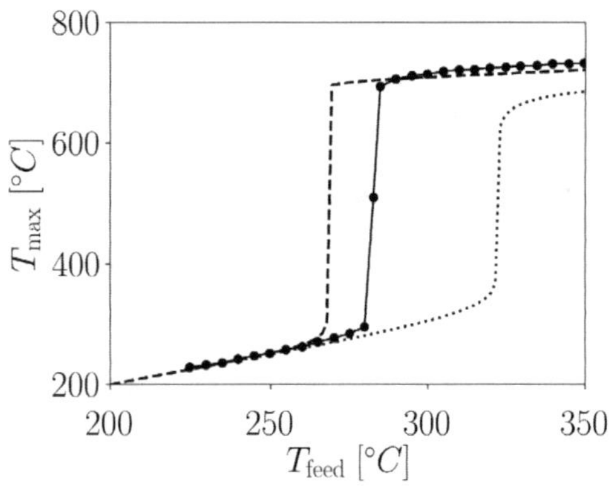

Figure 1. Maximum temperature in the PFR for a simulated reactor time-scale t_τ, obtained as a function of the feed (initial or inlet) temperature T_{feed}; see input parameters in [6]. The pseudo-homogeneous reactor model data from [6] is shown with a continuous line (——) and markers (●). The short tube Nu model described in this work is shown with a dashed line (– – –), while the long tube Nu model is shown with a dotted line (······).

3. Three-dimensional structured catalyst model

We now detail some considerations for a 3-D homogeneous structured catalyst model. For the 3-D model, we have resorted to the use of the open source library OpenFoam (v8) [11]. Specifically, we resort to the use of the reactingFoam solver, which solves a system of equations for the conservation of chemical species (mass fractions), mass (in the form of density), momentum, and total energy (formulated in terms of enthalpy). Assuming homogeneous media, i.e., uniform porosity, these equations are adapted from the homogeneous model in [2], that is,

$$\epsilon\frac{\partial(\rho Y_k)}{\partial t} + \epsilon\nabla\cdot(\rho\underline{V}Y_k) = -\epsilon\nabla\cdot(\rho\underline{V}_kY_k) + \overline{\dot\omega}_kW_k \tag{7}$$

$$\frac{\partial\rho}{\partial t} + \nabla\cdot(\rho\underline{V}) = 0 \tag{8}$$

$$\frac{\partial(\rho\underline{V})}{\partial t} + \nabla\cdot(\rho\underline{V}\circ\underline{V}) = -\nabla P + \nabla\cdot\underline{\underline{\tau}} + \underline{S} \tag{9}$$

$$\frac{\partial(\rho_{\mathrm{fs}}h)}{\partial t} + \epsilon\left[\nabla\cdot(\rho\underline{V}h) + \frac{\partial(\rho e_K)}{\partial t} + \nabla\cdot(\rho\underline{V}e_K) - \frac{\partial P}{\partial t}\right] = \nabla\cdot\underline{q}_{\mathrm{eff}} - \rho_{\mathrm{cat}}\sum_j\overline{\dot\omega}_{\mathrm{R},j}\overline{\Delta H}_{\mathrm{R},0} \tag{10}$$

Here, ∇ is the traditional Nabla operator, with $\cdot$ and $\circ$ symbolizing the scalar and dyadic products. Additionally, $\underline{V}$ is the velocity field, while $\underline{V}_k$ is a Fickian diffusion velocity field for species k. Likewise, P is the thermodynamic pressure, T is the temperature, $\underline{\underline{\tau}}$ is the shear stress tensor, $e_K = |\underline{V}|^2/2$ is the specific kinetic energy, and $\underline{q}_{\text{eff}} = \lambda_{\text{eff}} \nabla T$ is an effective heat flux, where λ_{eff} is an effective thermal conductivity. This is assumed to have a uniform value in axial direction $\lambda_{\text{eff},z}$, and a different uniform value in radial direction $\lambda_{\text{eff},r}$. The values $\lambda_{\text{eff},z}$ and $\lambda_{\text{eff},r}$ considered for this work are those mentioned in [5]; see [4] for related details. Also, ρ_{fs} is a fluid-solid weighted density, similar to the treatment in [2], i.e., $\rho_{\text{fs}} = \left[\epsilon\rho c_p + (1-\epsilon)\rho_s c_{p,\text{s}}\right]/c_p$, where ρ_{s} and $c_{p,\text{s}}$ are the density and specific heat capacities of the solid matrix.

The enforcement of the zero-net species mass-diffusion flux is achieved by the offset of errors into inert N_2 [12], i.e., $Y_{\text{N}_2} = 1 - \sum_{k \neq \text{N}_2} Y_k$. The chemical kinetics mechanism used for the 3-D homogeneous reactor model is the same one considered for the 0-D homogeneous reactor model. Unlike in the 0-D model, Cantera is not used for the processing of the reaction rates. Instead, OpenFoam itself is used. It is noted that the calculation of the equilibrium constant in OpenFoam differs from the calculation method used in Cantera, thus yielding different reaction rates (see the online OpenFoam v8 documentation).

The solution of equations (7-10) considers uniform initial conditions. The initial composition considered is that of a pure H_2 flow at the specified reactor feed temperature. This is similar to the reactor heating situation for catalyst pre-treatment in [7]. Likewise, zero Neumann boundary conditions (BCs) are considered for the chemical species at the reactor walls and reactor outlet, while fixed Dirichlet BCs are considered for the inlet gas composition. The velocity field considers no-slip BCs at the reactor walls, and inlet-outlet BCs. Finally, the temperature field considers zero Neumann BCs at the reactor outlet, fixed inlet gas temperature Dirichlet BCs, and fixed reactor (cooling) wall Dirichlet BCs, T_{w}.

3.1. Porous media modeling

The modeling of the porous media enters the momentum equation (9) via the additional drag term $\underline{S}$. Following the assumptions for homogeneous porous media, $\underline{S}$ is modeled following the Darcy-Forchheimer equation [1]. In practical terms, this is,

$$\underline{S} = -\frac{\mu}{K} u \hat{e}_z - \frac{F\epsilon}{\sqrt{K}} \rho |u| u \hat{e}_z \tag{11}$$

Here, u is the streamwise component of the velocity field, whereas $\hat{e}_z$ is the canonical streamwise vector, and μ is the dynamic viscosity. The permeability of the media K is estimated as in [13], using the hydraulic radius model and the Carman-Kozeny equation,

$$K = \frac{\epsilon^3 d^2}{36 k_K (1-\epsilon)^2} \tag{12}$$

where d is an equivalent spherical diameter obtained for the porous media, according to the specific surface area of the catalyst based on the solid volume, see [13]. Furthermore, k_K is the Kozeny constant, approximated as 5. Finally, F in equation (11) is obtained using the modified Ergun relation [13], such that

$$F = \frac{1.8}{\sqrt{180\epsilon^5}} \tag{13}$$

4. Case configuration and results

Dimensional parameters corresponding to the PC-HC and the G-3D structured catalyst reactors introduced in [5] are shown in table 1. In tables 2 and 3, we detail the operating and feed conditions for both the 0-D homogeneous reactor simulations, and the 3-D homogeneous reactor simulations. These are also taken from [5].

Table 1. Dimensional parameters for PC-HC and G-3D structured catalyst reactors.

Parameter	Dimension	Value PC-HC	Value G-3D
Monolith diameter (D)	cm	1.6	1.6
Monolith or reactor length (L)	cm	3.0	3.0
Catalyst surface area (S)	cm^2	540.0	504.0
Porosity (ϵ)	-	0.88	0.79
Estimated hydraulic diameter (D_h)	mm	2.66	1.58
Effective radial thermal conductivity ($\lambda_{\text{eff},r}$)	W/ (m K)	1.23	2.17
Effective axial thermal conductivity ($\lambda_{\text{eff},z}$)	W/ (m K)	2.31	3.91
Solid matrix specific heat capacity ($c_{p,\text{s}}$)	J/ (kg K)	466.0	466.0
Solid matrix density (ρ_s)	kg/m^3	7850.0	7850.0

Table 2. Operating conditions evaluated in the simulations.

Case	Condition	Dimension	Value
All cases	Weight hourly space velocity (WHSV)	m^3/ (kg$_\text{cat}$ s)	0.0167
	Gas volume flow rate (Q)	m^3/s	3.3333×10^{-6}
	Catalyst mass ($m_\text{cat} = Q/\text{WHSV}$)	kg	2.0×10^{-4}
	Catalyst mass per reactor volume (ρ_cat)	kg/m^3	1989.44
	Reactor (constant) pressure	kPa	101.325

Table 3. Feed compositions evaluated in the simulations.

Case	Condition
Case I (0-D)	Feed stream molar composition: 15% CO_2 + 60% H_2 + 25% N_2
	Inlet and reactor wall temperature: 473.15 – 673.15 K
Case II (0-D)	Feed stream molar composition: 8% CO_2 + 32% H_2+ 12% CH_4+ 48% N_2
	Inlet and reactor wall temperature: 473.15 – 673.15 K
Case III (3-D)	Feed stream molar composition: 15% CO_2 + 60% H_2 + 25% N_2
	Inlet and reactor wall temperature: 573.15 K

4.1. Zero-dimensional homogeneous reactor

The newly developed PC-HC and G-3D structured catalysts from [5], are evaluated in this section using a 0-D homogeneous reactor long tube Nu model. This evaluation corresponds to the numerical modeling of Cases I and II in table 3, using the operating conditions from table 2. In figure 2(a), the experimental results for CO_2 conversion for a Case I feed composition are shown and compared to the long 0-D reactor numerical model detailed in this work. The CO_2 conversion is defined based on the mole fractions of CO_2, X_{CO_2}, at the inlet and outlet of the reactor (or start or end of the simulated reactor time-scale),

$$x_{CO_2} = \frac{X_{CO_2,\text{inlet or start}} - X_{CO_2,\text{outlet or end}}}{X_{CO_2,\text{inlet or start}}} \tag{14}$$

The numerical results observe a large deviation with respect to the experimental results. First of all, we note that the chemical kinetics developed by [7], which are used in the numerical model, yield a lower CO_2 conversion in comparison to the actual experimental measurements. The kinetics developed by [7] correspond to Ni/$MgAl_2O_4$ catalysts, while the catalyst used in [5] is doped with Ru, i.e., Ru-Ni/$MgAl_2O_4$. Additionally, it is possible that by neglecting intermediate chemical species in the simplified chemistry, the established chemical equilibrium differs from the real reactor equilibrium. We observe that the real reactor runaway to chemical equilibrium is seen to happen at much higher temperatures in comparison to the numerical 0-D model, even when using the long tube Nu model. Despite the differences between experiments and numerics, the superior performance of the G-3D catalyst is confirmed in this work. The superior performance is attributed to the larger catalyst surface area of the G-3D catalyst, as suggested in [5]. Figure 2(b) shows similar CO_2 conversion results for a simplified biogas feed composition (Case II).

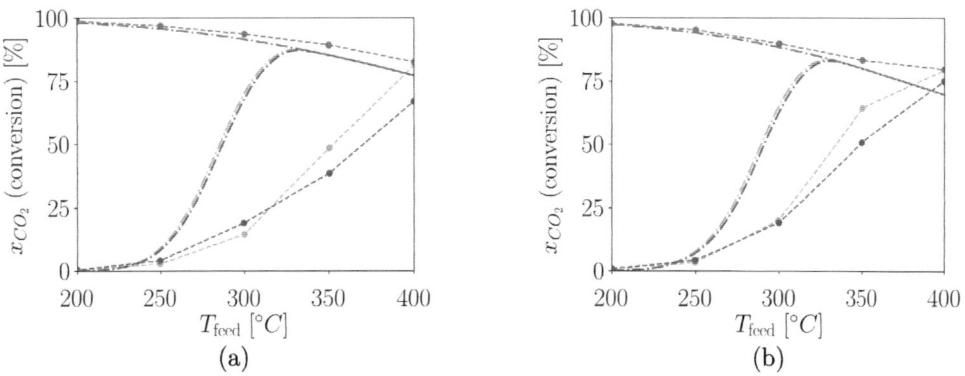

Figure 2. CO_2 conversion as a function of T_{feed}. (a) Feed composition Case I. (b) Feed composition Case II. Refer to tables 1-3 for operating conditions and feed compositions. Experimental data from [5] is shown with dashed lines (– – –) and markers (●), while simulation data is shown with dotted-dashed lines (— · —). Orange and magenta lines refer to the G-3D and PC-HC catalysts, respectively. The experimental chemical equilibrium, and the equilibrium obtained by the kinetics of [7], are shown with gray lines.

4.2. Three-dimensional homogeneous reactor

As an additional numerical tool for the evaluation and analysis of the experimental results in [5], we now perform 3-D homogeneous reactor simulations corresponding to the feed composition Case III (see table 3). Figure 3 shows the inlet, outlet and longitudinal fields for the CO_2 mass fraction in the simulated PC-HC catalyst, at half of the reactor time-scale. Results for the G-3D catalyst are very similar. There are no appreciable differences between the cases, and both yield CO_2 conversion values of approximately 90%, based on the cross-sectional averaged X_{CO_2} of the reactor inlet and outlet, see equation (14). The $\sim$ 90% value is of course larger than that obtained in the 0-D model for the same simulated inlet temperature (573.15 K). The reason for this are the higher flow temperatures in the regions away from the wall and the inlet, according to the temperature and composition BCs, in the 3-D model.

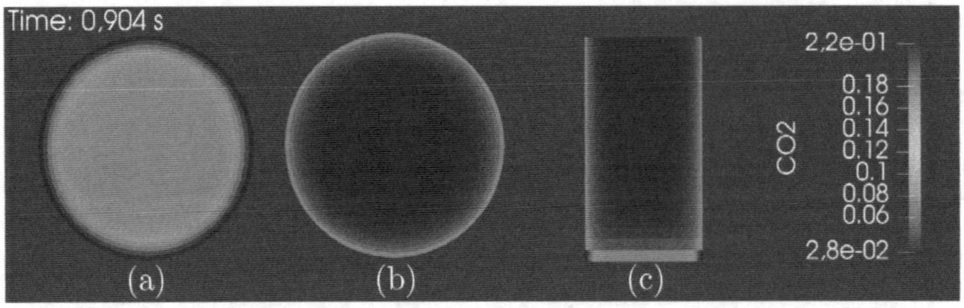

Figure 3. Snapshots of CO_2 mass fraction field at half of the reactor time-scale (feed composition Case III). (a) Inlet cross-section. (b) Outlet cross-section. (c) Longitudinal section.

5. Discussion and conclusions

We presented a framework for the homogeneous modeling of 0-D and 3-D structured catalysts. The modeling framework was applied for the numerical evaluation of the novel structured Ru-Ni/$MgAl_2O_4$ catalysts introduced in [5]. For that, chemical kinetics developed by [7] were used. Despite the numerical modeling being short on reproducing the experimental data, the numerical results confirm the superior performance of the G-3D catalyst in comparison to the PC-HC. This is solely due to the different porosity and catalyst surface area. The 3-D simulations were not conclusive in this work due to technical difficulties in the handling of the reaction rate calculation by OpenFoam, specifically regarding the calculation of the equilibrium constants, which differs from the method used in Cantera. In future work, we will use the Cantera library in OpenFoam for the calculation of the reaction rates, as a way to verify the 3-D results. The ultimate goal is to have a consistent comparison with the 0-D model, which simultaneously allows the adjustment of the chemical kinetics, and the evaluation of the monolith fluid-dynamic performance based on the porosity, drag, heat transfer, CO_2 conversion, among other variables. In this way, a reliable numerical tool can be developed for the future optimization of the catalyst design.

References

[1] Wood B D, He X and Apte S V 2020 *Annu. Rev. Fluid Mech.* **52** 171–203
[2] Liu B, Hayes R E, Yi Y, Mmbaga J, Checkel M D and Zheng M 2007 *Comput. Chem. Eng.* **31** 292–306
[3] Gao F and Goodman D W 2012 *Annu. Rev. Phys. Chem.* **63** 265–286
[4] Tronconi E, Groppi G and Visconti C G 2014 *Curr. Opin. Chem. Eng.* **5** 55–67
[5] Baena-Moreno F M, González-Castaño M, Navarro de Miguel J C, Miah K U M, Ossenbrink R, Odriozola J A and Arellano-García H 2021 *ACS Sustainable Chem. Eng.* **9** 8198–8206
[6] Schlereth D and Hinrichsen O 2014 *Chem. Eng. Res. Des.* **92** 702–712
[7] Xu J and Froment G F 1989 *AIChE J.* **35** 88–96
[8] Goodwin D G, Speth R L, Moffat H K and Weber B W 2018 Cantera: An object-oriented software toolkit for chemical kinetics, thermodynamics, and transport processes `https://www.cantera.org` version 2.4.0
[9] Gupta N and Balakotaiah V 2001 *Chem. Eng. Sci.* **56** 4771–4786
[10] Gundlapally S R and Balakotaiah V 2011 *Chem. Eng. Sci.* **66** 1879–1892
[11] Weller H G, Tabor G, Jasak H and Fureby C 1998 *Comput. Phys.* **12** 620–631
[12] Poinsot T and Veynante D 2014 *Theoretical and Numerical Combustion* (Aquaprint)
[13] Kaviany M 2012 *Principles of heat transfer in porous media* (Springer Science & Business Media)

Numerical analysis of influence of various bluff-body shapes on diffusion flame dynamics

Robert Kantoch[1], **Agnieszka Wawrzak**[1] **and Artur Tyliszczak**[1]

[1] Czestochowa University of Technology, Faculty of Mechanical Engineering and Computer Science, Al. Armii Krajowej 21, 42-201 Czestochowa, Poland

E-mail: robert.kantoch@pcz.pl, agnieszka.wawrzak@pcz.pl, atrur.tyliszczak@pcz.pl

Abstract. In this paper, various bluff-body shapes (cylindrical, square, star) and two different surface topologies (smooth, wavy) are applied as passive tools for controlling a non-premixed hydrogen flame in a combustion chamber. We focus on the dynamics of the flame and its time-averaged characteristics in the close vicinity of an injection system within formed recirculation zones and also in a far-field. The research is performed with the help of large-eddy simulations (LES) method using the ANSYS Fluent software and a high-order academic code SAILOR. Flame behaviour is found to be strongly dependent on the geometry of the bluff-body whereas its wall topology affects the flame characteristics only slightly. In the cases with the square and star bluff-body, small vortical structures originating at the corners deform large vortical structures created by the Kelvin-Helmholtz instability mechanism. This intensifies the mixing and combustion process and, in the configuration with the square shape bluff-body, translates to the shortening of the recirculation zone by 15% of the equivalent bluff-body diameter and the flame, which in the axis develops closer to the bluff-body. The star shape leads to the most uniform flame at the radial border or the recirculation zone.

1. Introduction

The development of efficient, safe and low-emission industrial devices (burners, combustion chambers, engines, etc.) is conditioned by the application of efficient passive or active flow and flame control methods. When properly selected, they ensure an optimal mixing process, reduce pollutant emissions, and prevent local extinctions or thermo-acoustic flame oscillations that could potentially lead to complete flame extinction.

In combustor chambers, one of the most commonly used methods of passive control relies on applying a bluff-body as a part of injection systems. It creates a recirculation zone, which stabilizes the flame position and makes the combustion process more homogenous, which, among others, allows lean mixtures to be burned in an efficient way. The dimension and inner structure of the recirculation zone and the flame formed in its region significantly depend on the bluff-body shape and flow parameters. Tyliszczak et al. [1] showed that a non-premixed flame stabilized in a central recirculation zone produced by a conical/circular bluff-body may extinguish following even a small change of the system parameters. Masri et al. [2] investigated the instantaneous mixing process and reactive scalar fields of non-premixed methane and hydrogen-enriched methane flames. They demonstrated that the combustion is most intense in the inner region of the recirculation zone. Both the passive and active control methods were applied jointly by Kypraiou et al. [3] in a combustion chamber with an axial and radial injection system. The active control relied on the acoustic excitation of the oxidizer stream. The performed analyses revealed a strong dependence of the flame on the amplitude and frequency of the excitations.

As it was recently found by Kuban et al. [4] for jet type flames, the level of control that can be achieved by a nozzle shaping is higher than previously expected. In general, it turns out that the jets coming out from sharp-edged orifices are more energetic compared to the jets generated by smoothly contoured nozzles, and therefore they are characterised by a faster axial velocity decay [5]. In contrast, recent research performed by Tyliszczak et al. [6] showed a different trend concerning the dynamics of the jets coming out from shaped orifices. The shorter potential core and faster velocity decay occurred rather for a circular jet than for a non-circular one, e.g., issuing from a triangular nozzle.

Undoubtedly, the investigations of the flow and flame response to changes at the inlet parameters often rely on a trial-and-error approach and require advanced research. Surprisingly, the literature related to these issues regarding the bluff-body stabilized flames seem to be rather poor. The purpose of this paper is to better understand the dynamics of non-premixed hydrogen flame behind three shapes of bluff-bodies (cylindrical, square, star) with flat or wavy surfaces. We apply the large-eddy simulations (LES) method that provides deep insight into the mechanism of the flame response on the modification of mixing intensity associated with particular bluff-body shapes.

2. Model formulation

2.1. Governing equations

We consider a low Mach number reacting flow for which the continuity, and transport equations of momentum and scalars within the framework of the LES method [8] are solved:

$$\partial_t \overline{\rho} + \nabla \cdot (\overline{\rho}\tilde{\mathbf{u}}) = 0 \tag{1}$$

$$\overline{\rho}\partial_t \tilde{\mathbf{u}} + (\overline{\rho}\tilde{\mathbf{u}} \cdot \nabla)\tilde{\mathbf{u}} + \nabla \cdot \overline{p} = \nabla \cdot (\tau + \tau^{SGS}) \tag{2}$$

$$\overline{\rho}\partial_t \tilde{Y_\alpha} + (\nabla \cdot \overline{\rho}\tilde{\mathbf{u}})\tilde{Y_\alpha} = \nabla \cdot (\overline{\rho}(D_\alpha + D_\alpha^{SGS})\nabla \tilde{Y_\alpha}) + \overline{\dot{\omega}}_\alpha \tag{3}$$

$$\overline{\rho}\partial_t \tilde{h} + (\nabla \cdot \overline{\rho}\tilde{\mathbf{u}})\tilde{h} = \nabla \cdot (\overline{\rho}(D + D^{DSG})\nabla h) \tag{4}$$

$$p_0 = \overline{\rho}R\tilde{T} \tag{5}$$

where the tilde and bar symbols denote filtered quantities, p_0 is the hydrodynamic pressure, ρ is the density, u_i are the velocity components and h stands for the total enthalpy. The symbols p_0 and R are the thermodynamic pressure and gas constant, respectively. The subscript α is the index of the species α = 1, . . . , N-species whereas the variables Y_α represent species mass fractions. An unresolved sub-grid stress tensor, resulting from the filtering of the non-linear advection terms is defined as $\tau^{SGS} = 2\nu_t S$, where S is the rate of strain tensor of the resolved velocity field and ν_t is the sub-grid viscosity computed according to the model proposed by Vreman [9]. The sub-grid diffusivities in equations 3, 4 are computed as $D^{SGS} = \nu_t/\sigma$ where σ is the turbulent Schmidt/Prandtl number assumed equal to 0.7.

Chemical source terms $\overline{\dot{w}_\alpha(\mathbf{Y}, h)}$ representing the filtering reaction rates of species α are computed directly from the Arrhenius formula and modelled as $\overline{\dot{w}_\alpha(\mathbf{Y}, h)} = \dot{w}_\alpha(\widetilde{\mathbf{Y}}, \widetilde{h})$ [10].

2.2. Numerical method

The computations are performed using ANSYS Fluent solver in the preliminary precursor simulations and an in-house SAILOR code as the main research tool. This code is based on the projection method for pressure-velocity coupling [7] combined with a predictor-corrector method (Adams-Bashforth / Adams-Moulton) applied for the time integration. The spatial discretization is performed on half-staggered meshes by the 6th order compact finite difference approximation for the momentum and continuity equations. The second-order TVD (Total Variation Diminishing) scheme with Koren limiter is used for the convective terms of the species mass fraction and enthalpy transport equations. Chemical reactions are modelled using CHEMKIN interpreter and a detailed mechanism proposed by Mueller et al. [11].

2.3. Computational setup

A typical combustion chamber with a conical bluff-body is shown in figure 1. It consists of a fuel pipe in the center and an outer section where from the bottom a co-flowing stream is provided through an inlet section with a cylindrical duct. Such a configuration has been used for instance in experimental [3] and numerical [1] investigations of the combustion process stability. Unlike in those works, in this paper,

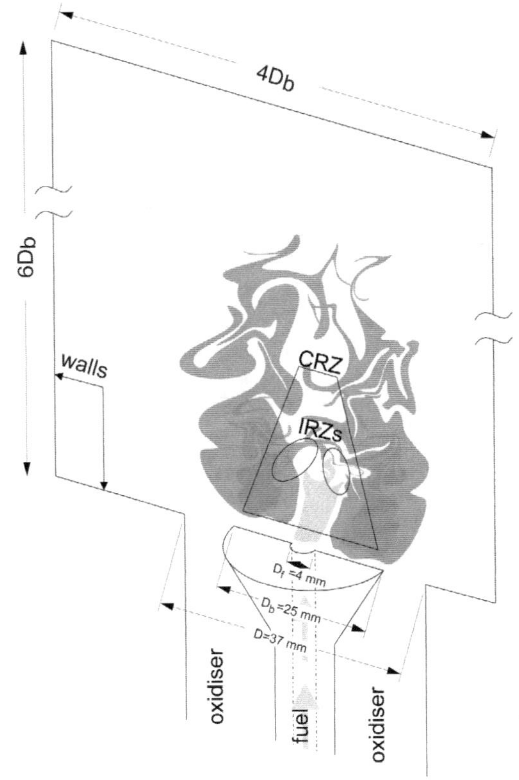

Figure 1: Sketch of the bluff-body combustion chamber.

besides the classical cylindrical bluff-body we consider its modified shape and assume it as a rectangular or a star. Additionally, in the configuration with the cylindrical bluff-body we analyze whether its wavy wall structure changes the flame behaviour compared to the case with the smooth wall. In all the cases the equivalent diameter of the bluff-bodies is $D_b = 25$ mm, and the outer duct diameter equals $D = 37$ mm. The fuel is a mixture of hydrogen and nitrogen with the mass fractions Y_{H_2} = 0.05 and Y_{H_2} = 0.95. The temperature of the fuel stream equals 300 K. It is injected into the combustion chamber through a thin pipe of diameter D_f = 4 mm with the axial velocity assumed equal to 10 m/s. Mixing of the fuel with the hot co-flowing air (1000 K) causes auto-ignition and the arising flame stabilizes as attached to the bluff-body plane in the region of the central and inner recirculation zones (CRZ, IRZs).

Figure 2 shows the computational domains and explains a two-part computational strategy. In the first part, the ANSYS Fluent solver with the WALE LES sub-grid model is used to model the flow inside the channel around the bluff-bodies. These precursor calculations are performed to acquire the instantaneous velocity fields at the end of the inlet section for a time period of $250D_b/U_b$ (U_b- bulk velocity of an oxidizer stream). The velocity profiles are collected at a distance of $0.04D_b$ upstream of the bluff-body upper surface (1-1 cross-section plane). This allows us to avoid errors caused by flow instabilities generated by sharp bluff-body edges. In the second part of the simulations, a high-order low Mach number LES code SAILOR [7] is used. The acquired velocity fields are superimposed onto the inlet plane of the main computational domain, which represents only the outer section of the combustion burner presented in figure 1. It is a rectangular domain with the length $6D_b$ in the axial direction ('y') and the width $4D_b \times 4D_b$ in the 'x-z' plane. Time-averaged results from the preliminary calculations, performed only for the air stream flowing in the duct around the cylindrical bluff-body, were used to verify the solution strategy. A comparison of the axial velocity profiles at three different heights above the bluff-body reveals that the imposed boundary conditions lead to similar velocity evolution downstream of the bluff-body. One can notice that the results obtained using the SAILOR code seem to be more

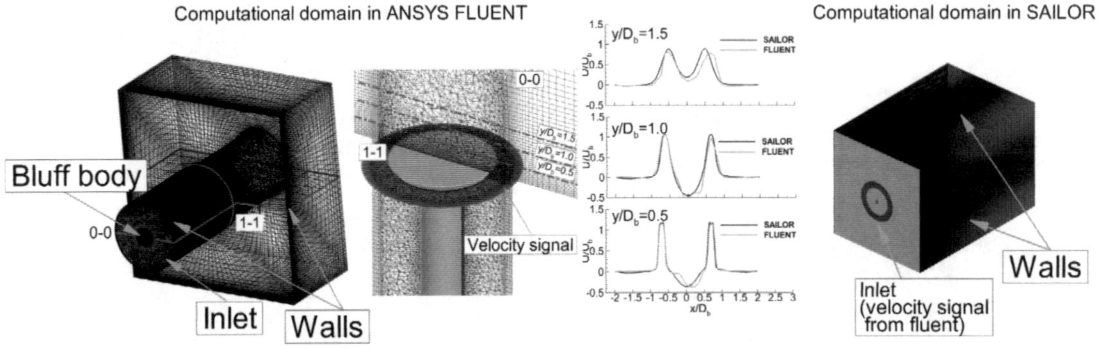

Figure 2: Computational configuration.

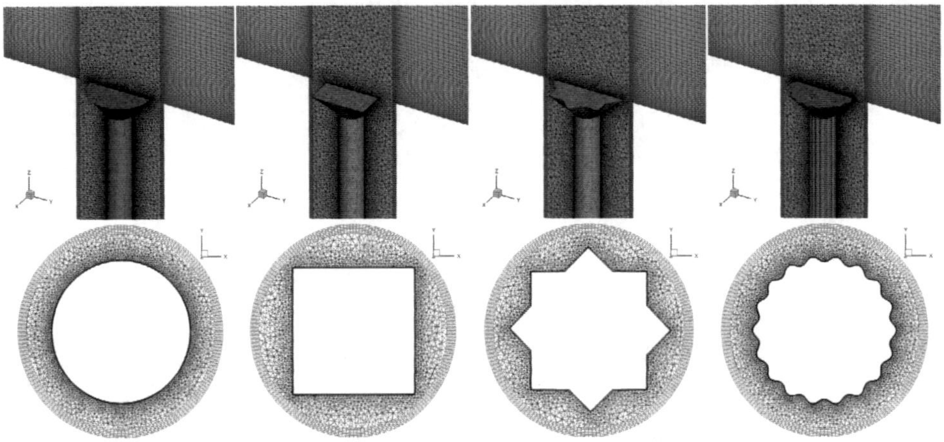

Figure 3: Computational meshes for various bluff-bodies. The bottom row presents the mesh points distribution at 1-1 cross-section plane.

accurate, at least qualitatively, as they reflect the symmetric shape of the bluff-body.

Figure 3 shows the cylindrical, square and star bluff-body and their flat or wavy surfaces. Note, that the grooves are oriented along the main flow direction. The numerical meshes were generated in the ANSYS Meshing module. The upper row shows them in the 0-0 plane and on the bluff-bodies surfaces whereas the bottom row displays the view on the meshes in 1-1 cross-section where the inlet velocity signals are gathered. The meshes have a block structure combining the hexahedral and triangular elements. This combination allows for a very precise mesh alignment to the complex bluff-body shapes. In all the cases, the near-wall mesh densities ensure a proper resolution of the boundary layer region with $y+ \approx 1$. The overall numbers of cells in the inlet duct are equal to 3.6×10^5 and 3.7×10^5 for the cylindrical bluff-body with the smooth and wavy wall, 3.5×10^5 for the rectangular bluff-body, and 3.6×10^5 for the star.

In the calculations carried out using the SAILOR code, the impact of the mesh density on the simulation results is assessed based on the time-averaged results obtained for 'cold' flow conditions (no chemical reactions). Five meshes with different numbers of nodes in the radial and axial directions are taken into account. The nodes are axially and radially compacted in such a way that a nearly uniform distribution of grid nodes is ensured in the inlet section. Figure 4 presents the axial profiles of the mean axial velocity in the simulations performed for the cases with the cylindrical (black line) and square (red lines) bluff-bodies. It can be seen that the use of the mesh counting of 144×192 nodes

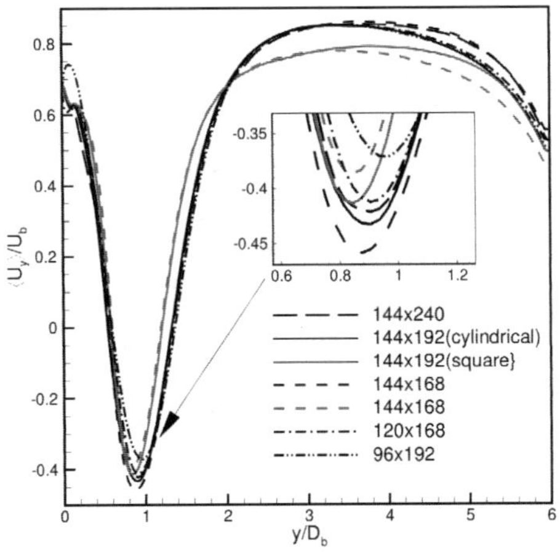

Figure 4: Profiles of the time averaged axial velocity (mean) along the axis obtained for different meshes and different bluff-bodies (black lines - cylindrical bluff-body, red lines - square bluff-body).

provides a nearly grid-independent solution. A further refinement in the axial direction (240 nodes) does not lead to a significantly different solution and only small quantitative differences are observed inside the recirculation zone. Therefore, taking into account the computational costs, the mesh consisting of $N_x \times N_z \times N_y = 144 \times 144 \times 192$ nodes is used for further comparative studies. In this case, the maximum cell volume is equal to 1.7×10^{-9} m^3, which corresponds to the minimum cell volume of the block-structured grids used in the computations performed with the ANSYS Fluent software.

3. Results

Figures 5-8 show instantaneous results for the cylindrical and square bluff-body shapes. First, the flow structure is discussed based on isosurfaces of Q-parameter presented in figures 5 and 6. They visualise toroidal coherent structures resulting from the Kelvin-Helmholtz instability. A shape of the vortex generated in close vicinity of the inlet plane is respective to the bluff-body shape, as can be seen in the zoomed views. The presence of small-scale vortices generated behind the sharp corners causes deformation of large vortical structures, which in the case with the star shape bluff-body are almost completely destroyed (not shown). This leads to an increase of the axial velocity fluctuations amplifying the mixing intensity, which form the point of view of the combustion process stability is very desirable. Instantaneous temperature distributions in the flames are presented in figures 7 and 8. It shows the temperature isosurfaces and contours in the (y-z) cross-section along the axial direction and in two different axial locations in the (x-z) planes. One can see that the flame dynamic is pronouncedly affected by the bluff-body shape both in the near field, where the flame propagates in the shear layer, as well as in the far-field. Here, the flame structure seems to be more complex for the square bluff-body due to the interaction of different scales vortices.

Despite of visible impact of the bluff-body shape on the instantaneous flame evolution its effective susceptibility to control can only be assessed by taking into account the time-averaged results. Figures 9 and 10 show the time-averaged mean axial velocity and its fluctuations along the axis. As can be seen, compared to the classical cylindrical shape the largest influence on the velocity distribution has the square bluff-body. In this case, in its close vicinity the axial velocity minimum is about U_y/U_b=0.1 larger and is shifted $15\%D_b$ towards the bluff-body. Concerning the differences in the fluctuations profiles it can be seen that for the case with the square bluff-body the maximum occurs earlier ($0.1D_b$) and achieves approximately 5% lower level. The shift and smaller value of the velocity minimum are the result of intensified mixing caused by the small vortices generated at the sharp corners. In effect, the fuel mass fraction in the axis drops the fastest, as can be seen in figure 11, and the temperature visibly rises, as can be seen in figure 12. In the localization $y/D_b = 1.6$ it is about 330 K higher than in the other cases and equals to approximately 1200 K. This means the presence of the flame, which causes a sudden drop of

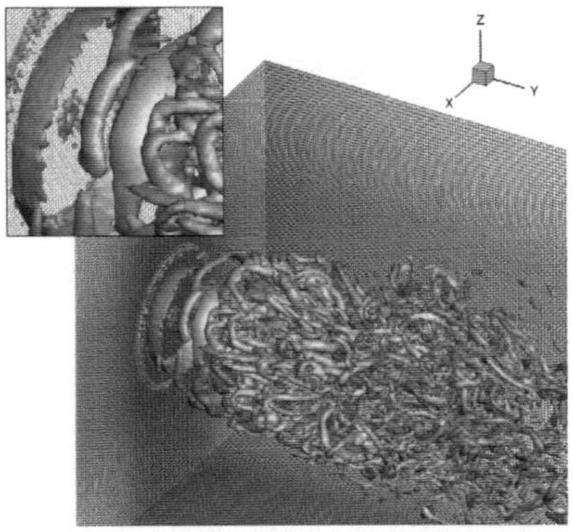

Figure 5: Isosurfaces of instantaneous Q-parameter. Results obtained for the case with circular bluff-body.

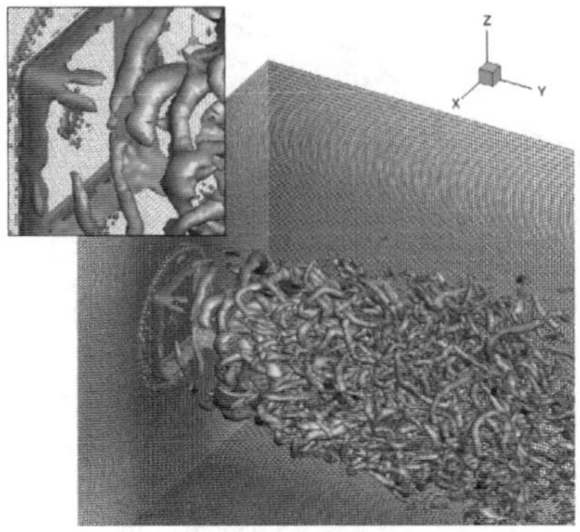

Figure 6: Isosurfaces of instantaneous Q-parameter. Results obtained for the case with square bluff-body.

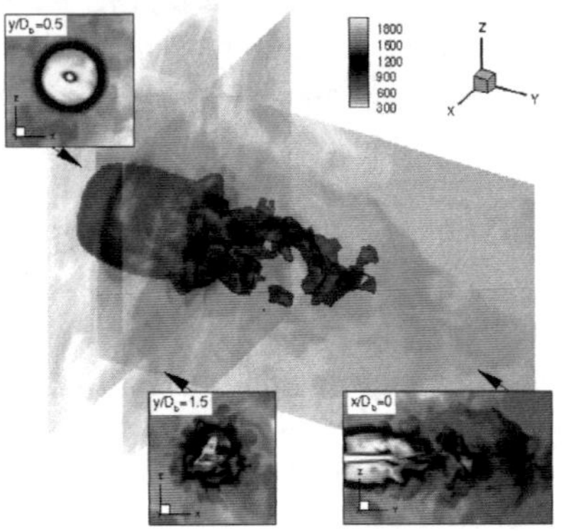

Figure 7: Isosurfaces and contours of instantaneous temperature. Results obtained for the case with circular bluff-body.

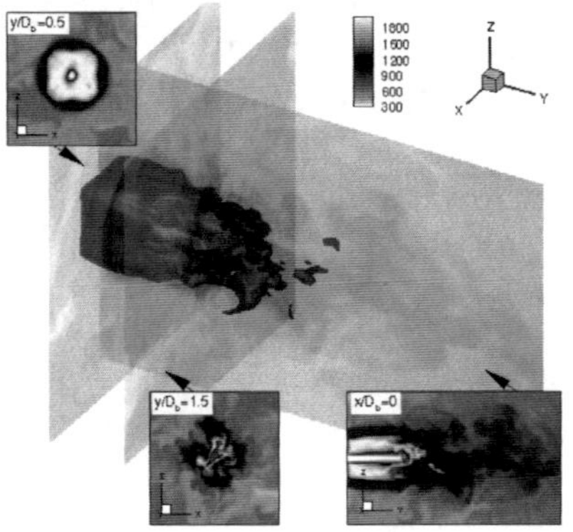

Figure 8: Isosurfaces and contours of instantaneous temperature. Results obtained for the case with square bluff-body.

the density that in the connection with the mass conservation law results in the flow acceleration. Indeed, from the localization $y/D_b \approx 1.6$ the velocity rapidly increases and its profile significantly diverges from the ones obtained for the remaining cases. In consequence, the flow dynamics in the far-field is pronouncedly different.

Figure 13 presents the localization of the point of occurrence of the maximum temperature in the radial direction. It can be regarded as a measure of the radial size of the flame. As can be seen, up to the distance $y/D_b = 1$ the profiles diverge from each other noticeably. Then, they start to overlap and moderately fluctuate ($\approx$5% of D_b). The profile obtained in the case with the star shape bluff-body is the smoothest that can be interpreted as an indicator of the most stable combustion process. A probable reason for such a situation is the occurrence of small vortices generated in the near corner regions. Compared to the square bluff-body, their large numbers cause almost complete destruction of the large Kelvin-Helmholtz like structures, as discussed at the beginning of this section. The lack of large vortices in the recirculation zone border prevents the natural flow oscillations, and thus, the flame shape becomes more regular.

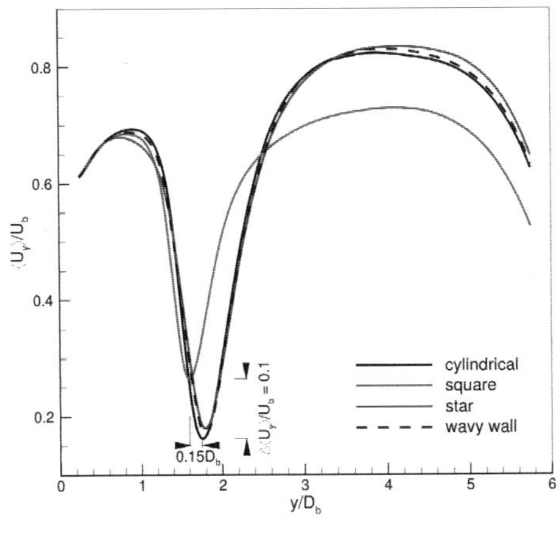

Figure 9: Time averaged mean profiles of axial velocity along the axis for different bluff-body shapes.

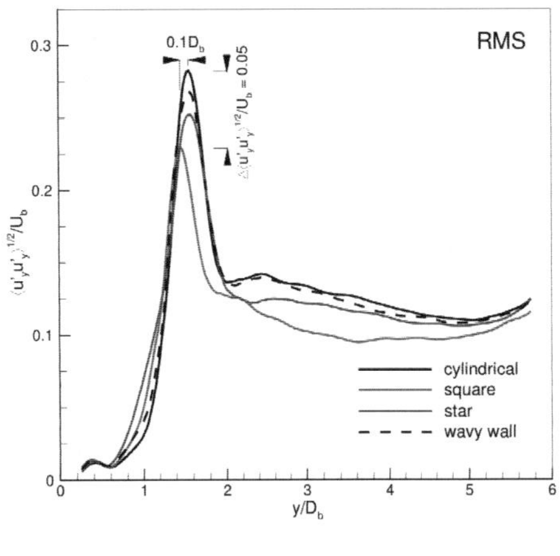

Figure 10: Time averaged RMS profiles of axial velocity along the axis for different bluff-body shapes.

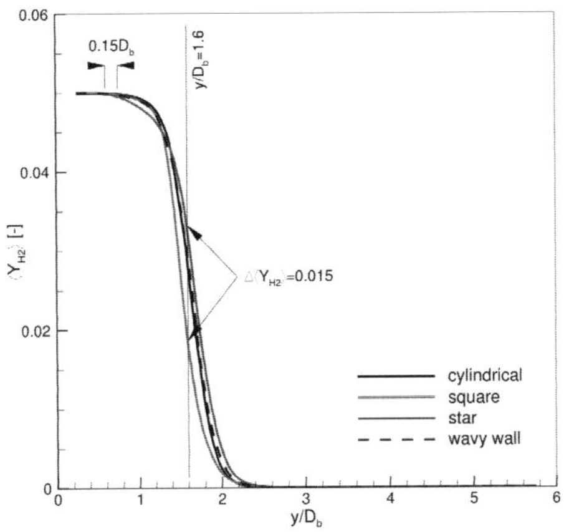

Figure 11: Time averaged mean profiles of H_2 mass fraction along the axis for different bluff-body shapes.

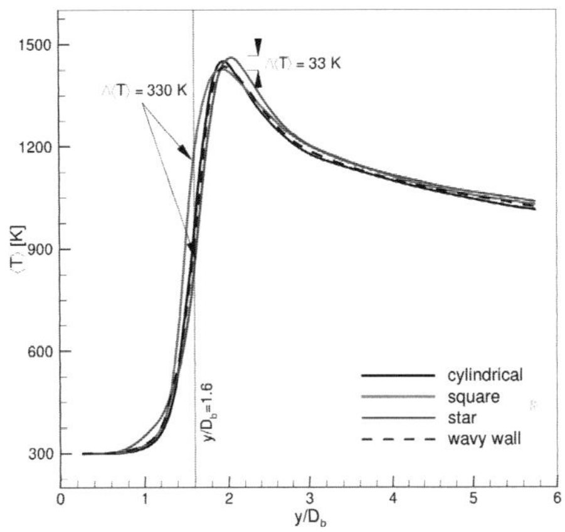

Figure 12: Time averaged mean profiles of temperature along the axis for different bluff-body shapes.

4. Conclusion

This research concentrated on a better understanding of the mixing process behind bluff-bodies and provided insight into the flame dynamics downstream of its various shapes, i.e., square, star and cylindrical, the latter with the smooth and wavy walls. The presented results showed that the flames adapt to the shape of the bluff-body in the initial region of the mixing layer, while the presence of sharp corners significantly affects the recirculation zone further downstream. The small vortical structures originating in the corners distort the large vortices and intensify the mixing process, and change the shape and structure of the flame. From the point of view of the reduction of the external flame border oscillations the most effective seems to be the star shape bluff-body.

In the configuration with the square bluff-body a shortened recirculation zone by 15% of the equivalent bluff-body diameter and the smallest axial velocity drop has been observed. Compared to the other cases, this caused a faster drop in the fuel concentration and a rapid temperature rise along the

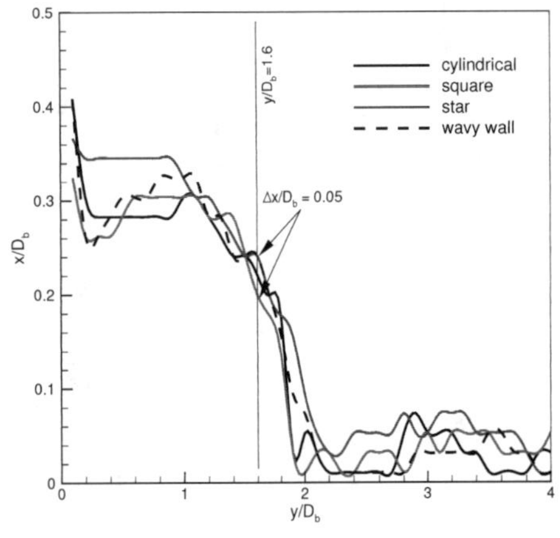

Figure 13: Radial location of maximum temperature along the axial direction.

axis. The wavy surface of the bluff-body affected the flame structure only slightly. Probably the impact of the waviness would be more pronounced if they have a larger amplitude but this issue requires further analysis.

Acknowledgments
This work was supported by the National Science Center in Poland (Grant 2020/39/B/ST8/02802) and statutory founds (BS/PB-1-100-3011/2022/P). The computations were carried out using the PL-Grid Infrastructure.

References
[1] Tyliszczak A, Cavaliere D E and Mastorakos E 2014 LES/CMC of blow-off in a liquid fueled swirl burner. *Flow Turbul. Combust.* **92** 237–267
[2] Masri A, Dally B, Barlow R and Carter C 1994 The structure of the recirculation zone of a bluff-body combustor *Symp. (Int.) Combust.* **25** (Amsterdam: Elsevier) pp 1301–1308
[3] Kypraiou A M, Allison P M, Giusti A and Mastorakos E 2018 Response of flames with different degrees of premixedness to acoustic oscillations *Combust. Sci. Technol.* **190** 1426–1441
[4] Kuban L, Stempka J and Tyliszczak A 2021 Numerical Analysis of the Combustion Dynamics of Passively Controlled Jets Issuing from Polygonal Nozzles *Energies* **14** 554
[5] Mi J and Nathan G J 2010 Statistical properties of turbulent free jets issuing from nine differently-shaped nozzles *Flow Turbul. Combust.* **84** 583–606
[6] Tyliszczak A, Kuban L and Stempka J 2022 Numerical analysis of non-excited and excited jets issuing from non-circular nozzles *Int. J. Heat. Fluid Flow* **94** 108944
[7] Tyliszczak A 2016 High-order compact difference algorithm on half-staggered meshes for low Mach number flows *Comput. Fluids* **127** 131–145
[8] Geurts B J 2004 Elements of Direct and Large-eddy Simulation (Edwards Publishing Philadelphia)
[9] Vreman A W 2004 An eddy-viscosity subgrid-scale model for turbulent shear flow: Algebraic theory and applications *Phys. Fluids* **16** 3670–3681
[10] Duwig C, Nogenmyr K J, Chan C K and Dunn M J 2011 Large eddy simulations of a piloted lean premix jet flame using finite-rate chemistry. *Combust. Theor. Model.* **15** 537–568
[11] Mueller M, Kim T, Yetter R and Dryer F 1999 Flow reactor studies and kinetic modeling of the H2/O2 reaction *Int. J. Chem. Kinet.* **31** 113–125

Strouhal number and vortices in far wake of cylinder

D Duda[1], V Yanovych[1,2], V Horáček[1] and V Uruba[1,2]

[1] University of West Bohemia in Pilsen, Faculty of Mechanical Engineering, Univerzitní 22, 306 14 Pilsen, Czech Republic

[2] Institute of Thermomechanics of the Czech Academy of Sciences, Dolejškova 5, 180 00 Prague, Czech Republic

E-mail: dudad@kke.zcu.cz

Abstract. The wake past a smooth cylinder perpendicular to the flow is studied $15D$ (cylinder diameters) downstream by using particle image velocimetry (PIV). Spatial shedding period is used to estimate the spatial Strouhal number (Sh) and to compare with time-resolved hot-wire anemometer (HWA). The vortices are detected within the spatial fluctuation field. The statistical vortex properties do not change within the explored range of Reynolds numbers (from 5 000 to 60 000), similarly the spectra and Sh.

1. Introduction

In 1878 Strouhal published his famous work about tones produced by a wire moving through static air [1]. By varying wire diameter and speed, he found that the tone follows ratio of these two parameters and the tone frequency $f \approx 0.2u/D$, where D is wire diameter and u its velocity. His conclusion remains valid in large range of speeds, dimensions and even object shapes. Nowadays, the ratio of fD/u is considered to be an useful parameter mainly in the problems of acoustic noise, induced vibrations or validation of numerical simulations. The direct measurement of frequency (typically by using hot wire anemometry or some microphone) has to deal the question, what is the velocity u – the local velocity or some averaged velocity or some artificial reference velocity?

Another question is, how far is enough to stabilize the wake. Jiang and Cheng [2] observed in a 2D numerical simulation at low Reynolds number a transition from classical von Kármán vortex street to, first *two-layered vortex street* and later to *secondary vortex street* with much longer shedding frequency. These transitions have been observed at downstream distance around 20 and 30 cylinder diameters respectively, which is much more far than our measurement at 15 cylinder diameters.

2. Experimental setup

The measurement has been performed in a open wind tunnel [3, 4] with speeds from 5m/s to 60m/s, the test-section has dimensions $750 \times 306 \times 204$mm. The cylinder of diameter $D = 15$mm is placed perpendicularly to the flow direction and its length fills the entire height of the test-section (i.e. 204mm). The Particle Image Velocimetry (PIV) [5] measurement plane is oriented perpendicularly to the cylinder axis in the middle of the test-section height. Field of view (FoV) of size $34.2 \times 34.2\text{mm}^2$ is illuminated by solid-state double-pulsed laser New Wave Solo and observed by a double-frame camera FlowSense MkII. The Safex fog generator produces the seeding particles. In the next, u means the stream-wise velocity component, v is the velocity component perpendicular to the stream and to the cylinder axis as well; the third component is not measured. Figure 1 shows the example of measured PIV data.

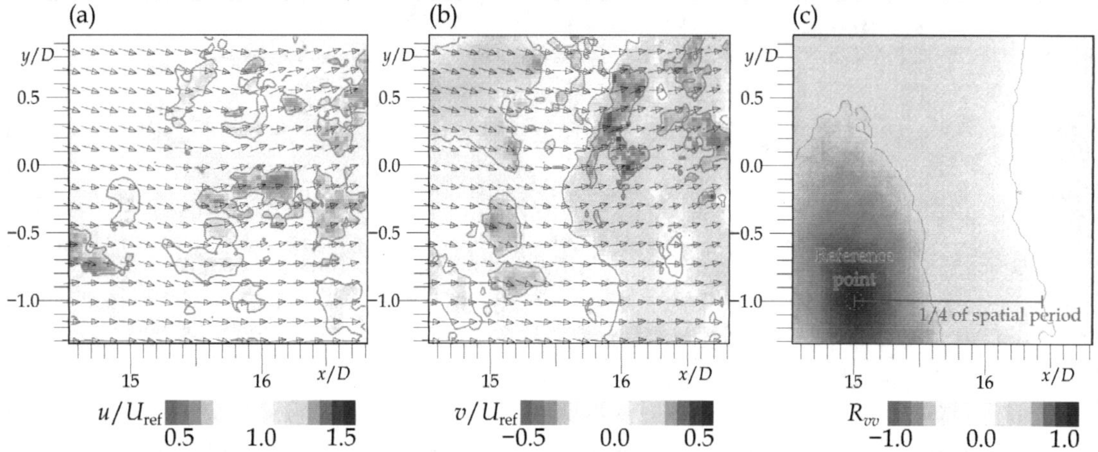

Figure 1. (a) Example of measured experimental data; colors correspond to stream-wise velocity component; only every 4th vector is plotted for sake of clarity. (b) The same snapshot with depicted velocity component v perpendicular to both: stream and cylinder axis. (c) Correlation function of cross-stream velocity v with reference point depicted by the circle (R_{vv}, equation (1)). The distance towards area with zero correlation can be understood as one quarter of Strouhal period; however, it is calculated by using the integral length-scale, which is an integral property less sensitive to noise.

Hot Wire Anemometry (HWA) [6] studies were performed using a miniature X-wire probe 55P63. This allowed us to investigate in more detail the two-dimensional structure (u and v components) of the turbulent flow [7]. The wire probes with tungsten platinum coating have a diameter of 5μm and an active length of 1.25mm. The ends of the wire are gilded with a thickness of 15μm to 20μm. Calibration of velocity sensors was performed by the StreamLine system. This system includes units for generating the flow and supply of dried air. Collis-Williams law was used to calibrate and estimate the instantaneous flow rate. This method was described in details in [8]. Hot-wire signals for all experiments were sampled at 70kHz. The measurement time at one point was 10s. The filtering of the received signal was also applied. For the High Pass filter the frequency was 10Hz, while for the Low Pass filter, the frequency was 30kHz.

3. Strouhal number estimation by using spatially resolved PIV data

Spatially resolved velocity data obtained e.g. by using Particle Image Velocimetry (PIV) [5] method offers a different approach: the Strouhal number $\mathrm{Sh} = fD/u$ can be seen as spatial wavelength λ, i.e. $\mathrm{Sh}_s = D/\lambda$ (subscript s like *spatial*). The wavelength can be read out of noisy turbulent data by using the Proper Orthogonal Decomposition [9] or by using simpler property of correlation function, which inherits the periodicity of the direct parameter [10]. Autocorrelation function [10] of fluctuation velocity component v (perpendicular to both: stream and cylinder axis) is calculated as

$$R_{vv}(\vec{x}) = \frac{\langle v'(\vec{x}) \cdot v'(\vec{x}_0) \rangle}{\sigma_v(\vec{x}) \cdot \sigma_v(\vec{x}_0)} \tag{1}$$

where $\vec{x}_0$ is some reference point depicted in figure 1c by circle and $v' = v - \langle v \rangle$ is fluctuation velocity; σ_v is standard deviation, $\sigma_v^2 = \langle v^2 \rangle - \langle v \rangle^2$. The autocorrelation function of velocity component v is displayed in figure 2a and compared with the HWA data. Figure 2b shows the comparison of stream-wise component u. The length of vortex shedding period is quantified by using *integral lengthscale* L_{vv} of autocorrelation function [11]

$$L_{vv} = \int_0^{X, R_{vv}(X)=0} R_{vv}(x) \mathrm{d}x \tag{2}$$

Integration continues up to a point, where R_{vv} reaches zero [11].Then, the wavelength is $\lambda = 2 \cdot 4 \cdot L_{vv}$ (2 because triangular shape of R_{vv}), thus $\mathrm{Sh}_s = \lambda / D$ reaching values between 0.20 and 0.25, see figure 3a. For comparison, the classical way of measuring Strouhal number is to normalize the frequency of dominating peak in power spectral density of HWA signal, see figure 3b. The frequency is normalized by using *local* average velocity, which reaches $80 - 90\%$ of the reference velocity, see figure 4a. Here we see, that the PIV spectral method is unable to detect Strouhal peak – the dynamical range of PIV is rather small limited by size of FoV on one side and by the size of one grid point on the other side.

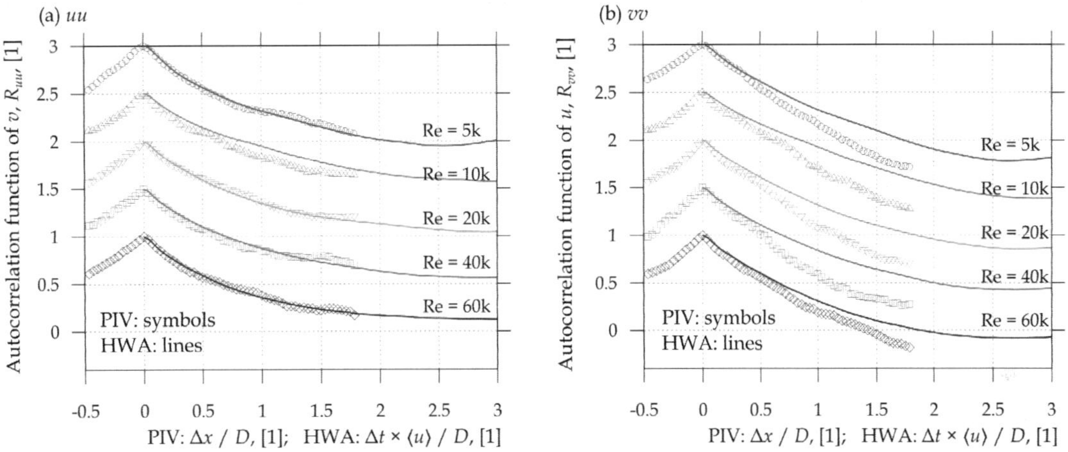

Figure 2. (a) Autocorrelation function R_{uu} of stream-wise velocity component as a function as dimensionless downstream distance or time delay respectively. (b) R_{vv} of cross-stream component. The y-value is shifted by 0.5 for each velocity consecutive to avoid chaos in similar data. Symbols represent the PIV data, while HWA data are plotted as solid lines. The range of PIV data is limited by the field of view, while the range of HWA data is limited only by the measuring time (10 s).

The velocity fluctuations are dominated by the cross-stream direction, because the von Kármán vortex street is more active in such direction. This effect is slightly stronger at lower velocities, where there is less smaller-scale turbulence and the dominance of vortex street is even stronger. Figure 4b shows the *degree of anisotropy* calculated according to [15] as

$$DA = \frac{\sigma_v - \sigma_u}{\sigma_v + \sigma_u} \tag{3}$$

where σ_u, σ_v are the standard deviations in stream-wise or cross-stream direction respectively. Note, this is not the only single definition of a quantity describing the anisotropy, other ways can be found in [14, 16, 17]. We are not able to explain the discrepancy between HWA and PIV measurements at middle Re.

Journal of Physics: Conference Series **2367** (2022) 012017 doi:10.1088/1742-6596/2367/1/012017

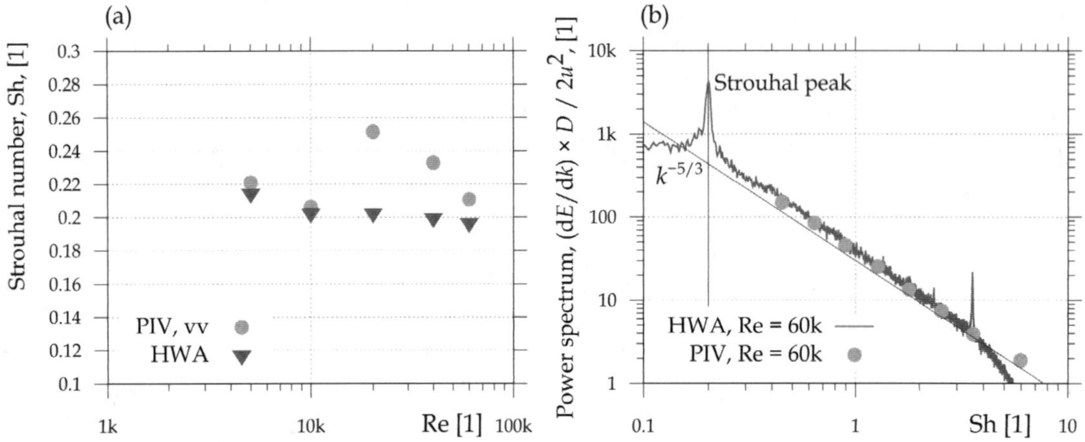

Figure 3. (a, green circles) Strouhal number measured by using integral lengthscale L_{vv} of correlation function of cross-stream velocity component measured by PIV. (a, red triangles) Strouhal number measured by using the dominating frequency in HWA data as a function of Reynolds number based on reference velocity and cylinder diameter. (b) Power spectral density obtained by PIV convoluted by set of band-pass filters [12] (green circles) and by Welch spectrum [13] of HWA time series (red line) respectively. Only data at $Re = 6 \cdot 10^4$ are shown. The Strouhal number (x-axis) for HWA data is calculated as $\mathrm{Sh}_{\mathrm{HWA}} = f \cdot D/u$, while for PIV data it is $\mathrm{Sh}_{\mathrm{PIV}} = D/(2\pi\sigma)$.

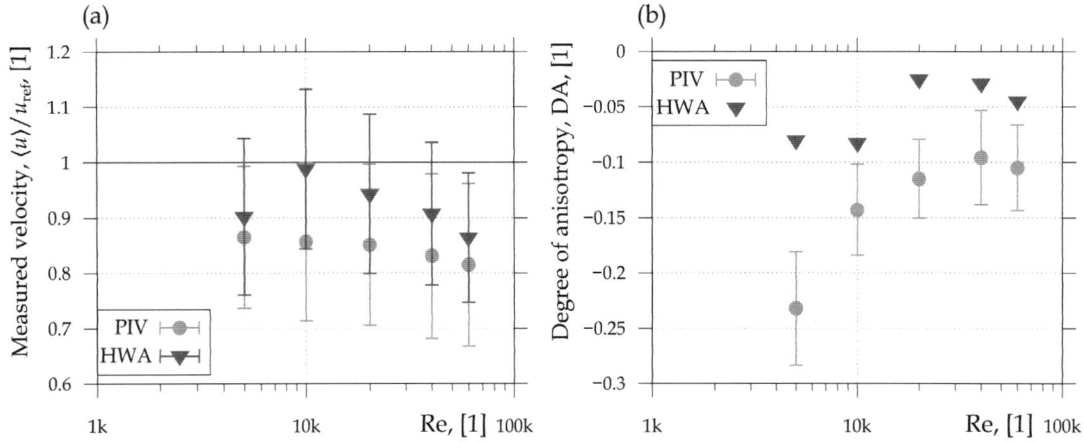

Figure 4. (a) Measured local average stream-wise velocity $\langle u \rangle / u_{ref}$ normalized by reference velocity. The errorbars represent the standard deviation of the signal (i.e. real turbulence + method noise). PIV data: green circles; HWA data: red triangles. (b) Degree of anisotropy $DA = (\sigma_u - \sigma_v) / (\sigma_u + \sigma_v)$ between fluctuations in stream and cross directions by both measuring methods, [14, 15].

4. Individual vortices of sub-von Kármán size

Strouhal shedding frequency of von Kármán vortex street describes the large structures of the wake. However, the dissipation, turbulent mixing and other phenomena take place at much smaller lengthscales highlighted by using the *Agrawal decomposition* [18, 19, 20]. Figure 5 shows example of instantaneous spatial structures of different lengthscales. The same snapshot as in figure 1a is used; i.e. at Re $= 6 \cdot 10^4$. The probed lengthscale is defined by the σ-parameter of corresponding Gaussian function[1] convoluted[2] with the measured velocity field; more details in [12, 20]. Vortices smaller or comparable to the PIV grid resolution are invisible for this method and even vortices covering 2 or 3 grid points are barely detectable in measured field containing noise. Therefore we select the lengthscale interval from $\sim 0.22D, (\sigma_L = 0.035D)$ to $\sim 1.6D, (\sigma_H = 0.25D)$. Example of instantaneous structures of these lengthscales is displayed in figure 5b The vortices at these scales are far from both: vortex street scale (represented by Strouhal) and dissipation scale (represented by Kolmogorov scale). At the same time, this lengthscale interval contains interesting and complex turbulent structures.

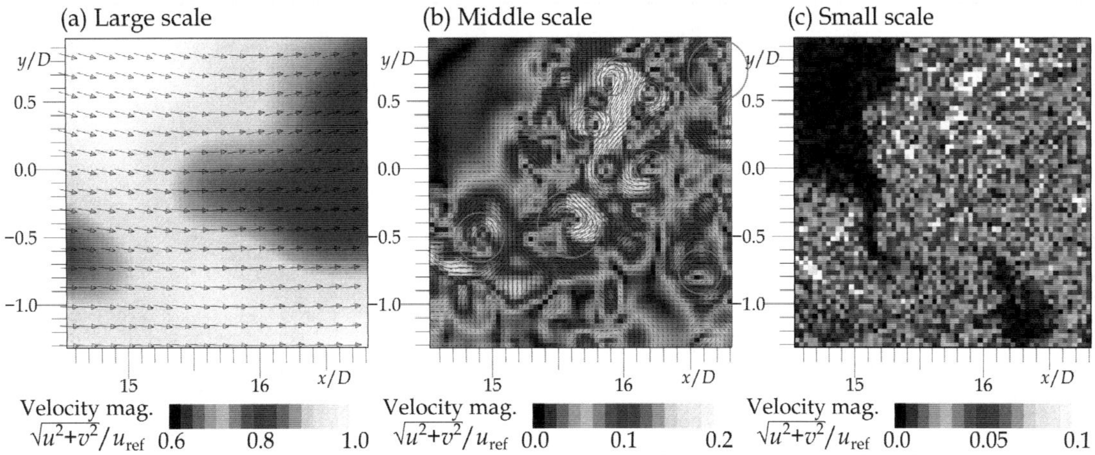

Figure 5. Example of instantaneous snapshot (see figure 1a) convoluted with (a) Gauss function of $\sigma_L = 0.25D$, (b) with the difference of two Gauss functions with $\sigma_L = 0.25D$ and $\sigma_s = 0.035D$ and (c) the rest after subtracting convolution with Gauss function of $\sigma_s = 0.035D$. The circles in panel (b) show positions of first 10 vortices algorithmically detected in this snapshot.

The algorithm of searching individual vortices is in depth described in a book chapter [21] and in conference contribution [22]. Figure 5b shows detected vortices in a single snapshot with highlighted middle-scale structures. Due to the lack of some well developed stop condition, the vortices are searched only up to certain artifitial limit, which is set to 10 in this case. Thus all following observations are valid only for the ten most dominant vortices additionally limited by the lengtscale described upper. Within these limits, it seems, that the size of vortices does not depend on the flow velocity and the population maximum lies around $0.1D$, see the probability density function in figure 6a.

The profiles of these vortices does not display some dependence on velocity as well (not plotted), instead, there can be observed a weak dependence on the vortex size – the vortex velocity profile displayed in figure 6b shows slightly faster velocity decay for larger vortices than for smaller ones. The threshold for declaring *small* or *large* vortex is the value of vortex core radius $R \approx 0.1D$, where the population is maximal. The observed profile does not follow the standard Oseen[3] vortex profile, because any larger structure is removed from the velocity field

[1] Gauss function $G[\sigma](\vec{x}) = \left(2\pi\sigma^2\right)^{-1} \cdot \exp\left(-x^2/2\sigma^2\right)$.

[2] convolution of two functions $f(x)$ and $g(\vec{x})$ is $(f * g)(\vec{x}) = \int\int f\left(\vec{x} - \vec{\xi}\right) \cdot g\left(\vec{\xi}\right) \cdot \mathrm{d}\vec{\xi}$.

[3] Oseen vortex tangential velocity profile: $u_\vartheta(r) = \frac{\Gamma}{2\pi R} \cdot \frac{R}{r} \cdot \left(1 - \exp\left(-\frac{r^2}{R^2}\right)\right)$, where r is the distance from vortex core and R is the vortex core radius.

by using the band-pass filtering shown in figure 5. The observed profile is closer to the Taylor[4] vortex, but, in respect to that profile, we observe a small, but clear area of velocity orbiting counter the orbital sense of vortex core. This effect is slightly stronger for larger vortices (blue line in figure 6b).

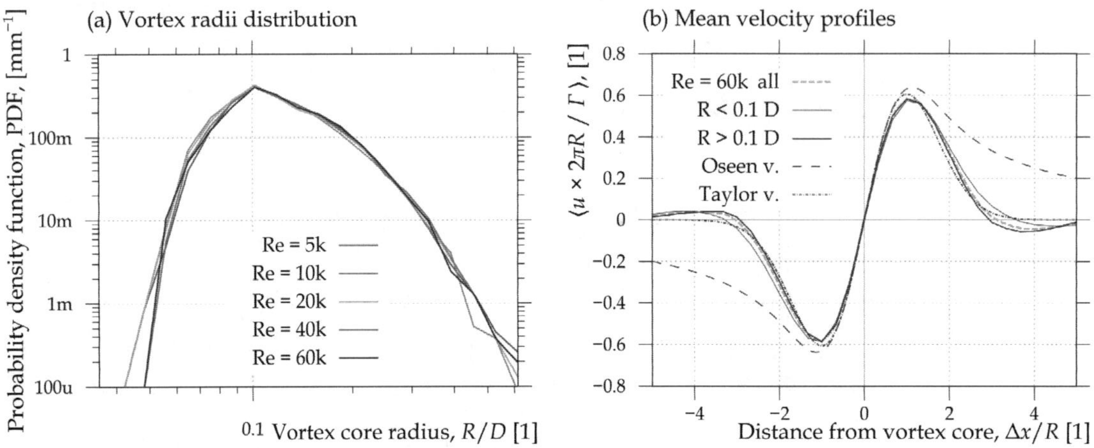

Figure 6. (a) Probability density function of vortex core radius R. Note that the distribution has to be naturally limited by the used band-pass filter. (b) Ensemble average of measured PIV velocity in normalized distance from each vortex position; the distance is normalized by detected vortex radius R; the velocity is normalized by the orbital velocity of each vortex $G = \Gamma/2\pi R$ before averaging. The data only for $\mathrm{Re} = 6 \cdot 10^4$, others are similar; the population is divided into small and large vortices. For comparison, two standard vortex models are plotted: the Oseen vortex with hyperbolic envelope and the Taylor vortex with Gaussian envelope, which fits better due to the Gauss-based filtering of the velocity field.

The distribution of normalized amplitude of orbital velocity $G = \Gamma/2\pi R$ is similar at all explored velocities, however, there is a very small decrease with increasing velocity, see figure 7a. The dependence on R (vortex core radius) shows clear decreasing tendency with R (see figure 7b) since larger R, where the number of vortices decreases. The orbital velocity G of vortices of size around $R/D \sim 0.1$ remains constant.

5. Conclusion

Both methods, PIV and HWA, are a standard tool for turbulence investigation. Each of them offer different advantages: PIV naturally maps the flow structures allowing deeper understanding of the fine physical mechanisms. However, its dynamical range of probed lengthscales is rather small (size of field of view to grid size – typically 1:64). Hot wire anemometry has much richer history of turbulence analysis [6], its astonishing range of probed timescales allows exploration from low-frequency wind tunnel stability up to electrical noise far behind the Kolmogorov scale. On the other hand, it is only a point method and the interpretation of measured signal is not straight forward.

The Strouhal number is measured by using PIV data within FoV smaller than the spatial Strouhal frequency by using the periodicity of autocorrelation function. But, the correspondence

[4] Taylor vortex is a second solution of Legendre expansion of Navier-Stokes equations; the tangential velocity u_ϑ depends on distance as $u_\vartheta(r) = \frac{\Gamma}{2\pi R} \cdot \frac{R}{r} \cdot \exp\left(-\frac{r^2}{2R^2}\right)$, where r is the distance from vortex core and R is the vortex core radius.

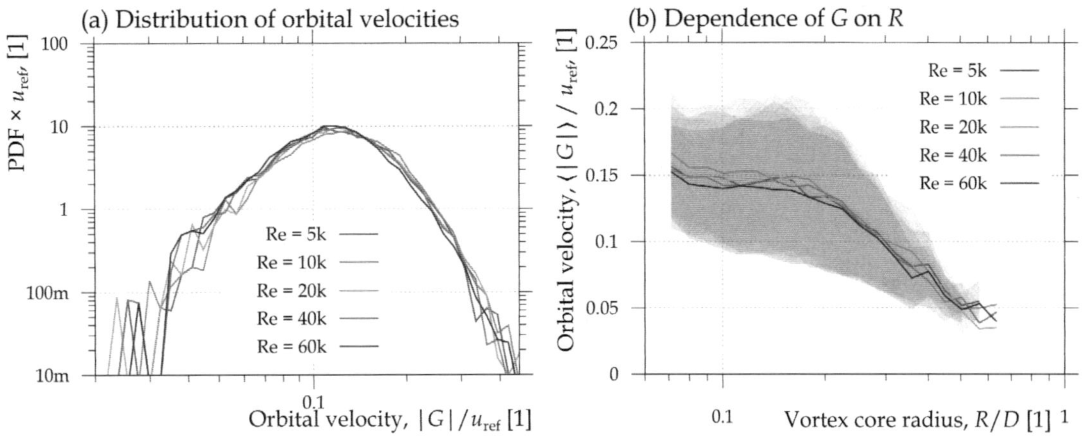

Figure 7. (a) Probability density function of the absolute value of vortex core orbital velocity $G = \Gamma/2\pi R$. (b) Dependence of average absolute value of G on the vortex core radius R. The shadowed areas represent the standard deviation of $|G|$ of each subset at given R.

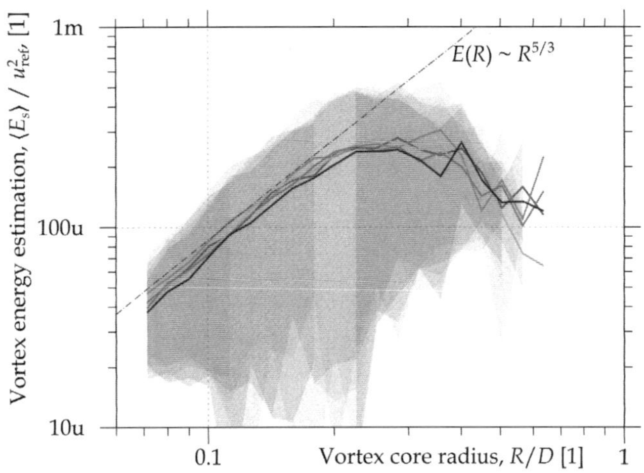

r[ht]

Figure 8. Energy of vortices of given radius (x-axis). The shadow area represents the standard deviation of the ensemble of each subset at given R (note the area cannot be plotted in points with standard deviation is larger than the average). The energy is estimated as the difference between original velocity field and the velocity field with subtracted vortex, more details in [22].

with HWA data is not satisfactory (figure 3a). On the other hand, the energy spectra show a reasonable correspondence (figure 3b). Therefore our algorithm [12] of spectral separation can be considered to be correct.

The inertial range vortices are directly observable by using PIV, their properties can be found and studied statistically, which opens a new way of turbulence exploration inaccessible for HWA. E.g., the observation, that large vortices rotate less than smaller ones, seems contradicting the the mechanism of Richardson cascade [23]. However, it applies only for the larger vortices, whose amount is small (note logarithmic plot in figure 6a); vortices of size around $R/D \sim 0.1$ display orbital velocity independent on radii and therefore their energy scales as expected – see the Figure 5, which shows the average dependence of vortex energy as a function of core radius; the dash-dotted line represents the scaling $E(R) \sim R^{+5/3}$.

Acknowledgments
The research was supported from ERDF under project LoStr No. CZ.02.1.01/0.0/0.0/16_026/0008389.

References
[1] Strouhal V 1878 *Annalen der Physik* **241**(10) 216 – 251
[2] Jiang H and Cheng L 2019 *Journal of Fluid Mechanics* **867** 691–722
[3] Yanovych V, Duda D, Horáček V and Uruba V 2019 *AIP Conference Proceedings* **2189**
[4] Yanovych V and Duda D 2020 *Strojnicky Casopis* **70** 181–196 doi: 10.2478/scjme-2020-0030
[5] Tropea C, Yarin A and Foss J F 2007 *Springer Handbook of Experimental Fluid Mechanics* (Springer, Heidelberg, DE)
[6] King L V 1914 *Philosophical Transactions of the Royal Society of London. Series A, Containing Papers of a Mathematical or Physical Character* **214**(509-522) 373–432
[7] Jonas P and Rehak V 1991 *Strojnicky Casopis* **42** 451–458
[8] Yanovych V, Duda D, Uruba V and Antoš P 2021 *SN Appl. Sci.* **3** 885
[9] Uruba V 2010 *Transactions of the VŠB – Technical University of Ostrava, Mechanical Series* **56**(3) 157–162
[10] Schulz-DuBois E O and Rehberg I 1981 *Applied Physics* **24** 323–329
[11] Azevedo R, Roja-Solórzano L R and Bento Leal J 2017 *Flow Measurement and Instrumentation* **57** 10 – 19 ISSN 0955-5986
[12] Duda D and Uruba V 2019 *ASME J of Nuclear Rad Sci.* **5**(3)
[13] Welch P 1967 *IEEE Transactions on Audio and Electroacoustics* **15**(2) 70–73
[14] Solís-Gallego I, Meana-Femández A, Femández Oro J M F, Argüelles Díaz K M A and Velarde-Suárez S 2017 *Journal of Applied Fluid Mechanics* **10** 1013–1027
[15] Porreca L, Hollenstein M, Kalfas A I and Abhari R S 2007 *Journal of Propulsion and Power* **23** 227–234
[16] Romano G P 2020 *International Journal of Heat and Fluid Flow* **83**
[17] Duda D, Yanovych V, Tsymbalyuk V and Uruba V 2022 *Energies* **15**(3) 1227
[18] Agrawal A and Prasad A 2002 *Experiments in Fluids* **33** 565–577 ISSN 07234864
[19] Agrawal A 2005 *Experiments in Fluids* **39** 836–840 ISSN 07234864
[20] Duda D and Uruba V 2018 *AIP conference proceedings* vol 2000 p 020005
[21] Duda D 2021 *Vortex Dynamics – From Physical to Mathematical Aspects [Working Title]* ed Bakırtaş I and Antar N (IntechOpen)
[22] Duda D 2020 *In Proceedings Topical Problems of Fluid Mechanics 2020* ed Šimurda D and Bodnár T pp 56–63
[23] Richardson L 1922 *Weather prediction by numerical process* (Cambridge, The University press) ISBN 9780511618291

Journal of Physics: Conference Series **2367** (2022) 012018 doi:10.1088/1742-6596/2367/1/012018

Experimental study of laminar-to-turbulent transition in an adverse pressure gradient flow

K Gumowski[1] **and S Kubacki**[1]

[1]Warsaw University of Technology, Faculty of Power and Aeronautical Engineering, Institute of Aeronautics and Applied Mechanics, Nowowiejska 24, 00-665, Warsaw, Poland

E-mail: konrad.gumowski@pw.edu.pl

Abstract. Experimental results on laminar-to-turbulent transition in a separated laminar boundary layer formed over a flat plate in an adverse pressure gradient are provided. Experiments have been performed for a range of Reynolds numbers ($1.4 - 2.4 \cdot 10^5$) and for two freestream turbulence levels (Tu =3.6 and 5.3%). The measurements of mean and fluctuating velocity profiles have been carried out at inlet to the test section using Constant Temperature Anemometry (CTA). The data might be useful for validation and development of transition models. In order to characterize the dynamics leading to the transition process in the separated boundary layer on the plate surface, the Particle Image Velocimetry (PIV) measurements have been performed. The results show that the transition process in the separated boundary layer is influenced by breakdown of the Klebanoff streaks.

1. Introduction

Despite the vast amount of studies on transitional flow approaches during the last 4 decades, understanding of the transition mechanism is still incomplete. Proper understanding of the laminar-to-turbulent transition is crucial for development of transition models. We discuss the mechanism for transition in a separated boundary layer developing along the plate subjected to the adverse pressure gradient and high freestream turbulence level. The present work also aims at providing the inlet conditions for numerical simulations. The test case is, therefore, useful for validation and development of the RANS-based laminar-to-turbulent transition models.

2. Experimental setup

The experiments have been carried out in closed-loop wind tunnel. Dimensions of test section are: $1m$ long, $0.45m$ high and $0.25m$ wide. The required pressure gradient is achieved by a contoured upper profile [1]. The settling chamber with two meshes and turbulence grid is placed in front of the test section to better control the inlet turbulence. The test section is presented in figure 1. The nominal length of the plate is $L = 0.5m$. The flat plate surface is followed by flap. The flap has been employed to control the small separation bubble at the plate leading edge. The coordinates of the test section are summarized in table 2.

The velocity field above the plate was measured with the 2D PIV system from LaVision. Images were captured by sCMOS camera equipped with a $50mm - Nikkor$ lens where the magnification of the optical system was 13. The resolution of the camera is 2560×2048 pixels and typical scaling accuracy is 2%. Illumination was provided by double-pulse laser Nd:YAG

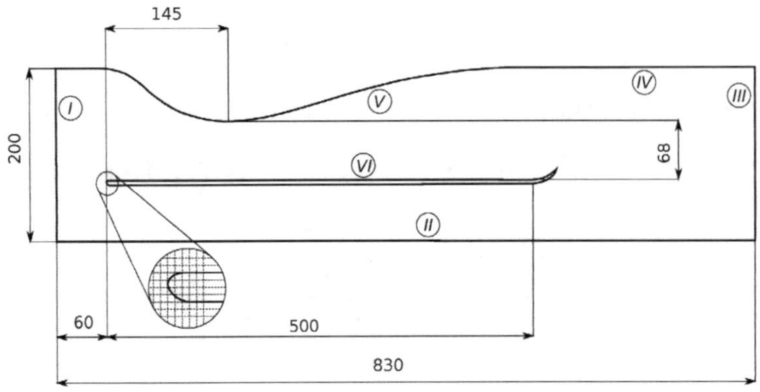

Figure 1. Schematic of the test section. The flow is from left to right. Dimensions are in [mm].

Spectra-Physics PIV-200 laser $200mJ$ per pulse. The number of pairs of images captured for averaging was equal to 1000 samples for each case with a frequency $15Hz$. The flow was seeded by typical DEHS oil droplets $(5-10\mu m)$ generated by Laskin nozzles in the PIVTEC seeding generator. The light sheet was placed in the symmetry plane of the test section (mid-span) and the axes were parallel to the surface of the plate to avoid undesirable reflections. The camera was set up perpendicular to the light sheet with a dedicated curtain to cut off the reflected light from the plate and upper wall. The calculations were done by Davis 8.3. software and final correlation window are 4×4 pixels with 50% overlap. The velocity field was measured with a spatial resolution of $12.13pix/mm$ and spacing of velocity field $0.165mm/vector$. The mean and fluctuating velocities have been measured at the channel inlet by StreamLine CTA anemometer from DantecDynamics with dual-channel X-probe 55P61. The signal was acquired by the National Instrument card PCIe-6320 with sampling $20kS/s$, $16bit$ resolution and $10sec$ recording time per point. The CTA probe was calibrated by the Automatic Calibration unit with the accuracy 1% and obtained calibration error was 0.05%.

3. Discussion of experimental results

3.1. Inlet conditions for numerical simulation and pressure distribution

Measurements of the mean and fluctuating velocity components have been carried out at the wind tunnel inlet plane using the CTA technique. As discussed later, the data will be needed for setting-up the numerical simulations. The measurements have been performed along a vertical line at mid-span (whole channel height) placed $60mm$ upstream of the plate leading edge (I symbol in figure 1). The inlet values are available for the three Reynolds numbers, $Re = 1.4 - 2.4 \cdot 10^5$, and for the two freestream turbulence levels (Tu=3.6 and 5.3%). The Reynolds number is based on the plate length, L, and the mean velocity at the plate leading edge (boundary layer edge). The turbulent intensity is based on the rms of fluctuating velocity at the leading edge ($30mm$ above the plate surface). Figure 2 shows the mean x- and y-velocity components. The results denoted by lines correspond to the moderate ($Tu = 3.6\%$) while the symbols correspond to the high ($Tu = 5.3\%$) turbulence levels. The symbols and black line correspond to the CTA measurements while the coloured lines correspond to the PIV results ($Tu = 3.6\%$). The agreement is good which shows feasibility of both measuring techniques. The reduced x-velocity in the middle part of the channel ($50 < y < 150mm$) is caused by enhanced turbulence mixing behind ribs of the turbulence grid (stronger velocity defect). The ribs are approximately located at distances: $y = 50$ and $150mm$. Figure 3 shows an evolution of the freestream turbulence intensity, $Tu = u'_{rms}/U_{loc}$ (in [%]), at distance 30mm above the plate surface for $Re = 1.4-1.8 \cdot 10^5$ and $Tu = 3.6\%$ and 5.3%. The symbols u'_{rms} and U_{loc} denote the

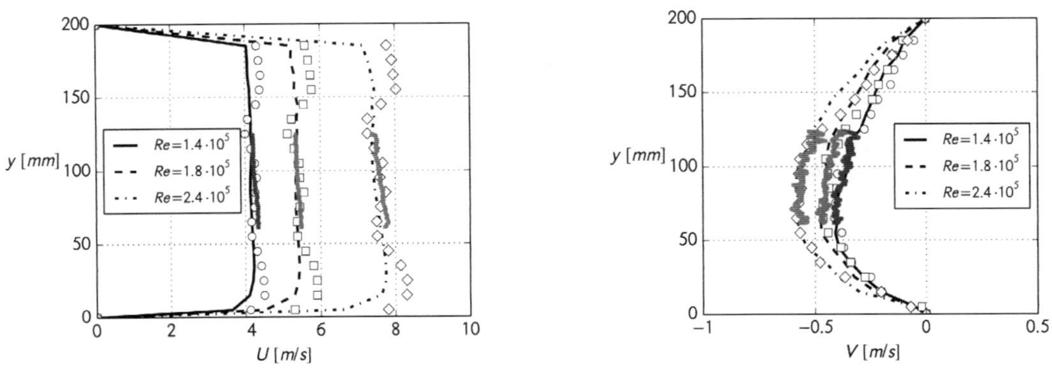

Figure 2. The mean velocity profiles at the channel inlet (line-segment No I in figure 1) obtained using the PIV and CTA techniques.

rms of fluctuating and the mean velocities, respectively. The turbulence grid was not employed in the measurements with the lower Tu level, $Tu = 3.6\%$, and it was only used in the higher turbulence case. All the Tu-profiles exhibit a characteristic decay of turbulence downstream the inlet plane. From distance $x/L = 0.312$ on, the profiles increase due to enhanced mixing at the edge of separated boundary layer and partly due to scaling (U_{loc} decreases downstream the throat). The pressure distributions on the plate surface (mid-span) are presented in Figure 4. The solid lines and dots show the results obtained with the pressure taps spacing equal to 10 and $40mm$, respectively. One can clearly see the separated boundary layer region for $x/L \in (0.45 - 0.5)$. Supplementary measurements of the static pressure have been performed along the line shifted with respect to the mid-span line by a small distance in z-direction. No differences were reported ensuring the flow two-dimensionality (results not showed).

The simulation of 2D flow over the flat plate (see discussion later) required supplementary measurements of the static pressure at the bottom surface of the wind tunnel below the plate (see II symbol in figure 1). The elements of measuring devices located below the flat plate perturbed the flow there and caused supplementary pressure losses. This has a small, but meaningful, influence on the flow characteristics above the plate. The pressure distributions were obtained at two Reynolds numbers and two freestream Tu levels (figure 5). Note a significant effect of mean flow velocity (pressure reduction is proportional to square of the velocity) and small effect of the freestream turbulence on the pressure distribution at the bottom wall of the wind tunnel.

3.2. Energy spectra

Figure 6 (left) shows the fluctuating velocity spectra at position $(x, y) = (-60, 30)mm$ (upstream the leading edge of the plate). The data are presented for the flows at the two Reynolds numbers and the two freestream turbulence levels. The energy content of the fluctuating velocity components becomes higher with an increased Reynolds number and with the increased freestream turbulence level. The power spectra shows a good correspondence with the Kolmogorov $-5/3$ law in the $f = 30 - 400Hz$ range. Similar quality results have been reported for the y-velocity signal (not showed). It means that the freestream conditions are representative of the fully turbulent flows in all cases analysed here. Figure6 (right) shows the power spectra at streamwise distance $x/L = 0.49$ (past the boundary layer separation point) for $Re = 1.41 \cdot 10^5$, $Tu = 3.6\%$. The spectra are analysed at four wall-normal distances: outside of the boundary layer at $10mm$ above the plate (denoted by $10mm$), at the boundary layer edge (δ), along the mean x-velocity inflection point (Inf. P.) and close to the wall ($1mm$). The local

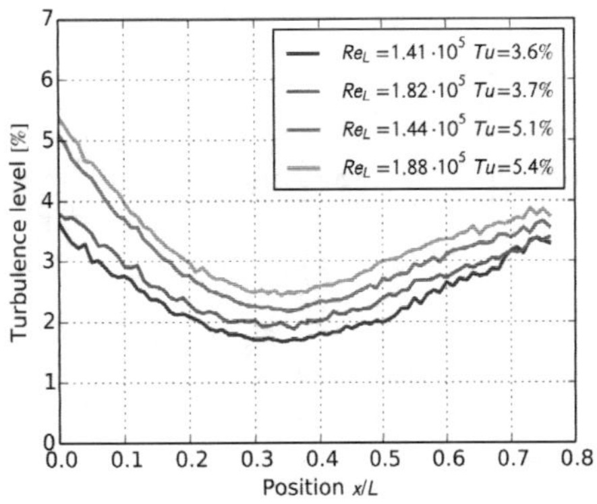

Figure 3. Freestream turbulence level at distance $30mm$ above the plate.

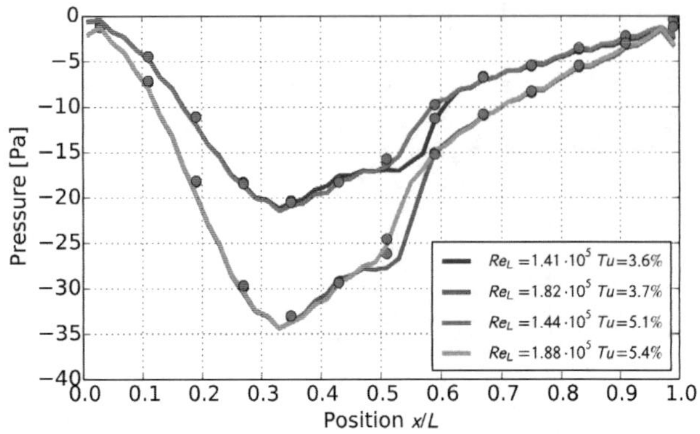

Figure 4. The pressure distribution on the flat plate surface.

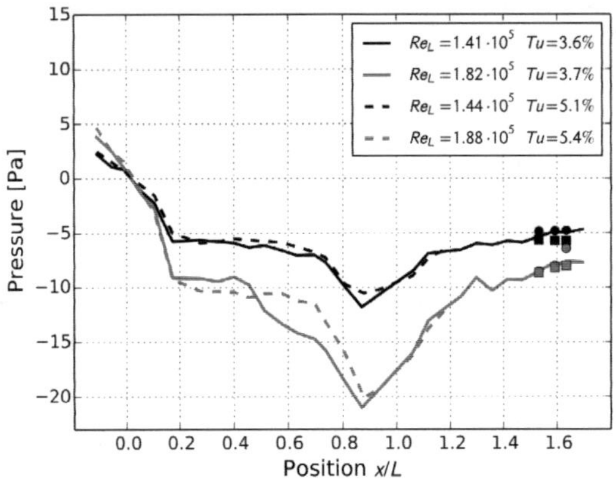

Figure 5. The pressure distribution at the bottom wall of the wind tunnel (line No II in figure 1). The symbols denote supplementary measuring points on the bottom side of the flat plate (profile No IV in figure 1).

freestream velocity spectra ($10mm$ and δ) show the higher fluctuating velocity amplitudes in the whole frequency range, compared to the amplitudes of the upstream flow (see figure 6 (left), $Re = 1.41 \cdot 10^5$, $Tu = 3.6\%$). This is due to the increased turbulence mixing at the edge of the separated boundary layer in the adverse pressure gradient flow. One can see an increased contribution of the low-frequency events ($f = 10 - 100Hz$) and damping of the high-frequency components ($f = 300 - 10,000Hz$) at the inflection point and near to the wall. This is caused by filtering out of the high-frequency components by the boundary layer shear, as observed in [2] and [3]. Only large-scale disturbances penetrate deeply into the separated laminar layer and induce the Klebanoff streaks there. Table 1 shows the flow and the boundary layer characteristics for the four cases analysed here. The last column shows the frequencies of presumed Klebanoff streaks according to the correlation by Lengani et al. [4]. For the $Re = 1.41 \cdot 10^5$, $Tu = 3.6\%$ case, the correlation gives $f = 187Hz$. This shows that the increased low-frequency amplitudes in figure 6 (right) (Inf. P. and $1mm$ above the plate) are due to Klebanoff streaks. Similar results were observed in the other cases analysed here (not showed). It means that for the elevated turbulence cases analysed here, the transition in the separated boundary layer is dominated by

Table 1. Summary of test cases. The mean velocity at boundary layer edge, U_0, and the turbulent intensity, Tu ($30mm$ above plate), are provided at the plate leading edge. Symbols s, δ^*, θ denote the separation point, displacement and momentum thicknesses, respectively. The Reynolds numbers Re_{xs} and $Re_{\theta s}$ are based on local velocity at boundary layer edge. Estimation of frequency of the Klebanoff streaks, $f_{Kleb.}$, is according to correlation [4].

Re	$Tu[\%]$	$U_0[m/s]$	Re_{xs}	$Re_{\theta s}$	$\delta^*_s[mm]$	$U_{\delta^*_s}[m/s]$	$f_{Kleb.}[Hz]$
$1.41 \cdot 10^5$	3.6	4.23	$1.03 \cdot 10^5$	221	1,88	3,69	187
$1.82 \cdot 10^5$	3.7	5.48	$1.34 \cdot 10^5$	265	1,79	4,75	253
$1.44 \cdot 10^5$	5.1	5.43	$1.41 \cdot 10^5$	260	2,01	3,79	180
$1.88 \cdot 10^5$	5.4	5.65	$1.49 \cdot 10^5$	308	1,88	4,95	251

the breakdown of the Klebanoff streaks.

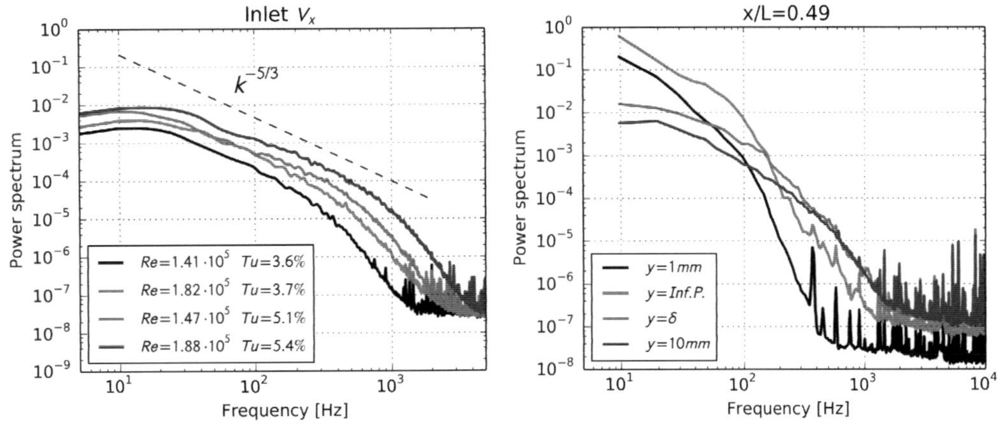

Figure 6. Power spectra of x-velocity component (left) in the freestream, $(x, y) = (-60, 30)mm$, and (right) close to the plate ($x/L = 0.49$) for $Re = 1.41 \cdot 10^5$, $Tu = 3.6\%$.

3.3. Flow characteristics along the plate

Figure 7 shows the boundary layer characteristics along the plate. One can see that the size of the separation bubble is reduced with increasing the Reynolds number and the freestream Tu level. The influence of the Klebanoff streaks becomes obvious in flows with varying turbulence levels.

4. Application of the data in numerical simulation

The transport equations of the $k-\omega$ model combined with the algebraic intermittency model by Kubacki and Dick [5] have been implemented with the user defined function (UDF) functionality. The specification of numerical details is as in [5]. A good quality grid was generated with 441,000 cells and with y^+ varying from 0.2 to 1.4 along the plate. The normalized residuals of the momentum and the transport equations were driven to below 10^{-6}.

The measured profiles of the mean x- and y-velocity components were directly imposed at the left inflow boundary of the computational domain (figure 8). Calculation of the inlet values of

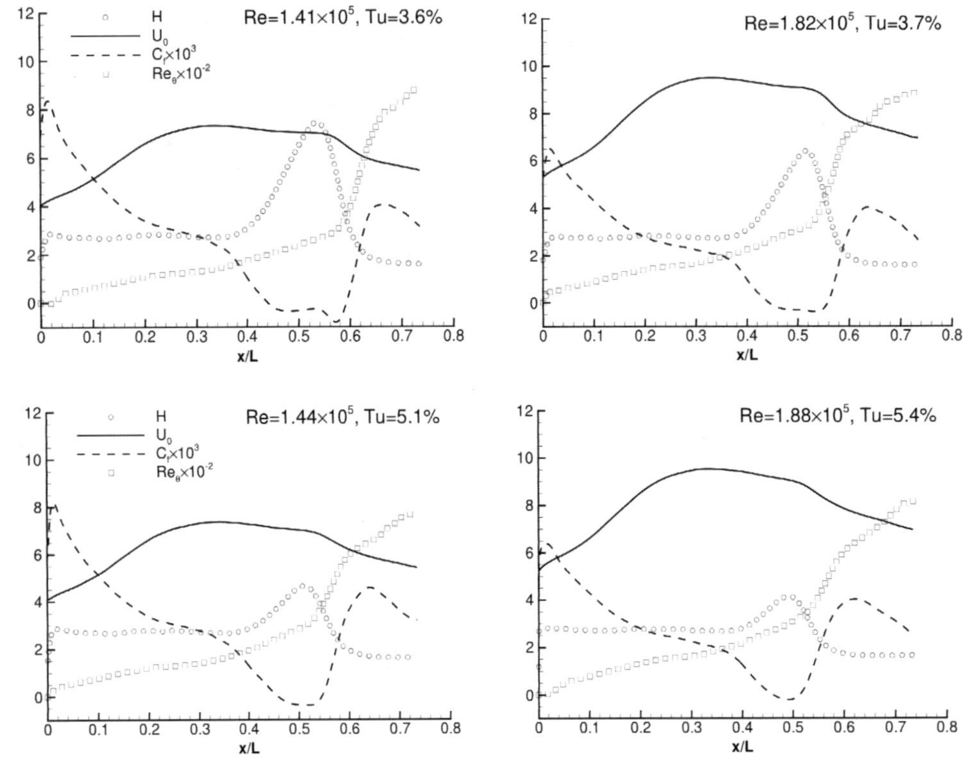

Figure 7. Evolution of the shape factor, H the mean velocity at the boundary layer edge U_0, the skin friction coefficient, C_f, and the momentum thickness Reynolds number, Re_θ along the flat plate.

the turbulent kinetic energy $k = 0.5[(u'_{rms})^2 + (v'_{rms})^2 + (w'_{rms})^2]$ required estimation of the rms of the spanwise fluctuating velocity component (w'_{rms}), because only the rms of the streamwise (u'_{rms}) and wall-normal (v'_{rms}) fluctuations are available from measurements. In present case $w'_{rms} = v'_{rms}$. The turbulent length scale, l_t, was not measured. In simulation, the turbulent length scale was estimated $(\omega = k^{1/2}/(\beta^* l_t))$ by matching the measured turbulent intensity profile at $30mm$ from the plate surface (figure 10 a). A simplified geometry, consists of the

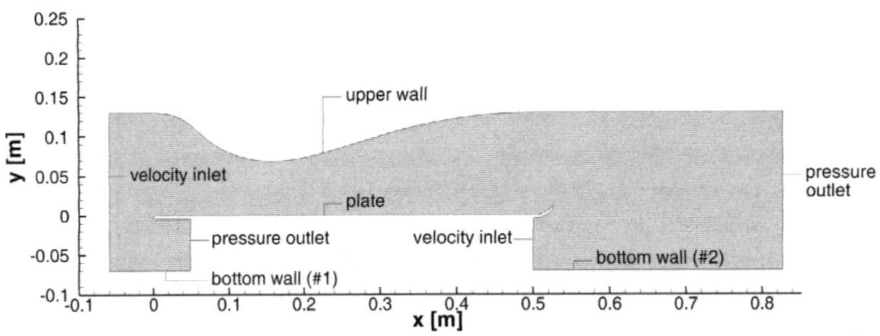

Figure 8. Scheme of the computational domain and types of the boundary conditions employed in numerical simulation of the flow along the flat plate.

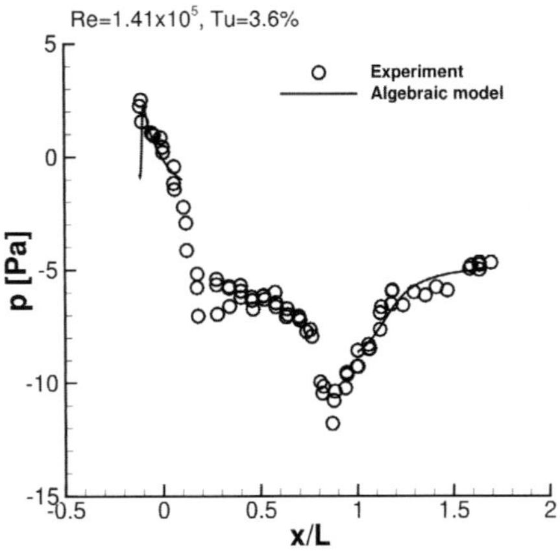

Figure 9. Flat plate at $Re = 1.41 \cdot 10^5$ and $Tu = 3.6\%$. The static pressure at the bottom wall of the wind tunnel. In simulation, the two bottom walls (figure 8) correspond to distances: $-0.12 < x/L < 0.1$ (bottom wall #1) and $1 < x/L < 1.654$ (bottom wall #2).

main test section (bounded by the plate and upper contoured wall) and the two parts one below the leading edge and one below the trailing edge of the plate, without the middle part (see figure 8), was considered. At supplementary outlet ($x = 0.05m$) and inlet ($x = 0.5m$) boundaries the boundary conditions have been imposed according to experiments. The static pressures equal to the constant values were imposed at the two outlets, placed at $x = 0.05m$ and $x = 0.827m$ (figure 8). The values of static pressures were deduced from measurements (see figure 5). At the second velocity inlet boundary, placed at $x = 0.5m$, the mean velocity, k and ω profiles were imposed. The profiles were obtained from a precursor simulation of 2D developing turbulent channel flow. The constant value of the mean velocity was imposed at the straight channel inlet. The turbulent intensity and the turbulent length scale were set to $Tu = 5\%$, $l_t = 0.07H_c$ (H_c is the channel height), respectively. The profiles of selected variables were extracted at the straight channel streamwise distance at which the same value of underpressure was reported as in the reference measurement, $x = 0.5m$ ($x/L = 1$) (figure 5). Figure 9 shows the comparison between measured and predicted (sections $-0.12 < x/L < 0.1$ and $1 < x/L < 1.654$) the static pressure at the bottom wall. The agreement is good. It shows that the proper conditions have been specified at the supplementary inlet and outlet boundaries (below the plate), resulting in a correct flow over the plate (figure 10 a).

Figure 10 (a) shows a comparison between measured and predicted using the algebraic intermittency model the mean velocity at the edge of the boundary layer and the turbulent intensity level at $30mm$ from the plate. Figure 10 (b) shows the shape factor profile along the plate. The model predicts the transition somewhat too early (figure 10 b). This results in somewhat too low value of the predicted mean velocity at $x/L = 0.5$ (figure 10 a). But the overall agreement is good. Similar quality results were obtained for simulation of the other cases (not showed). The results show that the experimental data can be used for assessment and development of the RANS-based transition models.

5. Summary

Experimental results for the flow over the flat plate subjected to the adverse pressure gradient and elevated freestream turbulence levels have been provided. The mean and fluctuating velocities have been measured at test section inlet. The data might be used for assessment and development of the laminar-to-turbulent transition models.

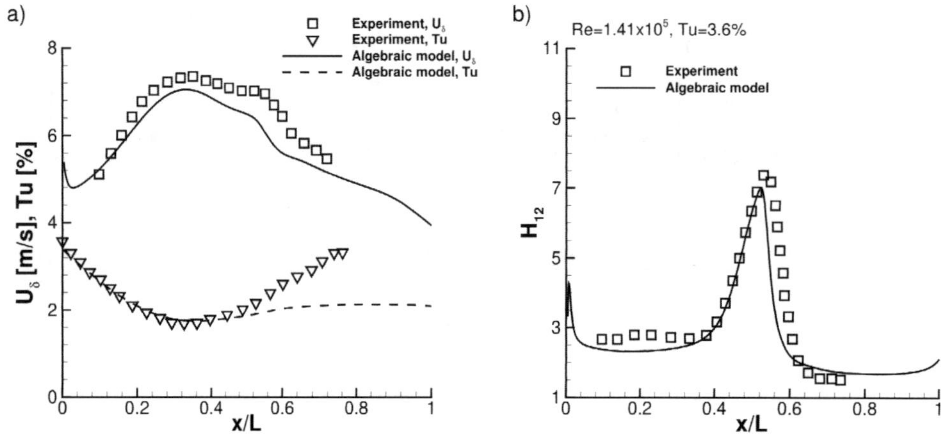

Figure 10. Flat plate at $Re = 1.41 \cdot 10^5$ and $Tu = 3.6\%$. Predicted by the algebraic model and measured a) the mean velocity at the boundary layer edge and the turbulence intensity at $30mm$ from the plate surface, b) the shape factor profile on the plate.

Table 2. Coordinates of the test section in $[mm]$.

Profile V				Lines I II III IV		Profile VI upper side		Profile VI lower side		Leading edge	
x	y	x	y	x	y	x	y	x	y	x	y
−12	130	185	70.75	−12	130	2	0	3	−4	2	0
10	129.54	210	75.46	−60	130	500	0	500	−4	1	−0.3
35	123.43	260	89.92	−60	−70	511, 5	2	511, 5	−2	0	−1.5
60	106.42	310	103.37	760	−70	516	4	516	0	1	−3.1
85	87.11	360	115.41	760	130	520	6	519.5	2	2	−3.75
110	75.69	410	123.88	510	130	522	8	522	4	3	−4
130	70.73	460	128.52			525	10	524.5	6		
156	68.57	510	130			527	11.5	527	10.5		

Acknowledgments

The authors acknowledge the Interdisciplinary Centre for Mathematical and Computational Modelling (ICM) at the University of Warsaw (Grant No. GB78-11) for granting the computational power for the numerical simulations.

This research was partially funded by IDUB against COVID-19 project granted by Warsaw University of Technology under the program Excellence Initiative: Research University (IDUB). The second author acknowledges the support from a research project funded by the Polish National Science Centre (Contract number DEC-2018/31/B/ST8/01717).

References

[1] Lou W and Hourmouziadis J 2000 *Journal of Turbomachinery* **122** 634
[2] Jacobs R G and Durbin P A 2001 *Journal of Fluid Mechanics* **428** 185–212
[3] McAuliffe B R and Yaras M I 2010 Transition mechanisms in separation bubbles under low- and elevated-freestream turbulence
[4] Lengani D, Simoni D, Ubaldi M, Zunino P and Bertini F 2017 *Experimental Thermal and Fluid Science* **84** 18–27
[5] Kubacki S and Dick E 2016 *International Journal of Heat and Fluid Flow* **62** 344—-361

Journal of Physics: Conference Series **2367** (2022) 012019 doi:10.1088/1742-6596/2367/1/012019

Wind tunnel investigations of aircraft airfoil in cruise conditions

P Kaczyński[1], R Szwaba[1], M Piotrowicz[1], P Flaszyński[1] and P Doerffer[1]

[1] Institute of Fluid-Flow Machinery Polish Academy of Sciences (IMP PAN), Fiszera 14, 80-231 Gdańsk, Poland

E-mail: piotr.kaczynski@imp.gda.pl

Abstract. Presented work is focused on analysis of the flow over the 2D model of Airbus A320 airfoil wing in cruise phase of the flight. For this purpose the measurement channel with an airfoil model was designed and assembled in a transonic wind tunnel to obtain a similar flow pattern as in the reference two-dimensional freestream flow. Experimental investigations were conducted in the IMP PAN transonic wind tunnel with a relative narrow test section which is a novel approach in terms of these type of research. The test section was designed using CFD simulations based on 2D freestream flow for the tested wing profile and in the next step the research were continued experimentally in transonic wind tunnel with a measurement chamber width of 100 mm. This paper presents the results of reference investigations on the A320 wing profile which combines experimental tests and CFD calculations. The obtained results show that the approach presented in the paper is appropriate and the obtained flow features in the tunnel do not differ much from the freestream conditions.

1. Introduction

One of the main challenges of the EU framework programme Horizon 2020 was to reach sustainable prosperity of efficient smart and green transport in Europe. The reduction of fuel consumption in aviation can lead to improved resource efficiency and diminish greenhouse gases emission [1]. The SMS (Smart Morphing and Sensing) project [2] was established based on those demands and was intended to obtain the optimum shape of the aircraft wing from Airbus A320 in order to increase aerodynamic performance and modify the lift and drag structure. Desired effects can be achieved e.g. by means of an electro-active morphing trailing edge which can lead to attenuate the flow separation and nuisance instabilities [3].

The IMP PAN contribution in SMS project was to investigate the reduced scale model of Airbus A320 airfoil in transonic regime which corresponds to the cruise condition of the aircraft flight. The first step in the research was to analyse flow structure around the reference wing profile, which was considered as the introduction to the second step, the investigations of the model with morphed trailing edge. Usually this type of research is carried out in large transonic wind tunnels (e.g. Institute of Aviation [4, 5]), where the model is mounted freely on the sting to ensure two-dimensional flow on most surfaces of the profile and at the same time to fulfil 2D freestream flow conditions for a given profile. Moreover, large scale models are also tested in cryogenic wind tunnels such as European Transonic Wind Tunnel (EWT) in range of low supersonic velocities [6]. The medium size transonic wind tunnels as ONERA S3MA, provides measurement campaigns which can be a complement to numerical calculation and optimisation of the design of airfoil [7]. Other example of measurement in medium size wind tunnel but in subsonic condition can be found in F. Grasso [8] which investigation was also intended for measurement of airfoil from wind turbine together with optimisation process and numerical calculation. However, wind tunnel measurements are sensitive on any errors thus one should put a great attention to details and procedures in order to avoid potential issues which can result from adequate model size regard to wind tunnel dimensions, proper model positioning or secondary flows [9]

It is also worth pointing out that the test campaigns in large transonic wind tunnels are expensive and the power needed for the their operation often reaches 1MW, therefore, it was taken the challenge to carry out this type of measurements in a smaller wind tunnel [10], in which the cost of testing is about one order lower.

The investigations of aircraft wing profile in such wind tunnel are very innovative undertaking and is itself a new achievement. It shows that the proper design of the measurement section meets with the requirements of transonic flow and such approach is feasible and may also reduce significantly the research costs.

The test section was designed using numerical simulations based on 2D freestream flow for the tested wing profile. When the measurement section design process of was completed, the first tests of the new channel were carried out. It appeared that the key element, which determines whether such a type of measurements can be conducted, is the sealing between the profile and the side walls. The available literature lacks this type of test examples and these are perhaps one of the first which shows that research of flow on aircraft profiles in the medium size transonic tunnels with a relatively narrow measurement space can be considered.

This paper presents the results of reference investigations on the A320 wing profile in a medium dimensions transonic tunnel with a measurement chamber width of 100 mm. The experimental tests and CFD calculations were performed and the obtained results show that the approach presented in the paper is appropriate and the obtained flow features in the tunnel do not differ much from the freestream conditions.

2. Test section

2.1. Experimental setup

The investigations of the scaled model of the Airbus A320 wing profile was carried out in the IMP PAN transonic wind tunnel (blowdown type). The internal dimensions of the test stand were 1800 mm long, 350 mm high and 100 mm wide. More details about experimental facility are included in Fomin [10].

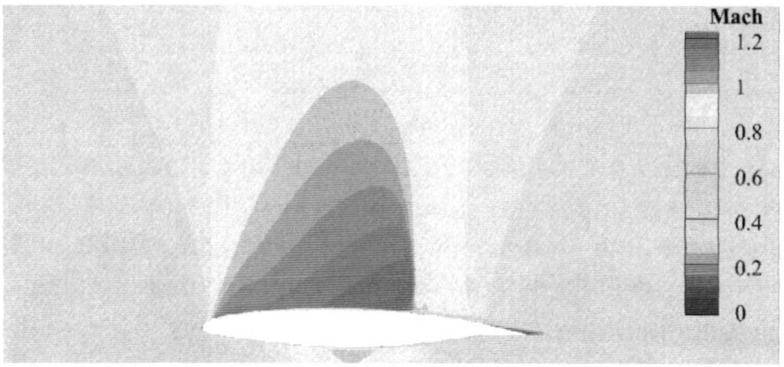

Figure 1. Numerical simulation of the model in freestream condition.

In order to investigate the aircraft airfoil in cruise conditions the dedicated test section was designed based on the reference flow data delivered by AIRBUS Toulouse. The first step of the design process was the numerical simulations for the flow over 2D airfoil in freestream (figure 1). One of the main criteria was the inlet Mach number and inflow uniformity which should be adequate to the real conditions. The Mach number upstream of the wing model corresponds to the typical aircraft cruise condition, M=0.78.

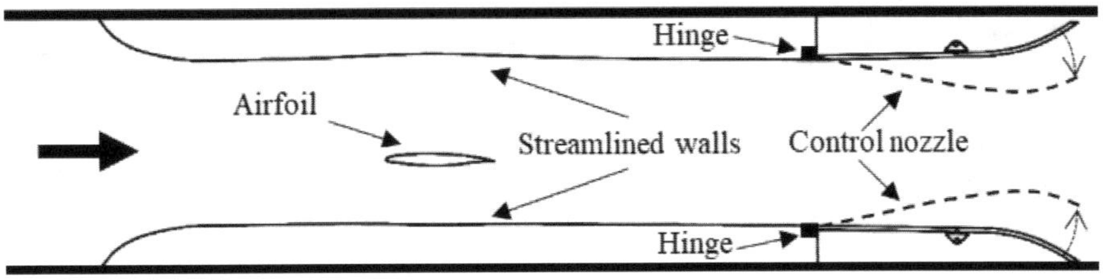

Figure 2. Scheme of the test section.

The second step of the design of the test section focused on modification of upper and lower walls in order to keep the similar flow structure like in freestream condition. The final shape of the upper and lower wall in the test section was extracted from streamlines from CFD calculation as well as relative position of the wing model. The sketch of the test section arrangement is shown in figure 2. The subsonic inlet conditions upstream of the profile were controlled by means of outlet nozzle. The nozzle was equipped with a hinges which allowed to alter the dimension of outlet plane.

The figure 3 shows the model of wing profile and measurement equipment assembled in the wind tunnel. The airfoil was attached to the side holders (red part, between $x/c = 0.14$ and 0.32) which pass through the window and then connected to the balance system, each on the one side. The airfoil model had 150 mm of chord and 100 mm of span. Moreover the whole side windows was able to rotate thus the angle of attack of the wing model could be differed in range from 1.8^0 to 2.4^0. The Mach number upstream of the airfoil, for all measured angles of attack, was 0.77 (179 mm upstream of profile's leading edge) and the Reynolds number based on its cord was 2×10^6. The Mach number upstream of the profile's shock wave was around 1.2.

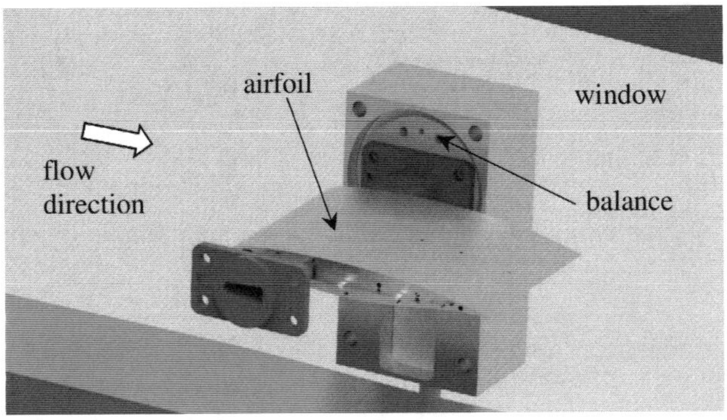

Figure 3. Airfoil in the wind tunnel, left - CAD view, right - assembled in test section.

The investigated model was equipped with 12 pressure taps on the suction side and 6 taps on the pressure side of the airfoil. What is more, two Kulite sensors for unsteady pressure measurements were installed on the suction side. In order to collect detailed information of an aerodynamic airfoil performance the two axis balance system was installed. The system of two balances on which the model was sustained was mounted outside the tunnel on transparent windows made of cast perspex.

Measurement campaign mainly consisted of pressure, velocity, lift and drag forces measurements and flow visualization. In details the following measurements, collected during the investigations concerning the reference flow over the aircraft wing model, can be listed:

- static pressure distribution along the centre line and around of the profile;
- schlieren photography of the flow structure and the shock wave topology;
- drag and lift forces by means of two axis balance system;
- the flow velocity distribution upstream of the model and downstream across the profile wake by means of LDA (Laser Doppler Anemometry);

2.2. Numerical model description

The numerical simulations were carried out by means of FINE/Turbo Numeca code with the two-equation nonlinear eddy viscosity turbulence model of the EARSM (Explicit Algebraic Reynolds Stress Model). The set of equations is closed by a perfect gas equation and Sutherland`s law for viscosity. It is applied 2nd order central scheme with scalar artificial dissipation for spatial discretization formulated by Jameson, Schmidt and Turkel [11].

The structured mesh was generated by means of IGG Numeca (Interactive Geometry Modeler and Multi-Block Structured Grid Generator). The multi-block topology for 3D test section model consists of 65 blocks and the total number of hexahedral cells is 13.7×10^6. The mesh resolution near the solid wall boundaries was assumed in such a way that y^+ was below 1. It assured proper modeling of the boundary layer.

Simulations for the adapted computational domain with upper and lower walls, inlet and outlet were done for 3D model in order to compare with experimental results. At the inlet the ambient conditions (as in the wind tunnel) were applied: total pressure 101 kPa and total temperature 293 K, while at the outlet, the static pressure (61 200Pa) were adjusted to reproduce the shock location (x/c = 0.6) predicted for the free-stream configuration and to retain inlet condition. Due to the presence of walls and increased dissipation effects in the channel, the outlet static pressure was varied with the chosen wall distance (y/c).

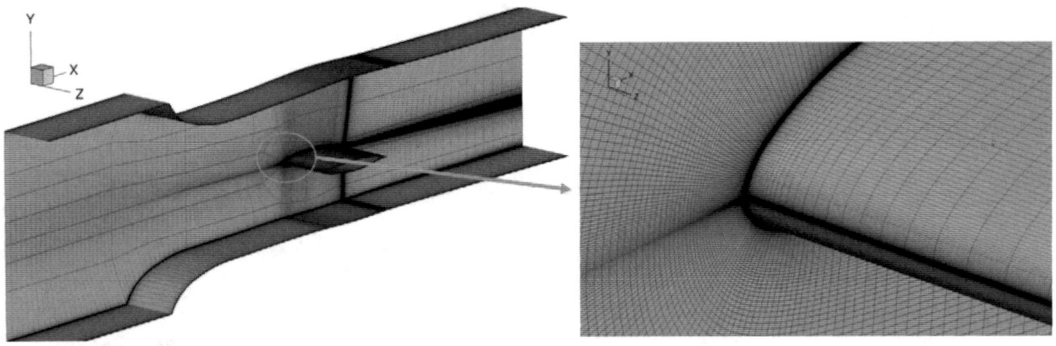

Figure 4. Domain and mesh for wind tunnel configuration (details on the leading edge).

3. Results and discussion

In preliminary phase of the tests of the the model of airfoil, after the test section had been assembled, it occurred that the selection of the proper sealing between the model and the windows would be crucial. As it was mentioned, the wing profile was attached to balances which were outside of the test section, while the airfoil itself was immersed in the side window in milled groove with the depth of 10 mm. The groove was larger than the profile's contours by 1 mm hence there was no direct contact between it and the window. It was significant to provide suitable sealing method parallelly with ensuring transverse and longitudinal leakproofness of the profile. The transverse leakproofness means lack of flow between the suction and pressure sides, while the longitudinal one is the blockage of the flow along the profile, e.g. due to the pressure difference between both sides of the shock wave.

Figure 5 shows a comparison of the Mach number distribution (for the airfoil center line) for two different applied seals S1, S2 and 3D numerical simulations. The S1 distribution was obtained with a rubber seal, while S2 shows the result for liquid (at the time of application) silicone. There are no significant differences in Mach number distribution between sealing S1 and S2, only one point for S1 distribution shows slight deviation (third point on suction side). On the pressure side, a lower velocity (higher pressure) for the S1 seal can be seen (figure 5). In the numerical simulation the airfoil is directly connected to the side walls (figure 4) thus there is no flow between sides of the profile. One can observe very good agreement between experimental and CFD data. Slight difference in the area of the shock wave (approx. x/c=0.7) results from its oscillations in experimental investigations. This affects the smoothing of pressure measurement in the shock vicinity, while calculations were carried out for stationary flow conditions.

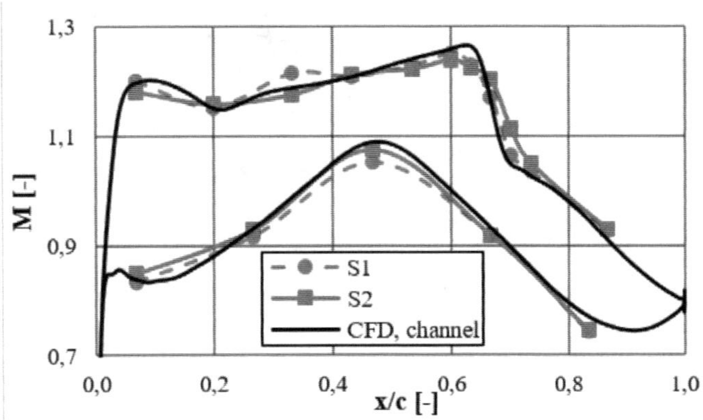

Figure 5. Mach number distribution on the airfoil for different sealing type.

In order to indicate the difference in the flow around the wing profile in freestream and transonic tunnel, the adequate Mach number distributions are compared in figure 6. The flow in freestream condition shows slightly different position of the shockwave (upstream) what implies in a lower flow velocity (higher pressures) on the airfoil pressure side.

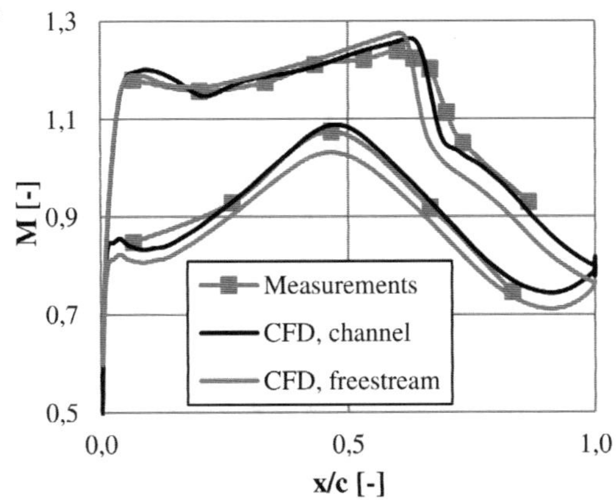

Figure 6. Comparison of Mach number distribution for channel flow and freestream.

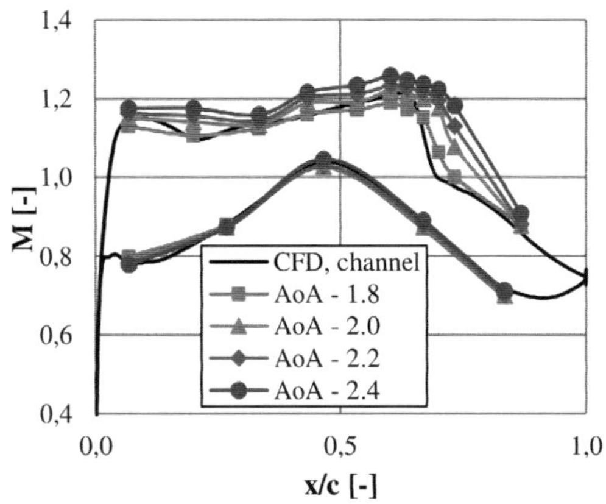

Figure 7. The Mach number distribution for various angles of attack.

Mach number distribution on the suction side of the airfoil is affected by angle attack of the profile (figure 7). As the AoA growth the flow velocity above profile also increases which results in shock wave displacement towards trailing edge. The pressure side is less influenced by the angle of attack modification, Mach number distribution shows slight differences.

Figures 8 and 9 show a schlieren photography from the experiment and numerical simulation. It can be seen in figure 8 that the shock wave reaches the wall without dividing, what indicates at this Mach number upstream of the shock (M = 1.2) on turbulent (shock wave with boundary layer) interaction. In the case of numerical simulation (figure 9), a very similar shock topology was obtained. On the pressure side due to the low local Mach number not much greater from unity any shock wave does not occur.

Figure 8. Schlieren visualization - Exp.

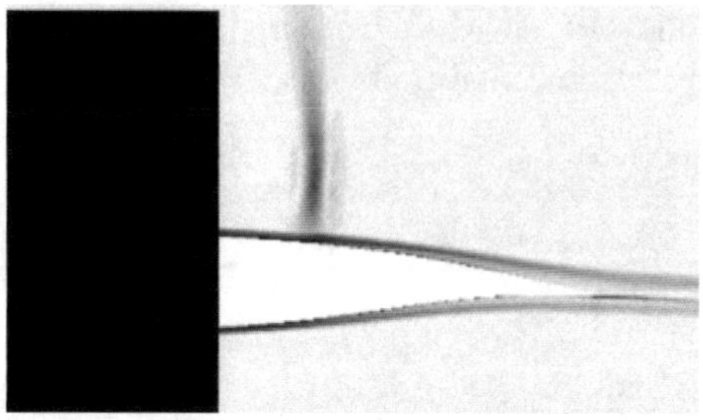

Figure 9. Schlieren visualization – CFD.

The above presented measurement and numerical simulation results show very good agreement, in particular the Mach number distribution on the profile resulting from the pressure distribution. When the pressure distribution is similar, it can be assumed with high probability that the flow structure on the profile surface will be also similar. Surface flow structure in the form of the surface friction coefficient and streamlines distributions on the suction (left side) and pressure (right) side of the airfoil surface is shown in figure 10. These results show that in the experimental measurements one can obtain two-dimensional flow on the majority of the profile. On the profile's suction side there is a small secondary flow area in the corners at the trailing edge and on the pressure side almost perfect two-dimensional flow is obtained. This is a strong indication which shows that transonic flows with airfoil can be also tested in a medium size wind tunnels.

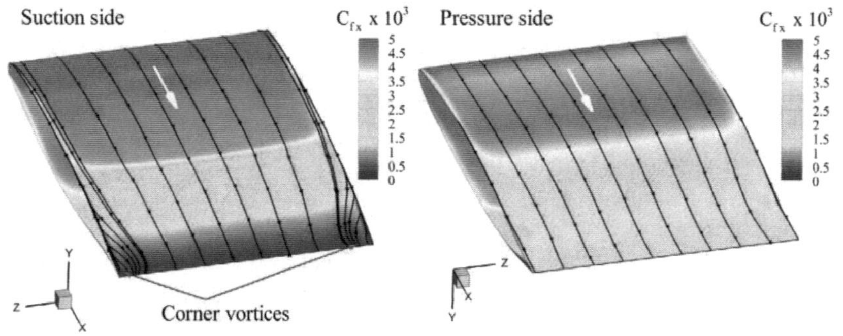

Figure 10. CFD - skin friction distribution.

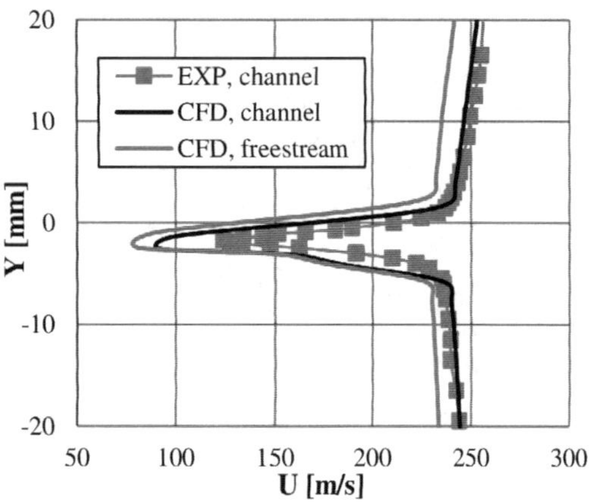

Figure 11. Velocity distribution across the wake for channel flow and freestream, AoA 1.8°.

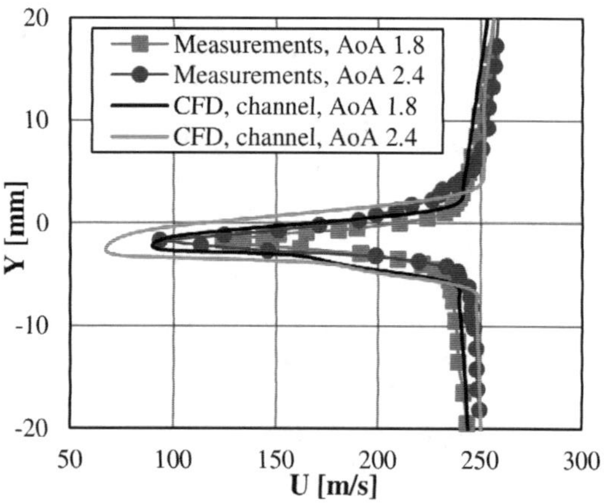

Figure 12. Velocity across the wake from numerical calculation and measurements for AoA 1.8°and 2.4°.

Plots in figures 11 and 12 show the results of the numerical simulation and the experiment concerning the flow in the profile wake. These velocity distributions measurements are located 7 mm downstream of the airfoil trailing edge and coordinate Y=0 is placed on the trailing edge. Comparing the wakes between the freestream flow and the channel, it can be seen that they are almost identical, except the velocity difference outside the wake. The airfoil in the wind tunnel is limited by the sidewalls thus additional secondary flow, as corner flow, results in narrowing non-disturbed main flow. In consequence, the main flow velocity is increased compere to the freestream condition.

The velocity profile in experimental investigations characterizes similar thickness as in numerical simulations, however in the centre of the wake a lower velocity was obtained as well for AoA of 1.8°and 2.4°. This is an expected result, because numerical simulations are quite often characterized by a larger velocity decrease [12] at this point.

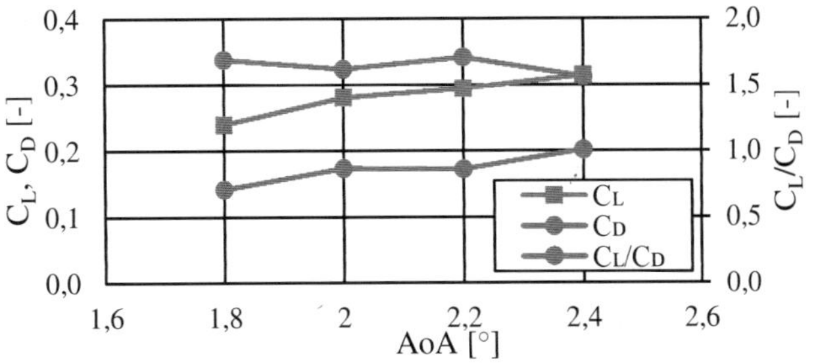

Figure 13. Results of drag and lift coefficient measurements.

Figure 13 shows the results of the drag and lift coefficient measurements determined based on forces measured using aerodynamic balances. Such measurements, can be commonly seen in the research in large transonic wind tunnels [13, 14]. As one can see, both coefficients increase as the angle of attack growths. It should be also emphasized that thanks to the mechanism of changing the angle of attack it is possible to precisely control its value with a resolution of 0.2°, this value in large wind tunnels is difficult to achieve. The C_L/C_D value, i.e. the lift to drag coefficient ratio due to the small variations in the angle of attack, practically does not change.

4. Conclusions

A new test section was developed, designed and implemented in transonic wind tunnel with a relative narrow measurement channel to obtain a similar flow pattern as in the reference two-dimensional freestream flow. An airfoil model of the aircraft wing was manufactured and equipped with two balances and necessary systems for aerodynamic measurements. The following conclusions can be drawn from the above described investigations:

- The flow field contains characteristic features of the flow over the reference two-dimensional airfoil in freestream, i.e. shock location and its topology;
- Velocity distributions both on the profile and in the wake show very good agreement between the numerical simulations and the experiment;
- Surface flow structure, surface friction coefficient and velocity distribution over the airfoil surface show that one has obtained two-dimensional flow on the majority area of the profile in the tunnel. This is an important indication which shows that transonic flows with airfoil can be also tested in a medium size wind tunnels. However it is essential that the measurement case should be properly selected and supported by CFD calculation;
- Lift and drag forces measurement using aerodynamic balances in innovative configuration outside the test section was succeeded what is an large achievement of this research. Such results up to now could be seen in the research in large transonic wind tunnels. It was applied the precise mechanism of changing the angle of attack which facilitate control this value with a resolution of 0.2°; such value in large wind tunnels is difficult to achieve.

The presented above results from experimental and numerical research show that a properly designed measurement channel with an airplane wing profile model along with the proper procedures at measurements commissioning (with particular importance on the sealing selection between the profile and the tunnel side walls) allow to obtain in transonic wind tunnel with a relatively narrow test section, the two-dimensional flow on the tested model. Obtaining such type of flow was one of the research objectives of the realized project [2], which was achieved, thus showing that a medium size transonic tunnels can be taken into account in case of testing the aircraft wing profile, thereby significantly reducing the research cost.

Acknowledgments

This study was realised under the H2020 European Research programme No 723402: SMS, ''Smart Morphing and Sensing" for aeronautical configurations", http://smartwing.org/SMS/EU/.

Numerical simulations were performed in the Computational Centre of TASK (Trójmiejska Akademicka Siec Komputerowa) and supported by PL-Grid Infrastructure.

5. References

[1] Di Lorenzo G, Frosina E, De Petrillo L, Lauria D, Senatore A, Curreri F, Saccone G, Mazuy M K and Pascarella C 2019 Design and development of hybrid-electric propulsion model for aeronautics *MATEC Web of Conf.*

[2] https://cordis.europa.eu/project/id/723402

[3] Jodin G, Scheller J, Duhayon E, Rouchon JF and Braza M 2017 Implementation of a hybrid electro-active actuated morphing wing in wind tunnel *Solid State Phenom.* **260**

[4] Wiśniowski W 2016. Wind tunnels in poland on the background of world tunnels *Trans. Inst. Aviat.* **242**

[5] Placek R and Ruchala P 2018 The flow separation development analysis in subsonic and transonic flow regime of the laminar airfoil *Transp. Res. Procedia* **29**

[6] Wang Y, Liu D, Xu X and Li G 2021 Investigation of reynolds number effects on aerodynamic characteristics of a transport aircraft *Aerosp.* **8** 177

[7] Zhang W, Sun J, Wang L, Wu J and He L 2020 Rotor airfoil aerodynamic design method and wind tunnel test verification, *Chin. J. Aeronaut.* **33** 8 p 2123-32,

[8] Grasso F 2016 ECN-G1-21 Airfoil: design and wind-tunnel testing, *J. Aircr.* **53** 5

[9] Placek R 2016 Errors and problems while conducting research studies in a wind tunnel -selected examples *Trans. Inst. Aviat.* **4** 245 pp 169-77

[10] Fomin N, Lavinskaya E, Doerffer P, Szumski JA, Szwaba R and Telega J 2009 Quantitative diagnostics of shock wave - boundary layer interaction by digital speckle photography *Shock Waves Conf. Proc.* **1**

[11] Jameson A, Schmidt W and Turkel E 1981 Numerical solutions of the euler equations by finite volume methods using runge-kutta time-stepping schemes *AIAA J.* **81**

[12] Flaszynski P, Doerffer P, Szwaba R, Piotrowicz M and Kaczynski P 2018 Laminar-turbulent transition tripped by step on transonic compressor profile *J. Therm. Sci.* **27** 1

[13] Hillenherms C, Schröder W and Limberg W 2004 Experimental investigation of a pitching airfoil in transonic flow *Aerosp. Sci. Technol.* **8**

[14] Gebbink R, Wang G and Zhong M 2018 High-speed wind tunnel test of the cae aerodynamic validation model *Chin. J. Aeronaut.* **31** 3

Microphysical measurements of hydrometeors using shadowgraph imaging technique

Moein Mohammadi [1], Jakub L Nowak [1] and Szymon P Malinowski [1]

[1] Institute of Geophysics, Faculty of Physics, University of Warsaw, Pasteura 5, 02-093, Warsaw, Poland

E-mail: moein.mohammadi@fuw.edu.pl

Abstract. The size distributions of raindrops and precipitation particles were measured in a series of in situ experiments during various weather events in Warsaw, Poland. To perform the measurements, a shadowgraph instrument, "VisiSize D30", was set up on the roof of the Institute of Geophysics, University of Warsaw, together with an "OTT Parsivel^2" laser disdrometer. The VisiSize D30, recently introduced to atmospheric research, is capable of accurately determining a wide range of particle sizes, from micrometre-scale cloud droplets to millimetre-scale precipitation particles, while directly measuring droplets by capturing shadow images of them. A comparison of the data collected by the two instruments during simultaneous measurements shows that combining the shadowgraph technique with the optical disdrometer can prevent droplet size distribution truncation and provide more accurate results.

1. Introduction

Cloud-related processes are the most unpredictable in weather forecasting and climate research, and continue to be a weak link in atmospheric circulation models [1]. The size distribution of cloud droplets plays a major role in the quantitative microphysical description of clouds and is critical for cloud lifetime, radiative effects, and precipitation formation [2]. As cloud droplets grow, they can form the raindrops and eventually fall to the ground as precipitation. The microstructure of precipitation is controlled by the intricate interplay between cloud microphysics and atmospheric dynamics [3]. A fundamental property of precipitation for the study of its microstructure is also the drop size distribution (DSD). All precipitation variables of interest can be derived as weighted moments of the DSD [4]. It is of paramount importance not only for remote sensing of precipitation, but also for numerical weather modeling, precipitation infiltration, soil erosion, and pollutant dispersal [5]. Consequently, accurate measurements of DSD are essential for the comprehensive study of hydrometeor microphysics.

There are two general approaches for precipitation measurement: in situ sampling and remote sensing. Although intrinsic difficulties are involved in both strategies, in situ measurements are generally considered fundamental since they provide instrumental access to individual precipitating particles within a sampling volume. Moreover, in situ data are utilized to generate and validate inversion methods implemented in remote sensing applications (e.g.[6]).

When it comes to in situ measurement of precipitation, disdrometers are considered the primary instruments. They measure the DSD of raindrops and precipitation particles at a given location. The first impact disdrometer was used by Joss and Waldvogel [7]. Later, the 2-Dimensional Video Disdrometers

(2DVD) [8] and optical disdrometers [9, 10] were developed. The optical disdrometers are relatively inexpensive, long-lasting, and reliable instruments that are particularly well suited for usage in networks to study the small-scale variability of DSD [5, 11].

However, according to recent studies, the commonly used disdrometers have some inaccuracy in measuring concentrations of very small droplets in the DSD. This can be due to either truncation of the DSD at the end of the small droplet or large uncertainties in these measurements. Consequently, the "drizzle mode" of precipitation, which consists of large concentrations of small drops and is usually detached from the main part of DSD by a shoulder region is not captured by many ground-based disdrometers [12]. The sampling uncertainty in the DSD measurements with disdrometers is investigated in [5, 12, 13].

On the other hand, shadowgraph imaging is known as an optical measurement technique capable of accurately sizing a wide range of particle sizes while performing direct measurement by capturing shadow images of the particles [14, 15]. Accordingly, the main objective of this study is to use a shadowgraph instrument, VisiSize D30, recently applied to atmospheric science, to perform an accurate DSD measurement of hydrometeors, especially small raindrops which are usually not captured or underestimated with disdrometers. Combining the shadowgraph technique with optical disdrometer, and performing measurements simultaneously can prevent DSD truncation and provide more accurate results. In addition, it could be considered as a reference for future studies of precipitation microphysics as well as flooding and soil erosion.

2. Instrumentation and methods

The following three subsections are going to briefly describe the apparatus and methodologies used in the measurement. The references for more detailed description are also provided in each part.

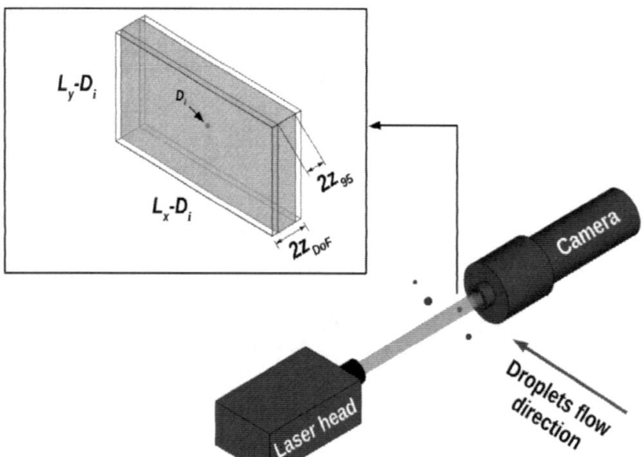

Figure 1. Schematic representation of the VisiSize setup and a magnified Sample Volume around a droplet, comparing the default and corrected depth of fields (z_{def} vs. z_{95}; Detailed description provided in [14])

2.1. VisiSize D30 shadowgraph system

The VisiSize D30 (hereafter called VisiSize) is a shadowgraph imaging instrument which is manufactured by Oxford Laser Ltd. and was originally designed to diagnose agricultural and industrial sprays. It works based on the particle/droplet imaging analysis (PDIA) method, which is first introduced and developed by Kashdan et al. [16, 17, 18]. A recent laboratory study on spray droplet size distribution using the PDIA shadowgraph together with the phase Doppler particle analysis (PDPA) method was conducted by Sijs et al. [15]. The results for droplets in range from 10 up to $\sim$ 400 microns correspond very well, where there is only 0.7% discrepancy between measured volume mean diameters by these two methods.

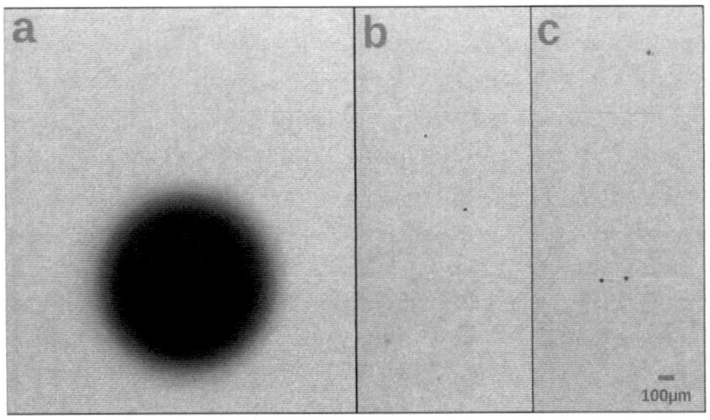

Figure 2. Sample shadowgrams captured during a mixed cloud-rain event at the UFS observatory on 13th of August 2019: (a) a captured rain drop; (b) and (c) a couple of cloud droplets from the same captured series of shadowgrams.

In the PDIA method implemented in VisiSize, the region of interest is illuminated from behind by a pulsed infrared laser while a high-resolution camera captures shadow images of droplets traversing the measurement volume at up to 30 frames or image pairs per second (See Figure 1). VisiSize provides a wide range of particle/droplet information, such as droplet size, velocity, shape, and direction, making it suitable for atmospheric research. Therefore, the applicability of VisiSize to measure cloud droplets has been investigated in previous studies [14, 19].

The details of droplet detection and sizing system of VisiSize together with essential corrections in built-in software algorithm to measure cloud droplets have been profoundly discussed in [14] through laboratory experiments. The VisiSize has been also employed for the first time to measure microphysical properties of droplets in atmospheric clouds at a mountain-based observatory, the environmental research station (Umweltforschungsstation Schneefernerhaus, UFS), located on the southern slope of Zugspitze in the Bavarian Alps [19]. Figure 2 includes three samples of cropped shadow images (shadowgrams) captured by the VisiSize during a cloud-rain mixed event at the UFS in which a rain drop of size $\sim$500 microns is detected (panel a) alongside a couple of cloud droplets $\sim$20 microns in diameter (panels b and c).

As mentioned, the VisiSize was successfully used to measure microphysical properties of cloud droplets in the previous studies. In this research, we have focused mainly on employing the VisiSize for capturing shadowgrams and post processing them for the mixed-phase and ice-phase precipitation while during the rain events the automatic live processing of the data can be used.

2.2. OTT Parsivel^2 disdrometer probe

In this study, we have used an optical disdrometer, OTT Parsivel^2, manufactured by OTT Hydromet GmbH. The Parsivel is an optical laser disdrometer that uses a sheet of light through which drops fall. The measuring surface is 180$\times$30 mm (54 cm^2). The diameter and velocity of a drop is then determined by measuring the shadow it casts and how long it casts that shadow [10]. The parsivels divide the drops into classes for velocity and diameter and record the number of drops measured per class over an integration time. The measurement range starts at 0.25 mm and goes up to 8 and 25 mm for liquid and solid precipitation, respectively. In previous studies, Parsivel has also been used to study snowfall measurements [20, 21, 22, 23] because of its ability to distinguish between solid and liquid precipitation. For more details on the Parsivel measurement technique, see [10].

2.3. *In situ measurements*

The VisiSize D30, a shadowgraph instrument recently applied to the atmospheric science, was utilized simultaneously alongside a laser-optical disdrometer, OTT Parsivel^2, in order to measure the size distribution of the hydrometeors (see Figure 3). The in situ measurements were performed on the rooftop of the Institute of Geophysics, Faculty of Physics, University of Warsaw (IGF-UW). While the Parsivel disdrometer is a standard, fully automated instrument and collects data continuously, VisiSize requires manual operation, presence of the investigator in the course of the measurements and initial calibration between each measurement series. Thus, data collection were limited to the precipitation events when such an operation was possible.

Figure 3. "VisiSize D30" shadowgraph instrument (front) and "OTT Parsivel^2" laser disdrometer (back) deployed next to each other at Meteo-station on the rooftop of Institute of Geophysics, Faculty of Physics, University of Warsaw (IGF-UW).

Figure 4 illustrates 10 samples of cropped shadowgrams of raindrops, ice particles and snowflakes captured with the VisiSize during the in situ experiments on the rooftop of IGF-UW. As it is shown, the VisiSize is not only capable of detecting micron-sized cloud droplets (see Figure 2), but it can also capture images of raindrops and precipitation particles up to 2.25 mm in diameter size.

3. Results

The probability size distributions of simultaneously collected data with the VisiSize and Parsivel are demonstrated in the following figures. Using probability size distribution allow us to merely compare the sizing performances between the two instruments without being influenced by their different sample volumes. Figure 5 shows data collected during a rain event in Warsaw on 18th of November 2018. The detected droplet counts and arithmetic mean values are also shown below the legend. The difference in sizing range and effect of drizzle drops on arithmetic mean values are apparent in these histograms. The total number of detected drops is different between the instruments due to different sampling volumes. Moreover, the arithmetic mean value for shadowgraph is lower than disdrometer. It must be noted that the purple color on the graph simply represents the area where the two histograms from the VisiSize (blue) and Parsivel (red) overlap each other.

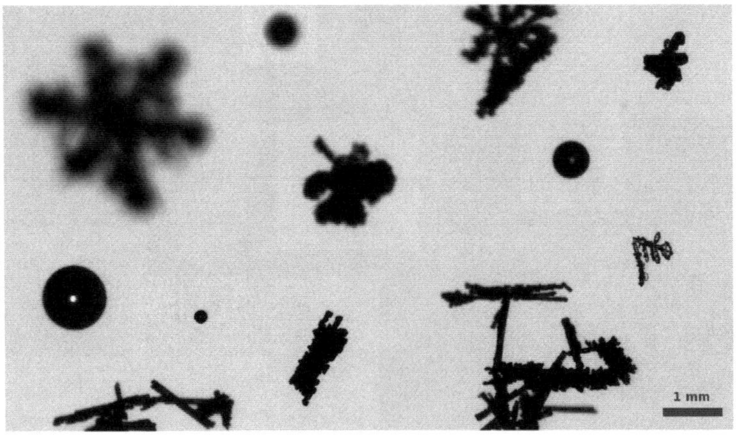

Figure 4. Sample cropped shadowgrams of of raindrops, ice particles and snowflakes captured during different weather events of Warsaw with the VisiSize deployed on the rooftop of IGF-UW.

In figure 6, the same measurement data are used only with analysing drops within the common sizing range of the two instruments (0.25-2.25 mm). Consequently, the arithmetic mean values and size distributions correspond better. However, discrepancy between the results in the smallest size bin (0.25-0.37 mm) does still exist, and the VisiSize data show almost three times higher value in this range which could be attributed to the inaccuracy of the Parsivel in drizzle mode.

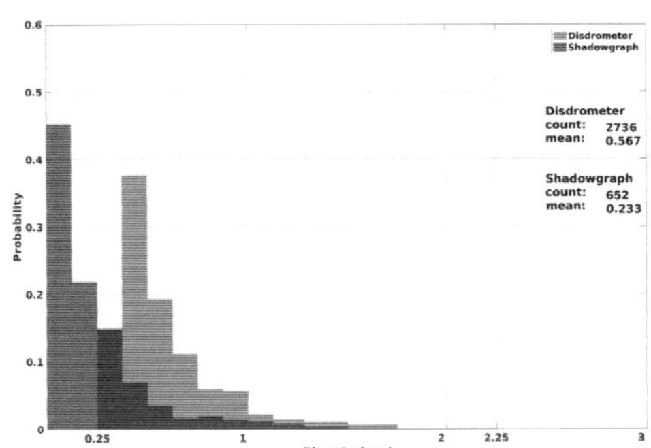

Figure 5. The DSD histogram of rain drops collected on the rooftop of IGF-UW on 18th of November 2018, 20:26'-23:54'.

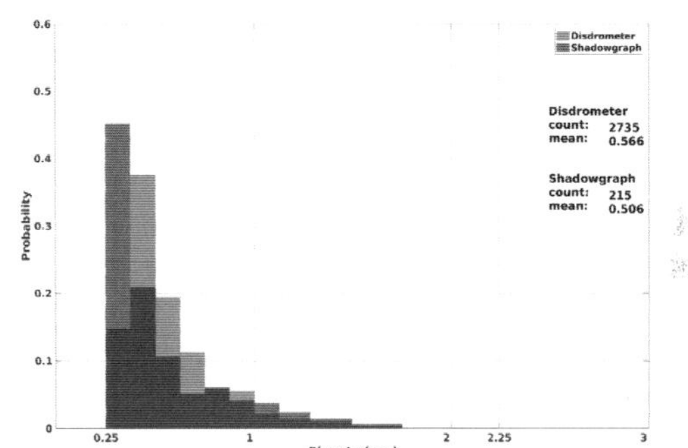

Figure 6. The DSD histogram of the same measurement as Figure 5, while drops out of common measuring range (0.25-2.25 mm) were excluded.

Figures 7 and 8 also show the DSD comparison between the two instruments before and after excluding particles out of common sizing range, respectively. The data were collected during a snow event in Warsaw on 24th of December 2018. The same trend as previous measurement case is also observed here after limiting particle size to the common measuring range. However, the agreement between the results seems better than the previous plots. It can be thanks to having more precipitation particles comparing to the drizzles and small particles in this case which not only increased the arithmetic value, but also slightly weakened the effect of drizzles for the DSD. Additional examples of simultaneous measurements with the VisiSize and the Parsivel are shown in Figures 9 and 10. The size distributions

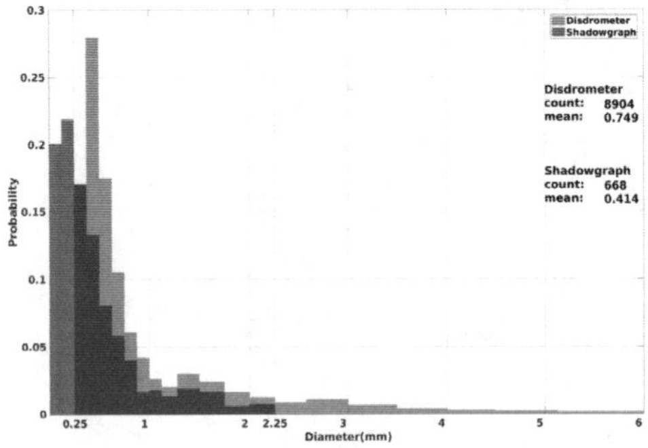

Figure 7. The DSD histogram of precipitation particles collected on the rooftop of IGF-UW on 24th of December 2018, 21:28'-24:00'.

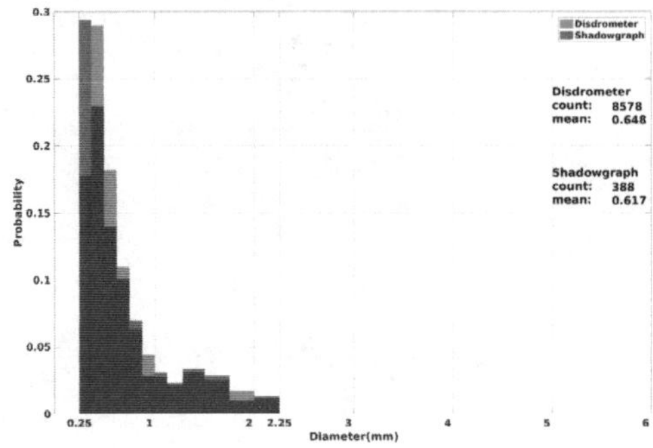

Figure 8. The DSD histogram the same measurement as Figure 7, while particles out of common measuring range (0.25-2.25 mm) were excluded.

of raindrops and precipitation particles collected with the two instruments compared with each other, excluding the drops out of common sizing range same as Figures 6 and 8.

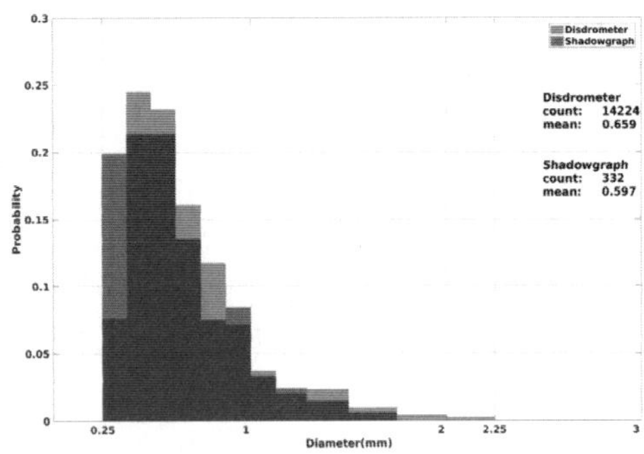

Figure 9. DSD histogram of rain drops collected on the rooftop of FUW on 13th November 2018, 19:27'-24:00', while drops out of common measuring range (0.25-2.25 mm) were excluded.

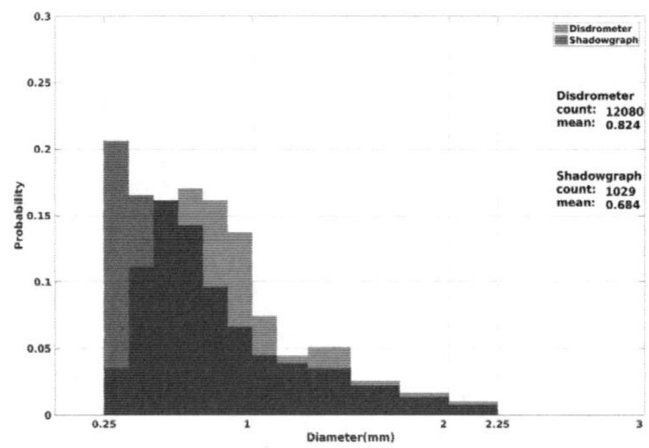

Figure 10. DSD histogram of precipitation particles collected on the rooftop of FUW on 31st January 2019, 11:10'-12:46', while particles out of common measuring range (0.25-2.25 mm) were excluded.

4. Conclusion

The drop size distribution of hydrometeors, collected in series of simultaneous measurements with the VisiSize D30 shadowgraph system and OTT Parsivel^2 laser disdrometer, have been investigated. The two instruments have been deployed on the rooftop of Institute of Geophysics of University of Warsaw. Each measurement series was planned based on the weather forecast and man resources availability. Eventually, a couple of case studies were selected including different weather events in which shadowgrams of raindrops and precipitation particles were captured by the VisiSize.

In conclusion, the VisiSize D30 shadowgraph instrument has been successfully applied to precipitation measurements. Furthermore, comparison between the VisiSize and Parsivel disdrometer showed a better correspondence in DSDs within the mid-sized range drops than smallest sizes (drizzles). Considering the capability of the VisiSize to accurately measure the drizzle mode, the discrepancies can be attributed to the difficulties in measuring drizzles with the Parsivel which were also observed in the past. Consequently, the synergy of these two techniques can lead us to more precise DSD measurements for the hydrometeors. However, the differences in measurement size ranges and sampling volumes of the two instruments for further investigations of kinetic energy of different types of precipitation must be carefully taken into account in future studies.

Acknowledgments The authors appreciate Wojciech Kumala for technical support during the in situ experiments, and Krzystof Markowicz who provided the recorded data of disdrometer probe. This research has been supported by the European Union's Horizon 2020 research and innovation programme under the Marie Skłodowska-Curie Actions, COMPLETE (grant agreement no. 675675).

References

[1] Bodenschatz E, Malinowski SP, Shaw RA, Stratmann F. Can We Understand Clouds Without Turbulence? *Science*. 2010; **327**:970-1.

[2] Devenish BJ, et al. Droplet growth in warm turbulent clouds. *Quarterly Journal of the Royal Meteorological Society*. 2012; **138**(667):1401-29.

[3] Testik FY, Barros AP. Toward elucidating the microstructure of warm rainfall: A survey. *Reviews of Geophysics*. 2007; **45**.

[4] Testud J, Oury S, Black RA, Amayenc P, Dou X. The concept of "normalized" distribution to describe raindrop spectra: A tool for cloud physics and cloud remote sensing. *Journal of Applied Meteorology*. 2001; **40**:1118-40.

[5] Jaffrain J, Berne A. Experimental Quantification of the Sampling Uncertainty Associated with Measurements from PARSIVEL Disdrometers. *Journal of Hydrometeorology*. 2011; **12**:352-70.

[6] Jakubiak B, Licznar P, Malinowski SP. Rainfall estimates from radar vs. raingauge measurements. Warsaw case study. *Environment Protection Engineering*. 2014; **40**:159-70.

[7] Joss J, Waldvogel A. Ein Spektrograph für Niederschlagstropfen mit automatischer Auswertung. *Pure and Applied Geophysics PAGEOPH*. 1967; **68**:240-6.

[8] Kruger A, Krajewski WF. Two-Dimensional Video Disdrometer: A Description. *Journal of Atmospheric and Oceanic Technology*. 2002; **19**:602-17.

[9] Hauser D, Amayenc P, Nutten B, Waldteufel P. A New Optical Instrument for Simultaneous Measurement of Raindrop Diameter and Fall Speed Distributions. *Journal of Atmospheric and Oceanic Technology*. 1984; **1**:256-69.

[10] Löffler-Mang M, Joss J. An Optical Disdrometer for Measuring Size and Velocity of Hydrometeors. *Journal of Atmospheric and Oceanic Technology*. 2000; **17**:130-9.

[11] Tapiador FJ, Checa R, De Castro M. An experiment to measure the spatial variability of rain drop size distribution using sixteen laser disdrometers. *Geophysical Research Letters*. 2010; **37**.

[12] Raupach TH, Thurai M, Bringi VN, Berne A. Reconstructing the Drizzle Mode of the Raindrop Size Distribution Using Double-Moment Normalization. *Journal of Applied Meteorology and Climatology*. 2019; **58**:145-64.

[13] Raupach TH, Berne A. Correction of raindrop size distributions measured by Parsivel disdrometers, using a two-dimensional video disdrometer as a reference. *Atmospheric Measurement Techniques*. 2015; **8**:343-65.

[14] Nowak JL, Mohammadi M, Malinowski SP. Applicability of the VisiSize D30 shadowgraph system for cloud microphysical measurements. *Atmospheric Measurement Techniques*. 2021; **14**:2615–2633.

[15] Sijs R, Kooij S, Holterman HJ, Van De Zande J, Bonn D. Drop size measurement techniques for sprays: Comparison of image analysis, phase Doppler particle analysis, and laser diffraction. *AIP Advances*. 2021; **11**:15315.

[16] Kashdan JT, Shrimpton JS, Whybrew A. Two-Phase Flow Characterization by Automated Digital Image Analysis. Part 1: Fundamental Principles and Calibration of the Technique. *Particle & Particle Systems Characterization*. 2003; **20**(6):387-97.

[17] Kashdan JT, Shrimpton JS, Whybrew A. Two-Phase Flow Characterization by Automated Digital Image Analysis. Part 2: Application of PDIA for Sizing Sprays. *Particle & Particle Systems Characterization*. 2004; **21**(1):15-23.

[18] Kashdan JT, Shrimpton JS, Whybrew A. A digital image analysis technique for quantitative characterisation of high-speed sprays. *Optics and Lasers in Engineering*. 2007; **45**:106-15.

[19] Mohammadi M, Nowak JL, Bertens J G Moláček, Kumala W, Malinowski SP. Cloud microphysical measurements at a mountain observatory: comparison between shadowgraph imaging and phase Doppler interferometry. *Atmospheric Measurement Techniques*. 2022; **15**:965-85.

[20] Egli L, Jonas T, Grünewald T, Schirmer M, Burlando P. Dynamics of snow ablation in a small Alpine catchment observed by repeated terrestrial laser scans. *Hydrological Processes*. 2012; **26**:1574-85.

[21] Löffler-Mang M, Blahak U. Estimation of the Equivalent Radar Reflectivity Factor from Measured Snow Size Spectra. *Journal of Applied Meteorology*. 2001; **40**:843-9.

[22] Tokay A, et al. Disdrometer derived ZS relations in south central ontario, canada; 2007. p. 6-10.

[23] Sandra EY, Kingsmill DE, Nance LB, Löffler-Mang M. Observations of Precipitation Size and Fall Speed Characteristics within Coexisting Rain and Wet Snow. *Journal of Applied Meteorology and Climatology*; **45**:1450-64.

Journal of Physics: Conference Series **2367** (2022) 012021 doi:10.1088/1742-6596/2367/1/012021

Experimental analysis of the impact of an acoustic excitation on a separated boundary layer behaviour

V Sokolenko[1], W Elsner [1], A Dróżdż[1], R Gnatowska[1], Z Rarata[2] and S Kubacki[2]

[1] Czestochowa University of Technology, Armii Krajowej 21 42-200, Czestochowa, Poland

[2] Warsaw University of Technology, Faculty of Power and Aeronautical Engineering, Institute of Aeronautics and Applied Mechanics, Nowowiejska 24, Warsaw, 00-665, Poland

E-mail: vasyl.sokolenko@pcz.pl

Abstract. The paper examines the effect of acoustic excitation at broadband frequency range (100 - 650 Hz) on separated boundary layer developing on a flat plate subjected to an adverse pressure gradient. The experiment was conducted with a low inlet turbulence intensity level (Tu < 1%) in order to provide a cleaner environment that magnifies the effects of the excitation frequency. It was shown that acoustic excitation at sound pressure levels: SPL=125 and 135 dB, can lead to a more rapid increase in flow instability followed by an earlier laminar-turbulent (l-t) transition. The boundary layer reattachment point was shifted upstream and the size of the separation bubble was reduced in the flow with the acoustic excitation active.

1. Introduction

Overall efficiency and performance of turbines and compressors in aero-engines heavily depend on viscous flow effects, where a substantial part of these effects are related to the development of boundary layers. It is well known that l-t transition is very sensitive to surface roughness as well as to the level of external disturbances like free-stream turbulence or incident wakes [1]. Additionally, in modern turbofan engines, the boundary layers developing on the surfaces of turbine blades are exposed to strong perturbations caused by acoustic waves. A high concentration of acoustic energy inside such an engine can have a major impact on the stability of the laminar boundary layer and can trigger an earlier laminar-to-turbulent transition [2].

On the other hand, when the boundary layer develops along an aerodynamic surface exposed to a strong adverse pressure gradient it may separate, and then l-t transition occurs in the separated shear layer. This is typically followed by the boundary layer reattachment and formation of a closed region of recirculated flow referred to as a laminar separation bubble (LSB). The transition process within the separated shear layer involves amplification of selective small-amplitude disturbances often attributed to an inviscid Kelvin-Helmholtz instability [3]. The results presented by Talan and Hourmouziadis [4] indicate that the instability correlates well within a relatively narrow range of Strouhal numbers defined by the momentum thickness at the separation point.

Most studies that investigate stability characteristics involve artificial forcing of the boundary layer by means of plasma actuation at a location upstream of the separation point [5]. Jones et al [6]

investigated by means of DNS the effect of volume forcing introduced in momentum equations on breakdown process in the separated boundary layer over NACA0012 airfoil. A number of previous studies have used controlled periodic forcing to investigate the sensitivity of transition in the shear layer of LSBs. Yarusevych and Kotsonis [5] demonstrate that forcing at the frequency of Kelvin–Helmholtz instability has a significant impact on the reduction of the size of existing LSB. Marxen et al. [7] observed that increasing the excitation amplitude leads to upstream shift in the mean reattachment locations, which however can lead to the mean flow deformation and a reduction in the frequency of most amplified disturbances. Kurelek et al. [3] studied the effect of tonal and broadband acoustic excitation on transition in the separated boundary layer formed over NACA0018 airfoil.

The present investigation is aimed to examine the impact of broadband acoustic excitation on l-t transition in the separated boundary layer. The inherent complexity of the problem is simplified by providing acoustic forcing from a precisely controlled external source (loudspeaker). The aim is to compare this case with the reference unexcited flow and to elucidate underlying physical mechanism of flow instability.

2. Experimental Setup

Experiments were conducted in the open-circuit wind tunnel with a 1.45 m long test section. The upper wall of the test section was shaped according to the assumed distribution of pressure gradient corresponding to the conditions encountered in the axial compressor blading allowed to impose adverse pressure gradient (APG) (Fig.1a). Measurements were carried out on the lower flat plate. Settling chamber with several screens and contraction ratio of 13:1 allowed for a low level of inlet turbulence intensity level ($Tu < 1\%$), which was necessary in order to provide a cleaner environment that magnifies the effects of the excitation frequency.

The velocity measurements were performed with the use of hot-wire anemometry (HWA) Dantec Dynamics Streamline Pro apparatus. The HWA probe type 55P05 with 1.25 mm in length and 5 μm in diameter was used. Data acquisition was performed with sampling frequency 25 kHz and with sampling time 60 s. The boundary layer was inspected in several traverses along the flow direction ranging from x=70 to 700 mm from the leading edge. Velocity signals in each traverse were recorded at several points, ranging from the wall to 20 mm in the normal direction.

Sound field measurements were performed with the use of advanced GRAS microphones system. The system consists of flush-mounted and free-field microphones. One free-field microphone was placed at the outlet of the test section during the measurement in order to measure the sound pressure level (SPL) generated by the speaker. Based on the instrument's accuracy for a frequency range of 10–4000 Hz, the uncertainty in the measured sound pressure levels (SPLs) is of ±0.1 dB, which is typical for such measuring devices [8]).

To accomplish the flow control a 15-inches 1-way speaker T115-800 (band: 100 Hz do 650 Hz) was located at a certain distance from the tunnel outlet and pointed upward in the flow. The speaker was connected to Powersoft K 10 DSP amplifier driven by a signal generator controlled by LabVIEW software and connected with a digital-analogue National Instruments converter, reproducing a noise in a specified frequency range.

Flow parameters determined in the core flow at the inlet plane to test section are the mean velocity $U_{e\_in} \approx 5$ m/s, which corresponds to the Reynolds number, $Re_x = 150\ 000$, based on the length of the plate at the separation point and the freestream turbulence level $Tu=u'/U_{e\_in}$ equal to 0.7% (u' is the rms of streamwise fluctuating velocity). In addition to the naturally developing flow (NE-no excitation), the flow was exposed to the pink noise characterized by the sound pressure levels: SPL = 125 dB and 135 dB at the frequency range 100 – 650 Hz. The pink noise is a random signal for which the spectrum density, i.e. narrow-band signal, varies as the inverse of frequency. The most important parameters at the separation point (x=600 mm) for all analysed cases are presented in Table 1, i.e. the mean velocity at the edge of boundary layer - U_e, the boundary layer thickness - δ, the shape factor - $H = \delta^*/\theta$, (where δ^* is displacement thickness and θ is momentum thickness) and the Reynolds numbers based on

Journal of Physics: Conference Series **2367** (2022) 012021 doi:10.1088/1742-6596/2367/1/012021

displacement - Re_{δ}^{*}, and momentum - Re_{θ}, thicknesses, respectively. The data show that as SPL increases, all values including the bubble size decrease.

Fig. 1b presents the noise power spectral density (PSD) measured by the free-field microphone, positioned 300 mm in front of the loudspeaker at the test section exit, for three considered cases. It can be seen that the activation of the broadband acoustic forcing in the 100 - 650 Hz range causes a strong increase in sound energy level above 80 Hz, with a characteristic for a pink noise decrease in sound power level above 100 Hz.

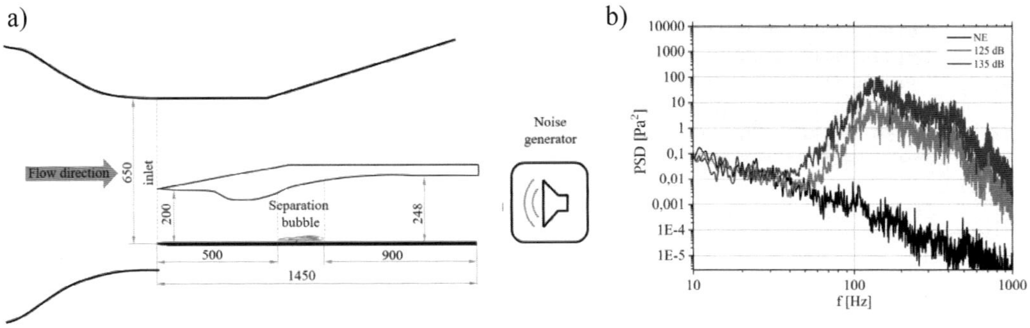

Figure 1. Schematic of the test section and speaker location (a), PSD noise spectra at the test section outlet (b).

Table 1. Parameters of the boundary layer (BL) at separation point, x=600 mm

Case	U_e [m/s]	δ [mm]	H [-]	$Re_{\delta*}$ [-]	Re_{θ} [-]
NE	5.25	9.47	6.83	1963	287
125dB	5.23	9.42	6.72	1934	288
135dB	5.18	8.72	5.92	1689	285

3. Results

The research described in this study aims at analysing a response of the separated laminar shear layer to the broadband acoustic excitation and investigate its effect on l-t transition. The presence of the LSB for the case without forcing is confirmed by smoke visualization presented in Figure 2. Separation onset likely occurs at a distance of 525 mm from the leading edge of the plate. The LSB is characterized by a distinct thickness 3 mm for the reference case (Figure 2a). Strong flow unsteadiness is visible at the edge of the separated shear layer starting at the distance x=600 mm and the boundary layer reattaches already at x=650 mm.

The introduction of an acoustic excitation results in a change in the flow topology, which, however, due to the strong dynamic impact of acoustic excitation is difficult to observe by visualization methods (see Figure 2b). This is also due to the fact that the images shown in Figure 2 illustrate only the instantaneous state of the near wall flow. However, for the disturbed case, a qualitative change can be observed, consisting mainly of a reduction in the height of the LSB as well as earlier attachment of the separated shear layer.

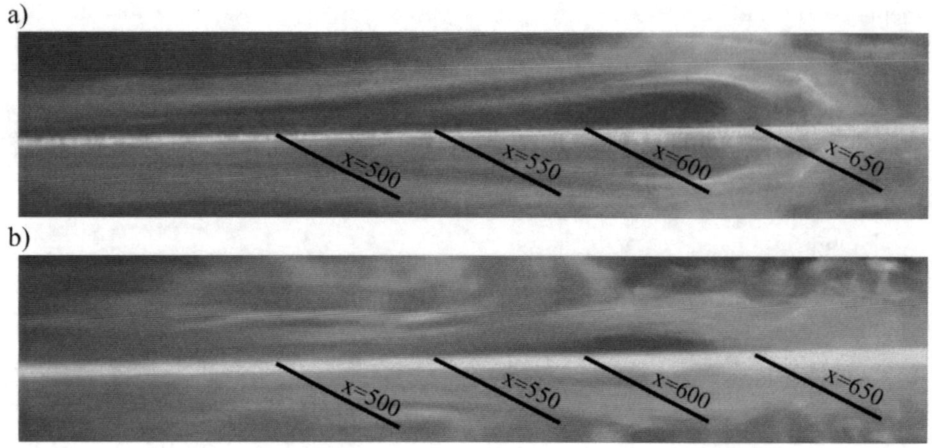

Figure 2. Visualization of separated bubble on the flat plate: without forcing (a) and with pink noise excitation at SPL=135 dB (b).

Figure 3a depicts the distribution of one of the basic boundary layer parameters, that is, the shape factor *H*. The shape factor is defined as the ratio of integral parameters, displacement thickness to the momentum thickness and takes the value of 2.59 for a laminar attached boundary layer (Blasius type) and 1.4 for the attached turbulent boundary layer. It is assumed that the shape factor beyond 4 indicates the boundary layer laminar separation [9]. At $x = 500$ mm, a clear increase in *H* is reported, which is due to the influence of APG. High values of shape parameter for undisturbed flow in the range x=525-650 mm confirm the presence of the LSB, as showed in Figure 2 (smoke visualization). The literature survey indicates that the maximum of the shape factor indicate locations of mean l-t transition and maximum bubble height [1,3]. It means that for the unexcited case (NE) the transition onset is located at x = 600 mm, while for acoustic forcing (denoted by 125 and 135 dB) it is shifted upstream. The analysis of the shape factor profile in Figure 3a shows, from a statistical point of view, that despite a strong level of acoustic excitation there in no effect visible on the laminar (x < 525 mm) and fully turbulent boundary layers ($x > 675$ mm). Confirmation of this finding is provided by the analysis of the time traces presented in Figure 4. Assuming, following Drela [9], that the shape factor at separation takes the value $H= 4.0$, and this value is reached for all cases at x= 525 mm, it can also be concluded that effect of acoustic excitation on separation point is hardly noticeable. The shrinking of the LSB with the acoustic excitation active might also be understood by analysing the distribution of the mean velocity at the boundary layer edge (Figure 3b). With the acoustic forcing active the mean velocity profiles approach the velocity distribution for the attached flow, denoted by the dashed line.

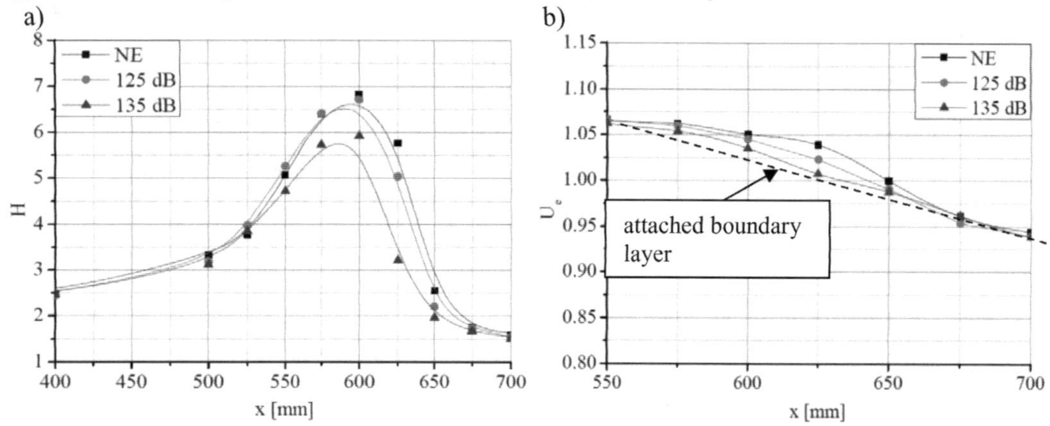

Figure 3. Shape factor distribution (a), edge velocity U_e distributions (b)

In order to gain better insight into the effect of acoustic excitation onto the transition process in the separated boundary layer the time traces are analysed in Figure 4 at four streamwise distances: x = 575, 600, 625 and 650 mm (recorded for wall-normal position y = 2mm). In the unexcited case (NE) only the large-scale unsteadiness is visible at x=575 mm and 600 mm. The amplitude of disturbances increases at distance x= 625 mm and the breakdown to turbulence is reported for the next traverse (650 mm). In the presence of acoustic forcing, the high-frequency perturbations superimposed on low-frequency unsteadiness are observed already at x=575 and 600 mm. The clear change in the shape of the time signal between cases, can be seen from the x= 625 mm. With SPL=135 dB the time traces at x=625 mm resemble the time traces for the unexcited flow at x=650 mm. It clearly shows earlier turbulence breakdown with the acoustic excitation active at SPL=135 dB. At distance 650 mm all the time traces show similar perturbed flow character, suggesting the transition is almost complete.

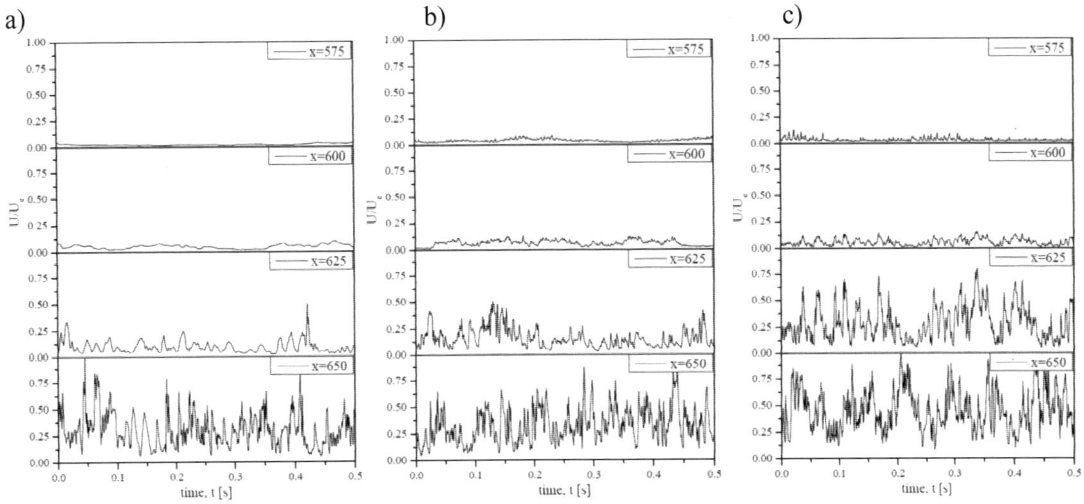

Figure 4. Sample hot-wire signals: without the acoustic excitation (NE) (a), with the acoustic excitation SPL = 125 dB (b) and SPL = 135 dB (c).

To clarify the phenomenon occurring in the separated boundary layer a spectral analysis of the velocity signal has been performed. Figure 5 presents the power spectral density (PSD) of the near-wall velocity at several streamwise locations (x=525-650mm). For the unexcited case in the shear layer (see Figure 4a), initially the energy is concentrated in a very narrow frequency range ~ 48 Hz, where the Strouhal number of the vortex shedding based on the momentum thickness and the velocity at the separation point equals $St_s = f\frac{\theta}{U_\delta} = 0.007$, what according to Talan and Hourmouziadis [4] suggest the presence of the Kelvin–Helmholtz instability. For further traverses from x = 600 mm there is an increase in energy in a wider bandwidth, but with a still visible maximum at around 48 Hz. In subsequent traverses a strong increase in the energy level of fine-scale is observed, characteristic of transitional flow. Although for the acoustic excitation cases (excitation is applied in the range of 100-650 Hz) there is also an effect for the low frequency range. In particular, in the early stages of LSB development (x=525 and 550 mm), frequencies around 64 Hz are amplified. This frequency corresponds to the Strouhal number around 0.009 and can also be associated with the Kelvin–Helmholtz instability, which is stimulated by broadband noise. Downstream the flow the peak is distorted, indicating a redistribution of the energy to adjacent scales and forcing an early loss of stability. Note that a similar effect was reported by Kurelek et al. [3], but deploying the broadband acoustic excitation with the frequency range including the frequency of most amplified disturbances of the unperturbed flow. So the present work shows that similar effect is observed with broadband acoustic forcing employed with the higher frequency range (f=100-650Hz) than the frequency of the unperturbed flow instability mechanism (f=48-64 Hz). The physical mechanism of this phenomenon is not entirely clear, but it can be assumed that there is coupling of the sub-harmonic forcing generated by acoustics to the natural frequency of the K-H instability.

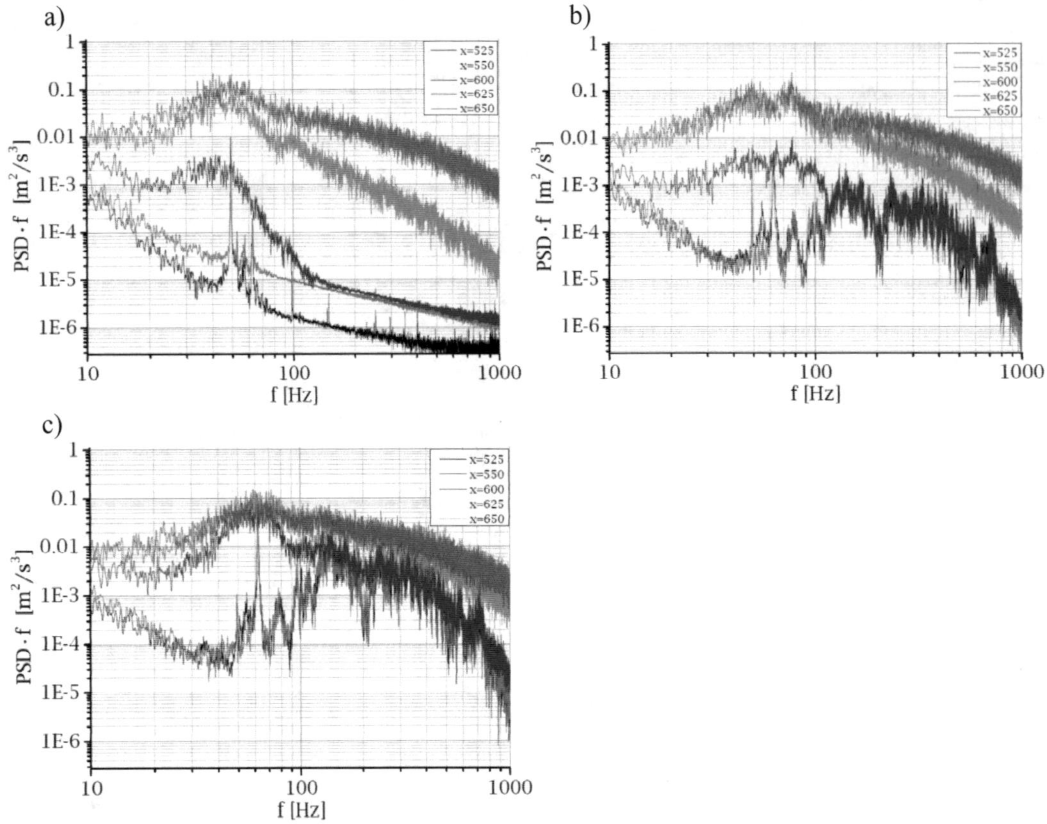

Figure 5. Premultiplied power spectra density: without the acoustic excitation (NE) (a), with the acoustic excitation SPL = 125 dB (b) and SPL = 135 dB (c).

4. Conclusions

It was shown that the acoustic excitation in the frequency range, f=100-650Hz, can lead to a more rapid increase in flow instability and advance the l-t transition. The effect of acoustic excitation on the separation point of the laminar boundary layer is hardly noticeable, while earlier reattachment point and decrease in the height of the bubble are observed. The decrease in the height of the bubble is confirmed not only by the drop of boundary layer thickness (see Table 1), but also by the analysing the distribution of the mean velocity at the boundary layer edge. The results indicate also a phenomenon, not yet observed in the literature, of amplification of flow instability (f=48-64 Hz) in the initial phase of the transition, forced by broadband pink noise generated by the higher frequency range (f=100-650Hz). It can be the result of coupling of the sub-harmonic forcing generated by acoustics to the natural frequency of the K-H instability.

Acknowledgments

The investigation was supported by National Science Centre under Grant No. 2018/31/B/ST8/01717. The author wishes to acknowledge Vaclav Uruba for some fruitful discussion performed within the project funded by the National Agency for Academic Exchange (NAWA) within the International Academic Partnerships Programme No. PPI/APM/2019/1/00062.

References

[1] Piotrowski W, Elsner W and Drobniak 2010 Transition Prediction on Turbine Blade Profile with Intermittency Transport Equation. J. Turbomach. 132 (1) 011020.

[2] Ezerskii A B 1985 Laminar-Turbulent Transition *Proc. of Symposium*, Novosibirsk.

[3] Kurelek W, Kotsonis M and Yarusevych S 2018 Transition in a separation bubble under tonal and broadband acoustic excitation. *J. Fluid Mech.* **853** 1–36.

[4] Talan M and Hourmouziadis J 2002 Characteristic Regimes of Transitional Separation Bubbles in Unsteady Flow *Flow, Turbulence and Combustion* **69** 207–227.

[5] Yarusevych S and Kotsonis M 2017 Effect of Local DBD Plasma Actuation on Transition in a Laminar Separation Bubble. *Flow Turbulence and Combustion* **98** 195–216.

[6] Jones LE, Sandberg RD and Sandham ND 2008 Direct numerical simulations of forced and unforced separation bubbles on an airfoil at incidence, *J. Fluid Mech.* **602** 175–207.

[7] Marxen O, Kotapati RB, Mittal R and Zaki T 2015 Stability analysis of separated flows subject to control by zero-net-mass-flux jet. *Phys. Fluids* 27 (2) 024107.

[8] Niegodajew P, Łukasiak K, Radomiak H, Musiał D, Zajemska M, Poskart A and Gruszka K 2018 Application of acoustic oscillations in quenching of gas burner flame. *Combust. Flame* 194, 245-249.

[9] Drela M 2014 Flight Vehicle Aerodynamics. Cambridge, MA: MIT Press.

Effect of the upstream cylinder shape on the flow around the downstream rectangular object in tandem configuration

Karolina Gajewska[1], Paweł Niegodajew[1], Renata Gnatowska[1] and Witold Elsner[1]

[1]Czestochowa University of Technology, Armii Krajowej 21 42-200, Czestochowa, Poland

E – mail: karolina.gajewska@pcz.pl

Abstract. The paper presents an experimental investigation of air flow around bluff bodies in tandem configurations. The first one concerns two square cylinders and in the second one a triangular cylinder was used as an upstream object. Experiment was performed for two different Reynolds number for the fixed distance between bluff bodies. To have an insight into the fluid flow, particle image velocimetry method was used. Particular attention was paid to examine the effect of the upstream cylinder shape on the flow around the downstream body.

1. Introduction

The flow around bluff bodies (BBs) is widely employed in a number of engineering and industrial applications. Flow control, enhancement of mixing processes, cooling of electronic components, thin wire probes or wind flows in urban areas are only a few examples of BB applications [1–4].

Whenever a flow passes a BB, such as a circular or rectangular cylinder (which are the most frequently studied BB shapes [5,6]), the vorticity created is directly associated with the formation of large-scale vortices that shed periodically into the wake region [7]. Since the vortices shed alternately, they generate an oscillatory force acting on the obstacle normal to a freestream. When the vortical regions are accompanied with low pressures, the pressure downstream the obstacle, will be low and so a pressure drag will be induced [8].

One should note that vortical structures oriented parallel to the BB axis are not the only ones that contribute to the instability of wake structure, streamwise vortices play an important role in this process as well. Consequently the wake structure is fully three-dimensional and non-stationary because structures appear randomly in time and space [9,10].

The aerodynamic characteristics of BBs are strongly dependent on the shear layers that separate from the leading edge. For instance, when concerning a flow around a square cylinder, the drag forces can be considerably reduced only by modification of corners' shapes [11]. The vortex dynamics can become even more complex due to the variation in Reynolds number (*Re*). The effect of *Re* is particularly important especially when considering such fundamental flow characteristics as separation, shear layers, vortex pattern, stagnation etc. [12–14].

From the practical point of view, even more important are configurations when the flow passes around multiple BBs. Early studies shown that force fluctuation and vortex shedding patterns for multiple BBs are completely different when comparing with those for a single BB for the same *Re*. Separated wakes generated from each BB interact mutually generating in turn the flow regimes that are different from those occurring when a flow passes a single BB [15]. Moreover, when gap spacing between objects and *Re* alter unexpected distinct flow regimes emerge [16,17]. When the flow passes multiple BBs even more complex phenomena may occur such as untypical vortex shedding, wake interactions or gap vortices [18,19].

Most of the research works on flow passing multiple BBs concerns circular and square cylinders. As pointed out in the review paper of Derakhshandeh and Alam [20], related to wakes generated by BBs of different shapes, not much attention has been devoted to triangular cylinders which generate stronger vortices than square or circular cylinders. Since the triangular cylinder produces a higher vorticity [20], one may expect that it will act differently on objects around (comparing to square or circular cylinder). Comprehensive fundamental knowledge about the effect of the wake downstream triangular cylinder on the other objects may be useful in practice in processes employing BBs to control the flow, especially where the selection of the proper object shape is crucial. Therefore, it is of interest to conduct such a study at least for a very fundamental configuration such as a tandem arrangement [21] where the triangular cylinder would be used as the upstream one and confront the results with classical configuration such as two square cylinders in tandem. The study also included an analysis of the effect of Reynolds number carried out for the subcritical regime.

2. Experimental setup

The experiment was conducted in an open-circuit wind tunnel with 2 m long test section with square cross-section (0.3 x 0.3 m). Two tandem configurations were studied, namely, the first one incudes two identical squares (each with the side length $D = 0.02$ m) and the second one includes triangle and square with dimensions (see Figure 1). Note that the height of each cylinder was the same as the channel height (0.3 m) so taking into account that both investigated objects were in tandem the value of the blockage ratio parameter was less than 7% which is satisfactory for such measurements. The so far literature indicate that in the case of two BBs arranged in tandem, the governing parameter is the inter-cylinder spacing ratio $S = x/D$, where x is the streamwise coordinate. In the present experiment, the distance S was chosen to be equalled 6 because for a smaller spacing vortex shedding from the upstream object is suppressed and so, has a low impact on the wake behind the downstream cylinder [22]. For the selected spacing S, the shear layers separated from the upstream cylinder alternately roll up into vortices and impinge upon the downstream one [23]. Two different Reynolds numbers ($Re = U_\infty D/\nu$, where U_∞ is the free stream velocity and ν the kinematic viscosity) were investigated for each configuration and the details about each investigated test case can be found in Table 1.

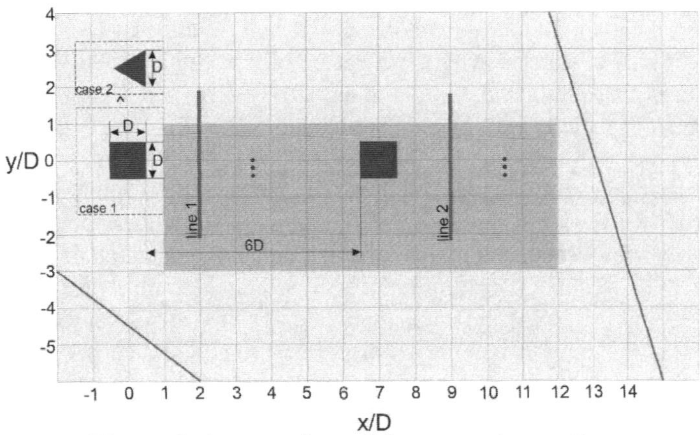

Figure 1. An overview of the measuring section.

A detailed insight into the flow field was accomplished by using the particle image velocimetry (PIV) method with SpeedSense VEO340 CCD camera and the flow was seeded with ~1 μm oil droplets. A double-pulsed laser emitting two pulses of 25 mJ at the wavelength of 527 nm with up to 20 kHz repetition rate was used for the illumination purpose. The approximate illuminated area is presented in

Figure 1 as the region between two red lines. Instantaneous velocity fields were investigated within the area, marked in Figure 1 with a grey region, lying in the plane passing through the centres of the cylinders, exactly at the half of the channel height. Moreover, averaged in time streamwise and transvers velocity distributions for both BBs configurations were measured at two selected lines behind each cylinder (marked in Figure 1 as 'line 1' and 'line 2' in blue).

Table 1. Tandem configurations investigated.

case	upstream cylinder	downstream cylinder	*Re*
SS-Re1	square	square	5000
SS-Re2	square	square	10000
TS-Re1	triangle	square	5000
TS-Re2	triangle	square	10000

3. Experimental results

Presentation of the results begins with a qualitative assessment of the evolution of instantaneous streamwise (U) velocity fields for cases SS-Re1 and TS-Re2 are presented in Figure 2a and Figure 2b, respectively. Note that the time interval between capturing subsequent images in each configuration was identical, i.e. 0.005 s. Streamlines are superimposed to highlight the flow structures especially in the gap region between BBs. At a first glance, the case SS-Re1 exhibits much wider wake which should be attributed to the flat front of the upstream square cylinder. The vortex forming process develops in the space between the cylinders (the case SS-Re1) and is well marked downstream of the second one. The other observation that can be made from Figure 2 is that the vortex shedding is different for each configuration. For the case SS-Re1 one can observe the periodical changes in the flow asymmetry in the gap. For the case TS-Re1 the flow intermittently slides off the trailing edge of the downwind cylinder and causes an alternate vortex shedding. In the case SS-Re1 the vortices shedding frequency is visibly lower downstream the first object comparing to the case TS-Re1. One can also see that there is an evident region with significantly reduced velocity very close to the front of the second cylinder for the case SS-Re1. This effect is not visible for the configuration with triangle cylinder. Similar conclusions can be drawn from the analysis of the spanwise component of the velocity field (Figure 3). In addition, the stronger variation of the transverse component behind the square cylinder is noted, which directly accounts for the wider wake.

The effect of each configuration on the flow downstream the second object also can be observed. In both cases, the vortex path behind the second object is heavily dumped which is clearly seen in Figure 3. This observation is also connected with the absence of a dominant frequency characteristic of the Karman vortex street in that region. There is a much stronger recirculation area for TS-Re1 which is manifested by presence of regions with negative values of U/U_∞ (Figure 2) which should considerably contribute to the drag force. Moreover, for TS-Re1, the vortex structures generated by the upstream object impinge directly the front of the downstream one, strongly modifying the generation of Karman vortices. In the case SS-Re1 with a wider wake behind the square cylinder, this effect is less pronounced because the upstream vortex structures pass the downstream cylinder much farther away.

The vortex shedding frequency or dimensionless Strouhal number (St) is one of characteristic parameters of the BB and important indicator that reflects the vortex dynamics [24]. In general, the shorter formation length or the wider wake width, the lower St and vice versa [25,26]. In the present experiment St was obtained based on the velocity signal collected in points marked in Figure 1 with circles downstream the first and second object for each configuration from Table 1. One should emphasize here that it was not possible to obtain St downstream the second object for any configuration as there was no distinctive peak visible in amplitude spectra. The values of St from the present experiment downstream the first object are summarized in Table 2. Note that for each point marked in Figure 1 identical values of St were observed in each configuration. Obtained St for both cases with upstream square cylinders is identical which suggests that it is independent of Re for its investigated range. Values of St behind the triangular cylinder are higher which is justified as this configuration exhibits narrower wake behind the first object. Worth noting is that obtained *St* for the configuration with square and triangular cylinders agrees with available literature data for similar *Re* (see Ref. [23]).

Table 2. Strouhal number behind the upstream cylinders for each configuration.

case	St
SS-Re1	0.12
SS-Re2	0.12
TS-Re1	0.18
TS-Re2	0.19

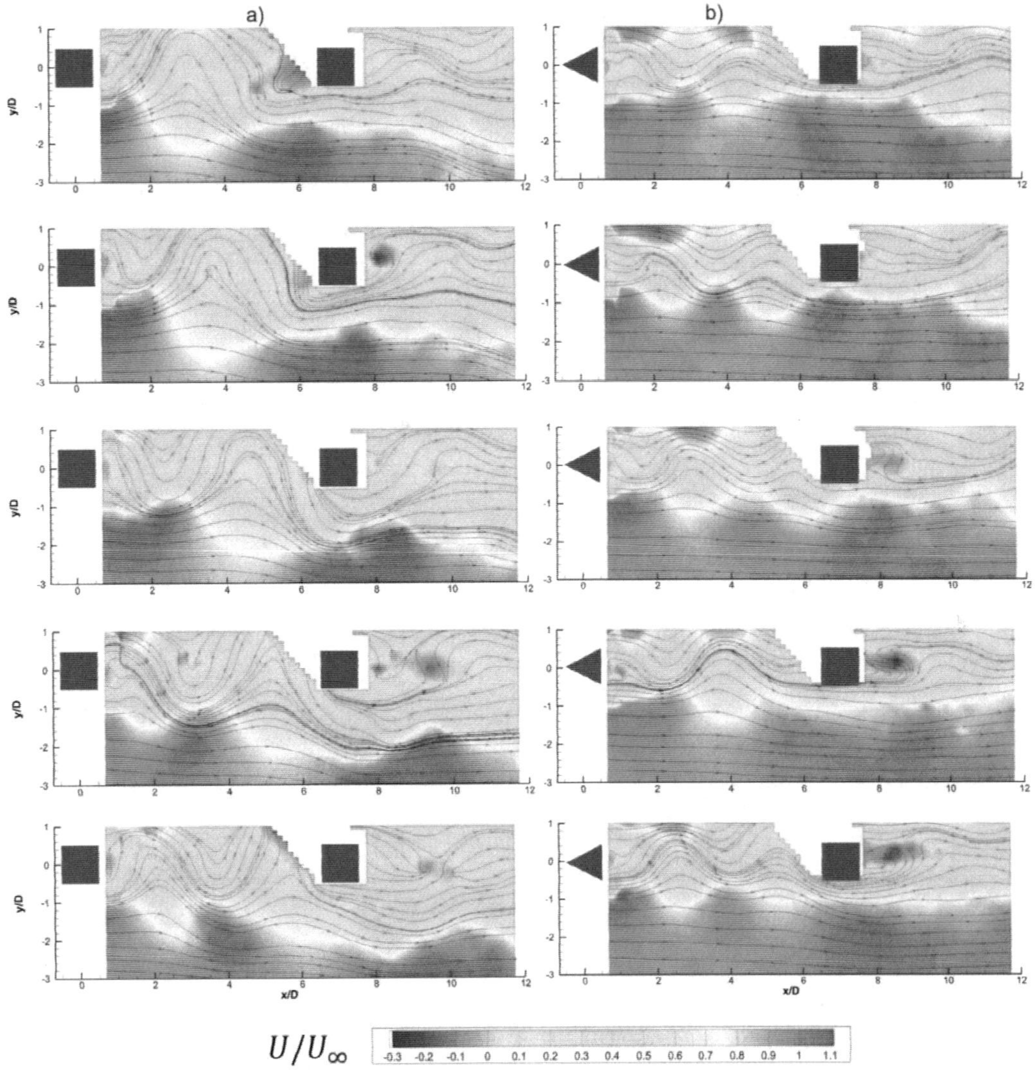

Figure 2. Evolution of instantaneous streamwise velocity fields with superimposed streamlines pattern for case SS-Re1 (a) and TS-Re1 (b).

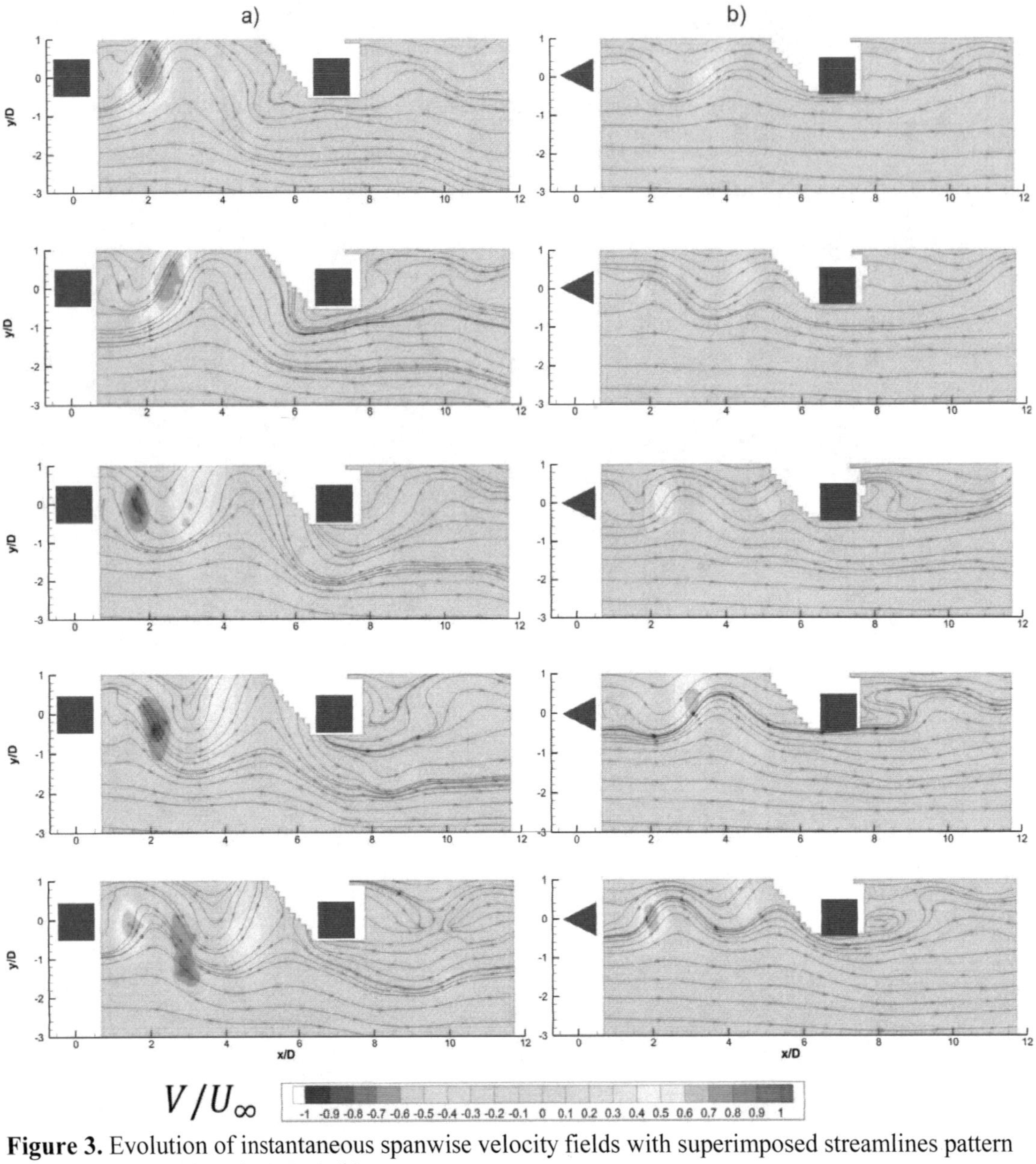

Figure 3. Evolution of instantaneous spanwise velocity fields with superimposed streamlines pattern for case SS-Re1 (a) and TS-Re1 (b).

More detailed insight into the flow downstream each investigated case is presented in the form of normalised streamwise and transvers (V) velocity distributions along line 1 and line 2 in Figure 4. The configuration with two squares exhibits much wider wake downstream the first object (about twice as wide as for the setup with triangle and square) which is represented by much wider streamwise velocity deficit in Figure 4a. Interestingly, the maximum streamwise velocity deficit behind the first BB (that can be read from line 1) is identical for each investigated configuration. The effect of *Re* on U/U_∞ behind the first object is marginal as the wake becomes only slightly wider when increasing *Re*. The effect of upstream cylinder shape is still visible behind the second object (line 2) which is manifested by a wider wake for the case with two square cylinders (see the U/U_∞ width deficit change in Figure 4b), however the effect is less pronounced than for the line 1. Triangular cylinder is responsible for a more profound velocity deficit behind the second object which confirms the observations from the velocity fields. The velocity distributions shown in Figure 4 together with the data in Table 2 confirm the finding of Derakhshandeh at al. [20] that for the subcritical regime the wake width and *St* depend on the geometrical shape of the bluff body, and both are almost insensitive to the Reynolds number.

Regarding the transvers velocity component, the configuration with two square cylinders is characterised much wider change in V/U_∞ (see Figures 4c and 4d). The effect of *Re* on V/U_∞ is more pronounced here especially for the case with two square cylinders. It is particularly evident in Figure 4d for two square cylinders as with increasing *Re* there is a notable reduction of the local minimum in V/U_∞.

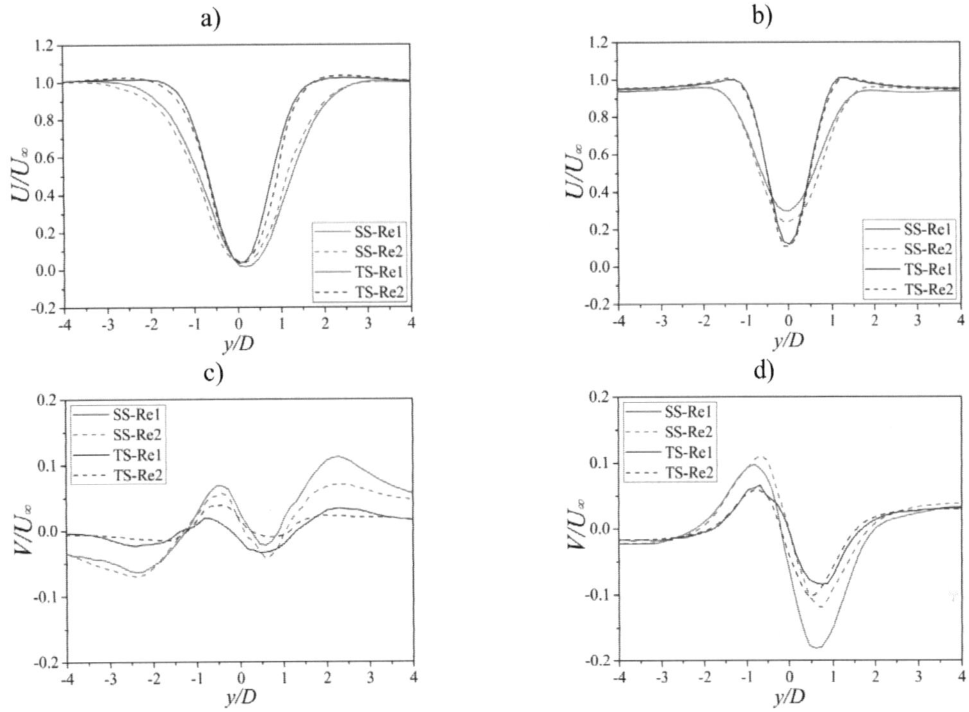

Figure 4. Normalised distributions of the streamwise velocity component for each investigated case for line 1 (a) and line 2 (b), and normalised distributions of transvers velocity component for each investigated case for line 1 (c) and line 2 (d).

4. Conclusions

The paper examines the effect of upstream cylinder shape on the air flow around the second square object, both arranged in tandem. Two different upstream cylinder geometries were investigated, namely square and triangular cylinder for fixed distance between objects and for two different Reynolds numbers.

The results shown that the configuration with two square cylinders exhibits much wider wake downstream both, the first and the second cylinder and so the effect of upstream cylinder shape is maintained even beyond the region between objects. The effect of each configuration on the flow downstream the second object also can be observed. In each case, the vortex path behind the second object is heavily dumped, which is also connected with the absence of a dominant frequency characteristic of the Karman vortex street in that region. For TS-Re1 case there is a much stronger recirculation area which should considerably contribute to the drag force. The maximum streamwise averaged velocity deficit behind the first BB is identical for each investigated configuration, however, behind the second object there is a notably smaller minimum in U/U_∞ for the case with triangle and square.

The effect of *Re* (for the considered range of change in *Re* in this study) on U/U_∞ was found to be marginal for each configuration, which confirm the finding of Derakhshandeh at al. [20] for the subcritical regime. On the other hand, the effect of *Re* on V/U_∞ was found to be noticeable especially for the case with two square cylinders.

Acknowledgements
This work has been supported by the statutory funds BS/PB-1-100-3016/2022/P the NAWA funds PPI/IAL/2018/1/00021. Special thanks are given to Konrad Gumowski from Warsaw University of Technology for his contribution in experimental work and post processing the data.

References

[1] Gao W, Yan Y, Huang L, Shen K, He Z and Gao B 2021 *Int. J. Hydrogen Energy* **46** 24382–94
[2] Blocken B 2015 *Build. Environ.* **91** 219–45
[3] Jafari M and Alipour A 2021 *Sustain. Cities Soc.* **72** 102979
[4] Dróżdż A, Niegodajew P, Romańczyk M, Sokolenko V and Elsner W 2021 *Exp. Therm. Fluid Sci.* **121** 110291
[5] Yanovych V, Duda D and Uruba V 2021 *Eur. J. Mech. B/Fluids* **85** 110–23
[6] Uruba V and Procházka P 2020 POD Spectrum of the Wake behind a Circular Cylinder *MATEC Web Conf.* **328** 05006
[7] Manickam Sureshkumar E, Arjomandi M, Cazzolato B S and Dally B B 2018 *Ocean Eng.* **164** 181–98
[8] Beaudoin J F, Cadot O, Aider J L and Wesfreid J E 2006 *J. Fluids Struct.* **22** 973–8
[9] Beaudoin J F, Cadot O, Aider J L and Wesfreid J E 2006 Drag reduction of a bluff body using adaptive control methods *Phys. Fluids* **18**
[10] Sobczyk J, Wodziak W, Gnatowska R, Stempka J and Niegodajew P 2018 *J. Wind Eng. Ind. Aerodyn.* **179** 184–9
[11] Tamura T and Miyagi T 1999 *J. Wind Eng. Ind. Aerodyn.* **83** 135–45
[12] Achenbach E and Heinecke E 1981 *J. Fluid Mech.* **109** 239–51
[13] Bai H and Alam M M 2018 *Phys. Fluids* **30** 015102
[14] Dróżdż A, Niegodajew P, Romańczyk M and Elsner W 2021 *Exp. Therm. Fluid Sci.* **125** 110377
[15] Zdravkovich M M 1977 *J. Fluids Eng. Trans. ASME* **99** 618–33
[16] Chatterjee D, Biswas G and Amiroudine S 2010 *Comput. Fluids* **39** 49–59
[17] Singha S, Nagarajan K K and Sinhamahapatra K P 2016 *Phys. Fluids* **28** 053603
[18] Nosek Š, Kukačka L, Kellnerová R, Jurčáková K and Jaňour Z 2016 *Boundary-Layer Meteorol.* **159** 259–84
[19] Du Y, Blocken B, Abbasi S and Pirker S 2021 *Environ. Model. Softw.* **145** 105172
[20] Derakhshandeh J F and Alam M M 2019 *Ocean Eng.* **182** 475–88
[21] Freidooni F, Sohankar A, Rastan M R and Shirani E 2021 *Build. Environ.* **201** 107985
[22] Sohankar A 2012 *Int. J. Numer. Methods Fluids* **70** 1244–1257
[23] Wang X K, Hao Z, Zhang J X and Tan S K 2014 *Exp. Fluids* **55**
[24] Okajima A 1982 *J. Fluid Mech.* **123** 379
[25] More B S, Dutta S, Chauhan M K and Gandhi B K 2015 *Exp. Therm. Fluid Sci.* **68** 339–58
[26] Yen S C, San K C and Chuang T H 2008 *Exp. Therm. Fluid Sci.* **32** 927–38

On Dynamics of Flow past a Stage Axial Air Turbine

V Uruba[1,2], D Duda[2] and V Yanovych[1,2]

[1] Institute of Thermomechanics of the Czech Academy of Sciences, Dolejškova 5, 182 00 Praha, Czech Republic

[2] Department of Power system engineering, Faculty of Mechanical Engineering, University of West Bohemia, Univerzitní 8, 301 00 Plzeň, Czech Republic

E-mail: `uruba@it.cas.cz`

Abstract. The air flow inside a single-stage test turbine is studied experimentally using the Particle Image Velocimetry technique. The measurements were performed at the axial-tangential plane just behind the rotor at the middle blade radius. The large vortical structures are identified in the shear layers between the blade wakes and channel jet. The phase-locked structures topology of random dynamical structures in the flow behind the stage are to be analysed using the Proper Orthogonal Decomposition method.

1. Introduction

The flow inside an axial turbine is complex and highly dynamical, consisting of an important periodical component generated by moving wakes behind the rotor blades, and random, turbulent component. These two components play very different roles in the flow dynamics, including mixing properties and shear flow stability. The effect of periodical fluctuations on the stability of a laminar boundary layer has been intensively studied in the past, mainly in simplified situations using both experimental and mathematical models, see e.g. [1]. In the presented study we will concentrate on study of truly turbulent behavior.

Classical approach to experiments in turbomachinery prefers pressure measurement strategy [2]. Various probes are used for indication of total and static pressures as well as the flow direction. However this approach relies on subsequent point measurements of the flow field, which allows for reconstruction of the fields of statistical characteristics based on averaging procedure. However this strategy could be applied only under assumption of the flow stationarity. The more sophisticated approaches rely on so called phase averaging, see e.g. the pioneering work [3].

Recently, some attempts of application of optical methods in the turbine experiments have been reported in literature. Both LDA and PIV methods has been used. As an example, we could refer to [4], where Stereo PIV method has been used. A limited number of other publications on this subject could be find during the last decade.

The presented work presents results of optical methods applied in a model air turbine. The model turbine stage geometry represents the typical 30%-reaction stage. The nominal velocity ratio of circumferential mid-span rotor velocity and the isentropic stage velocity was 0.55. The nominal stage isentropic Mach number was 0.4. The measuring plane was located close to the blade mid-span. More details on the experiment could be found in [5]. Several technical problems connected with optical access and seeding has been faced during the experiments, some of them are addressed in [6,7]. In the presented paper the advanced dynamical analysis of the same data is presented.

2. Experimental setup and methods

The experimental facility and methods of measurement and data evaluation are to be introduced.

2.1. Experimental facility

The measurements have been carried out in the single-stage experimental turbine at the Czech Aerospace Research Centre (VZLU). The facility is connected to the closed-loop wind tunnel with Reynolds and Mach numbers independent control. The static pressure is controlled by a system of vacuum pumps and the air mass flow rate is set using the twelve-stage radial compressor driven by a 1.3 MW engine. The quality of air, temperature and humidity, are controlled by interstage heat exchangers and condenser dryer, respectively. There is the feedback rotational speed system involving a hydraulic dynamometer. The facility allows changing the pressure ratio, Reynolds and Mach numbers and the rotational speed of the turbine independently. The external view on the turbine with the PIV measuring equipment is shown in Figure 1.

Figure 1. Turbine outside view.

The full-stage geometry is designed as a typical low-reaction stage with the rotor velocity to inlet velocity nominal ratio $u/c = 0.55$ and nominal isentropic Mach number 0.4. The other parameters of the used stator and rotor wheel are listed in Table 1.

Table 1. Parameters of the stator and rotor.

Parameter on tip	Units	Stator	Rotor
Hub diameter	mm	440	440
Blade height	mm	78.5	80
Blade pitch	mm	28.42	22.99
Blade chord	mm	44.61	26.00
Axial chord	mm	32.62	21.15
Blade number	1	66	82
Inlet angle	deg	0	25

Outlet angle	deg	74.5	71.9
Nominal *u/c*	1	0	0.55

Note: all relevant parameters are listed for the blade tip, angles are measured from the axial direction.

The same experimental setup was used in preliminary studies published recently [5,6,7].

2.2. *Measurement technique*

Particle Image Velocimetry (PIV) method was applied to explore the plane of measurement just behind the stator-rotor stage – see Figure 2. The phase of the rotor is locked during data acquisition using the synchronization with the rotor.

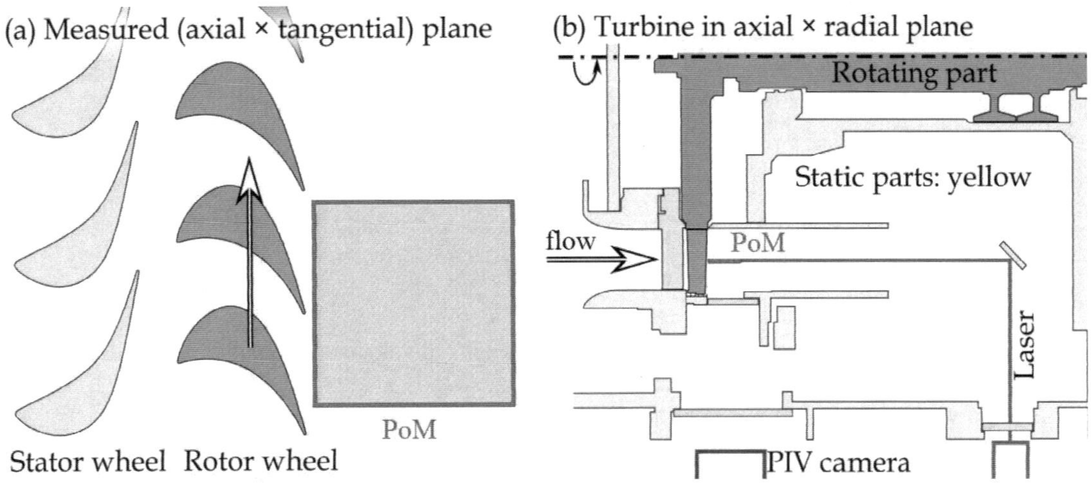

Figure 2. Turbine inside view (a) axial x tangential plane, (b) axial x radial plane. The Plane of Measurement PoM position is shown in red.

An adjustment of the test rig for the PIV setup was done. The special optical access to the space just downstream the stage was designed and manufactured both for the laser and camera. The mirror was placed at the diffuser outlet and the diffuser was equipped by a window for the camera, see Figure 2. The Plane of Measurement (PoM) position is shown in red.

The rotor phase position is detected using an optical gate, its signal is connected to the PIV system by Dantec to synchronize the images acquisition with the rotor angular position. The trigger signal is produced by the optical gate at the rotor shaft; therefore, the frame-pairs are synchronized with the rotor angle allowing it to perform phase averaging. The 675 measurements were acquired for statistical evaluation. Thus, the statistics of the obtained datasets reveals the random, turbulent part of the flow dynamics, the periodical dynamical component is filtered efficiently.

The camera FlowSense MkII with resolution 2048 × 2048 pixels in combination with the lens of focal distance 180 mm gives the physical size of the Plane of Measurement 29.2 × 29.2 mm^2. The seeding particles are illuminated by solid state double pulse laser New Wave Solo with cylindrical optics. The time between pulses during this measurement has been set to 5 μs. The SAFEX generator of oil droplets as a seeding particles with the typical diameter 1 μm has been used, the particles were introduced into the wind tunnel about 5 meters upstream the turbine.

2.3. *Evaluation methods and data analysis*

Standard evaluation procedure of the PIV measurement has been applied. The correlation method implemented in the Dantec Dynamic Studio software has been used, interrogation areas were 32 × 32 pixels.

Journal of Physics: Conference Series **2367** (2022) 012023 doi:10.1088/1742-6596/2367/1/012023

For the data analysis the standard statistical methods based on assemble averaging technique have been used.

The Proper Orthogonal Decomposition (hereinafter POD) has been used for the dynamical patterns detection in the velocity field. The POD provides an objective mathematical method for identifying deterministic topological objects from a turbulent flow-field. The method extracts the topology, which is best correlated in space, in a statistical sense, with the background velocity field. The method itself is of statistical type based on averaging technique. The topological modes are identified with the orthogonal eigenfunctions of the decomposition theorem of probability theory. The method could be used for identification of structures both in space and time, depending on the primary data character. The resulting modes poses specific features.

The POD method yields an orthonormal basis of decorrelated (i.e. orthogonal) modes maximizing the primary spatio-temporal data variance. For the velocity primary data, the maximized quantity physical interpretation is a total turbulent kinetic energy.

The details of the POD method are given e.g., in [8] including its application and interpretation.

The presented data are velocity components acquired as phase-locked manner and thus the turbulent, random part is not correlated. Thus, in our case only spatial modes are relevant.

3. Results

The results are to be presented in the form of vector and scalar quantities distributions. First, the basic statistical quantities are to be shown, next the POD analysis of dynamical structures.

3.1. Statistical results

The statistics of the velocity field will be presented. The mean velocities, velocity fluctuation variances and velocity components correlations have been evaluated.

In Figure 3, there are some representative results. In Figure 3a there is distribution of the mean velocity modulus. The strips of high velocity regions are shown. Then, the Turbulent Kinetic Energy (TKE) has been evaluated, however we have at our disposal only 2 velocity components in the PoM, velocity fluctuations u and v. Thus, the TKE was defined as follows:

$$TKE = 3/4\left(\overline{u^2} + \overline{v^2}\right), \tag{1}$$

where bar denotes averaging. The third, unresolved velocity component is assumed to be an average of the two resolved.

Figure 3c shows distribution of the correlation coefficient of the velocity components. Its negative value (in blue) denotes positive Reynolds stress component and thus production of turbulence. Those regions are located on the border of high-velocity strips in shear regions.

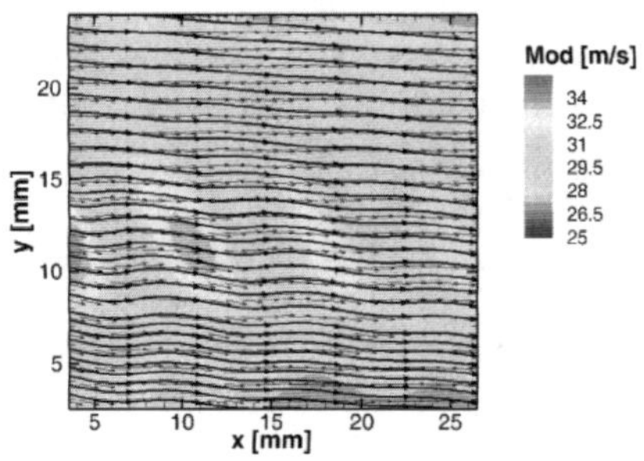

(a)

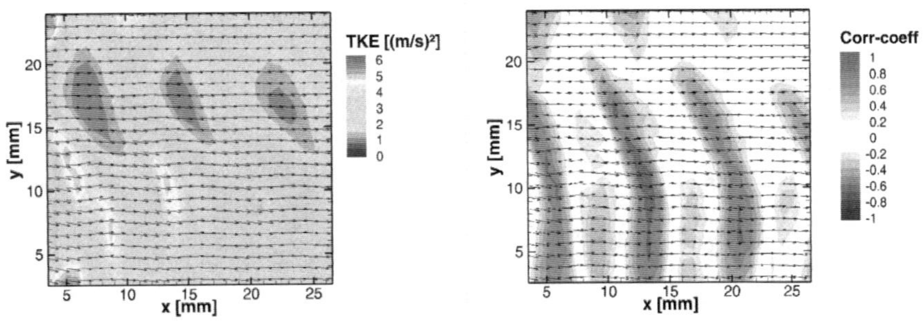

(b) (c)

Figure 3. The velocity field statistics distributions: (a) mean velocity modulus, Turbulent Kinetic Energy, (c) Correlation coefficient.

The strips, or wakes, topology in Figure 3 corresponds to the rotor output angle, see Table 1.

3.2. Proper Orthogonal Decomposition results

The results of the POD analysis are to be presented next. They characterize the dynamical structures in the flow-field. Please note, that the mean velocity field has been subtracted from the data prior to this analysis.

Out of 675 velocity fields the same number of POD modes with decreasing kinetic energy have been evaluated. The POD modes energies are shown the graph in Figure 4.

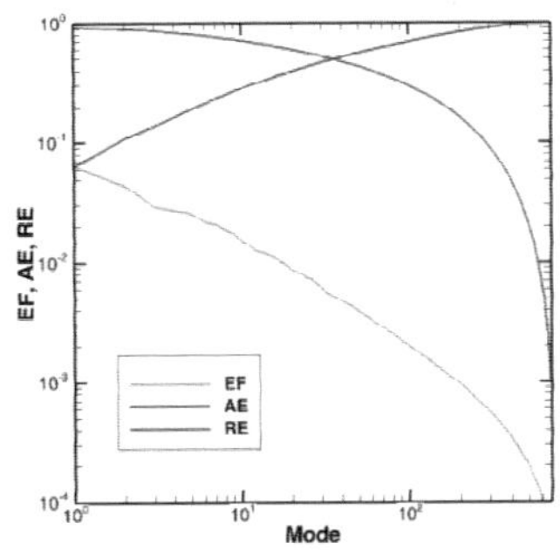

Figure 4. POD modes energies, Energy Fraction, Accumulated Energy and Residual Energy.

In Figure 4, there are Energy Fraction (EF), Accumulated Energy (AE) and Residual Energy (RE) shown, respectively, for all POD modes. Please note that both axes are logarithmic, and all quantities are dimensionless. The Energy Fraction represents the fraction of the kinetic energy assigned to a given POD mode. The first and second POD modes represent 0.058 and 0.052 (i.e., 5.8% and 5.2%) of the total turbulent kinetic energy, respectively.

The Accumulated Energy represents the contribution of the sum of energies of all lower order POD modes and the Residual Energy is the unexplained part of the total kinetic energy.

Next, the POD modes topologies are to be presented. In Figure 5, there are POD modes 1 and 2 topologies, respectively.

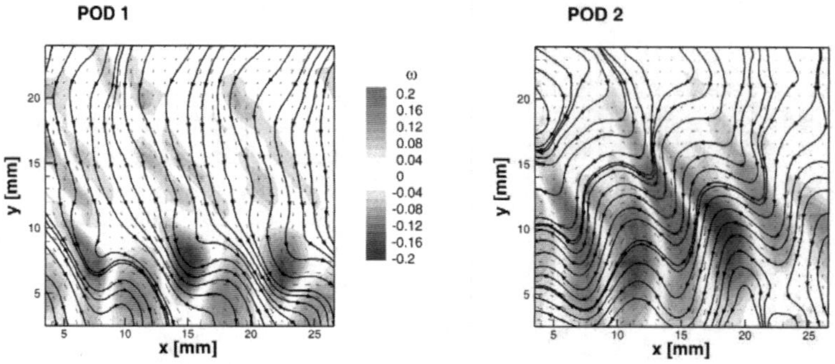

Figure 5. The first two POD modes topologies.

The POD modes topologies are shown as velocity-vector fields. To make these results clearer, the vector-lines are added arbitrarily, and vorticity is evaluated and shown in color. Please note that the POD modes are normalized, so all quantities are dimensionless, the only meaning represents the color: in red positive and in blue negative vorticity.

The most energetic dynamical POD modes 1 and 2 are represented by vertical (POD 1) and horizontal (POD 2) wavy patterns. Very distinct shear regions are represented by concentrated vorticity, they coincide with the blades wakes, compare with the mean velocity distribution in Figure 3a.

The higher POD modes with lower kinetic energy are more complex and less regular. The POD modes 3,4 and 5 contain weak vortical structures on the top of wavy patterns. They are shown in Figure 6.

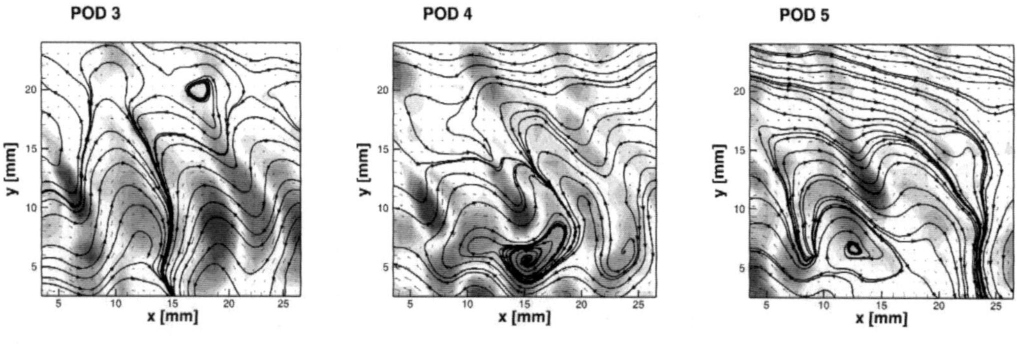

Figure 6. The POD modes 3,4 and 5 topologies.

The higher POD modes are even more complex as for the topology and random. In Figure 7 the POD modes 6-10 are shown. Those topologies contain both streamwise and spanwise regions and very distinct shear regions in the form of strips with predominant orientation defined by wakes behind the blades.

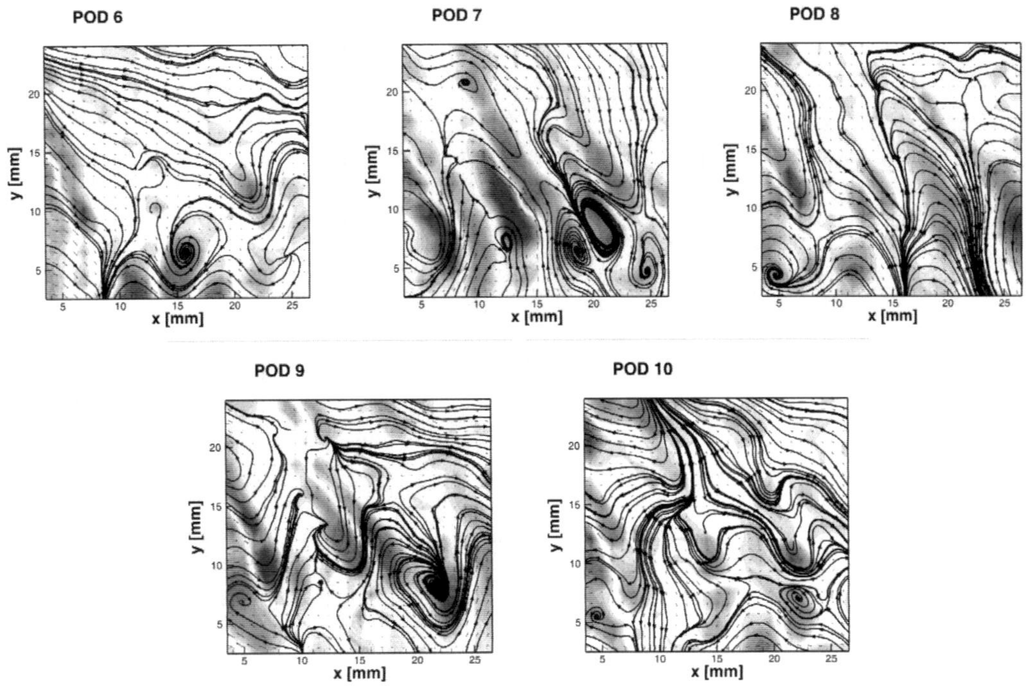

Figure 7. The POD modes 6,7,8,9 and 10 topologies.

The higher order POD modes contain some vortical structures, however the dominant vorticity concentration is in shear regions, which are linked to wakes behind the blades and/or with flow in channels between them.

4. Conclusions

The paper presents the POD analysis of experimental results obtained on air axial research turbine in nominal operational conditions using the PIV method.

The acquired data was used for instantaneous velocity fields evaluation. Phase-lock averaging was performed, and random statistical characteristics distribution was identified.

The dynamical properties of the random process were studied using POD method.

The topology of the POD modes shows the waves highlighting the wakes behind the individual blades. The 1st POD mode dominates in tangential fluctuations perpendicular to the machine axis, while the higher order modes are more complex, and fluctuations are oriented also perpendicularly to the wakes. Vorticity indicates important shears, but no distinct vortices are present in the dynamical topology.

References

[1] Michelassi, V.; Wissink, J.; Rodi, W. Analysis of DNS and LES of Flow in a Low Pressure Turbine Cascade with Incoming Wakes and Comparison with Experiments. Flow Turbul. Combust. 2002, 69, 295–330.Duda D., Uruba V., Spatial Spectrum From Particle Image Velocimetry Data, ASME J of Nuclear Rad Sci. 5, 030912 (2019).

[2] Tropea C., Yarin A.L., and Foss J.F. Springer handbook of experimental fluid mechanics (2007).

[3] Zaccaria, M.A.; Lakshminarayana, B. Unsteady flow field due to nozzle wake interaction with the rotor in an axial flow turbine: Part ii-rotor exit flow field. J. Turbomach. 1997, 119, 214–224.

[4] Porreca, L.; Yun, Y.I.; Kalfas, A.I.; Song, S.J.; Abhari, R.S. Investigation of 3D Unsteady Flows

in a Two Stage Shrouded Axial Turbine Using Stereoscopic PIV and FRAP: Part I—Interstage Flow Interactions. In Proceedings of the ASME Turbo Expo 2006, Power for Land, Sea, and Air, Barcelona, Spain, 8–11 May 2006; GT2006-90752; pp. 711–723.

[5] Duda, D.; Jelínek, T.; Milčák, P.; Němec, M.; Uruba, V.; Yanovych, V.; Žitek, P. Experimental Investigation of the Unsteady Stator/RotorWake Characteristics Downstream of an Axial Air Turbine. Int. J. Turbomach. Propuls. Power 2021, 6, 22. https://doi.org/10.3390/ijtpp6030022.

[6] Duda D., Jelínek T., Němec M., Uruba V., Yanovych V., and Žitek P. Observation of Flow Structure Past a Full-Stage Axial Air Turbine at the Nominal and Off-Design States, AIP Conference Proceedings 2323, 030004 (2021); https://doi.org/10.1063/5.0041491.

[7] Duda D., Jelínek T., Němec M., Uruba V., Yanovych V., and Žitek P. Particle Image Velocimetry Measurement inside Axial Air Test Turbine – Effect of Window. AIP Conference Proceedings 2323, 030005 (2021); https://doi.org/10.1063/5.0041492.

[8] Uruba V., Decomposition methods in turbulent research, EPJ Web of Conferences, vol. 25 01095 (2012).

Journal of Physics: Conference Series **2367** (2022) 012024 doi:10.1088/1742-6596/2367/1/012024

Least-order representation of control-oriented flow estimation exemplified for the fluidic pinball

Songqi Li[1], Wenpeng Li[1] and Bernd R Noack[1]

[1] School of Mechanical Engineering and Automation, Harbin Institute of Technology, 518055 Shenzhen, PR China

E-mail: songzi32@outlook.com

Abstract. Deep learning has been widely utilized to accurately estimate the flow state from the sparse sensor measurements. Yet there is still a lack of understanding of the actual dimension of this type of regression problem. In this study, we propose an Autoencoder (AE) based estimation method to tackle the control-oriented, sensor-based flow estimation problem. This method encodes the input information into a few-nodes bottleneck, then decodes the compressed information to the estimated the flow state. This network stands for the least order representation of the input-output relationship. We choose the fluidic pinball as the benchmark problem which contains multiple-input and multiple-output. The rotations of three cylinders in the flow generates rich flow physics. Consequently, strong non-linearity arises between input measurements and output flow state. For fair comparison the Deep Neural Network (DNN) is also employed to highlight the influence of the encoding-decoding process in AE. The results show that when the flow is periodic, the input vector can be effectively encoded into a five-dimensional subspace without a significant loss of estimation accuracy. However, the five-dimensional encoding will eliminate partial information from the input vector when the forced flow becomes chaotic, resulting in a lower estimation accuracy than DNN.

1. Introduction

The accurate estimation of spatial-temporal activities of coherent structures from sparse sensor signals in the turbulent flows remains a challenging engineering problem ([1]). This problem is desirably connected to the flow control, where the inference of the flow state could play a significant role in the development of control algorithm ([2]). Modern machine learning methods have come into the researches' field of vision. Several machine learning based estimation architectures ([3],[4],[5]) have been proposed to estimate full-state flow from sparsely placed sensor signals. As a continuation, this study discusses the least order representation of this type of non-linear regression problem. More specifically, we propose a data-driven, estimation method based on the autoencoder (AE, see Section 2.2). This architecture encodes the input information into a few-nodes bottleneck, then reconstructs the flow field using the subspace variables. The bottleneck layer represents the least-order information bridging up the inputs and the high dimensional flow. This AE-based estimation method is examined in the fluidic pinball plant (Section 2.1) aiming for control-oriented flow estimation. In this plant the flow is controlled by the steady rotations of three cylinders and the flow contains fruitful physics under different rotation speeds of three cylinders in the flow. The estimation results from AE will be compared to the ones from the Deep Neural Network (DNN) which DNN optimizes a non-linear mapping function with the maximum performance, and the results will be presented and discussed in Section 3.

2. The Fluidic Pinball Plant and Estimation Methods

2.1. The fluidic pinball plant

The fluidic pinball plant is a classical control problem which contains fruitful dynamical behaviors under different control parameters ([6]). This plant is composed of a two-dimensional, uniform flow passing a set of three cylinders of the same diameter D. Centers of the three cylinders form an equilateral triangle pointing towards the upstream direction, and the side length of the triangle is $1.5D$. The flow control is realized via the independent rotation of the three cylinders. This allows the control of the wake formed downstream, and is capable to replicate most of the known control mechanisms ([6, 7]).

The flow is depicted in a two-dimensional Cartesian coordinate system, where the origin is located on the midpoint of the two rear cylinders. The x axis represents the streamwise direction and the y axis points perpendicular to the freestream direction. Both axes are non-dimensionalized by the diameter of the cylinders. The freestream velocity is represented by U_∞. The velocity vector is denoted by $\boldsymbol{u} = (u, v)$ where u represents the streamwise velocity and v the transverse velocity. In this study we consider an incompressible Newtonian fluid with constant density ρ and kinematic viscosity ν. The Reynolds number based on the cylinder diameter D is $Re_D = U_\infty D/\nu = 100$. The computational domain Ω is bounded by a rectangular box $[-6, 20] \times [-6, 6]$ externally and the three cylinders internally. The centers of three cylinders are $[-3/2\cos(30°), 0]$,$[-3/4, 0]$, and $[3/4, 0]$, respectively.

The computational domain Ω is discretized by the unstructured mesh shown in Figure 1 This mesh contains 4225 triangles and 8633 vertices. Given the effect of the cylinder rotations, The flow adopts the circumferential velocities $\boldsymbol{b} = [b_1, b_2, b_3]^{\mathrm{T}} \in [-2, 2]^3$ on the boundary of the front, bottom and top cylinders. For each cylinder, the rotation speed on the boundary is within the range of [-2, 2]. Here a positive value corresponds to counter-clockwise rotation of the cylinders and vice versa. The numerical method follows the ones that are described in [7, 8]. In this study, flow snapshots are obtained from a total of 1,000 randomly generated control commands $\mathcal{B} = \{\boldsymbol{b}\}_{j=1}^{1000}$. Here only the post-transient data are collected which will be used to model the input-output relationship between the sparse sensors and the flow state. For every control command, 100 snapshots are recorded with a non-dimensionalized time interval $dt = 1$.

To monitor the flow state, a total of 9 virtual probes are placed downstream to measure the discrete velocity signal, either in streamwise or spanwise component. The velocity probes are deployed as a 3×3 rectangular grid symmetric about the centerline, where the upper and lower sensors measure the streamwise velocity and the middle ones on the centerline measure the spanwise velocity. The streamwise locations of the sensors are $x = 5, 6.5, 8$, and the spanwise locations are $y = -1.5, 0, 1.5$. A summary of the sensor placement can be found in Table 1, and a comprehensive visualization regarding to the flow state, control parameters, and the downstream sensing is presented in Figure 1.

2.2. Estimation Methods

The goal of this study is to discuss the least order representation of the mapping function which bridges up input measurements and the output flow state in the control-oriented flow estimation problem. The input vector $\boldsymbol{q} = [\boldsymbol{s}, \boldsymbol{b}]^{\mathrm{T}}$ comprises the sensor signal $\boldsymbol{s} \in \mathbb{R}^9$ and the steady control command $\boldsymbol{b} \in \mathbb{R}^3$ in the flow. The output is the estimated flow state. Here we propose a data-driven estimation method based on the autoencoder (AE), and this method will serve as a minimum representation of both the flow state and the measurement information. In general, the autoencoder is a specific type of the Neural Network (NN) in which a bottleneck is imposed in

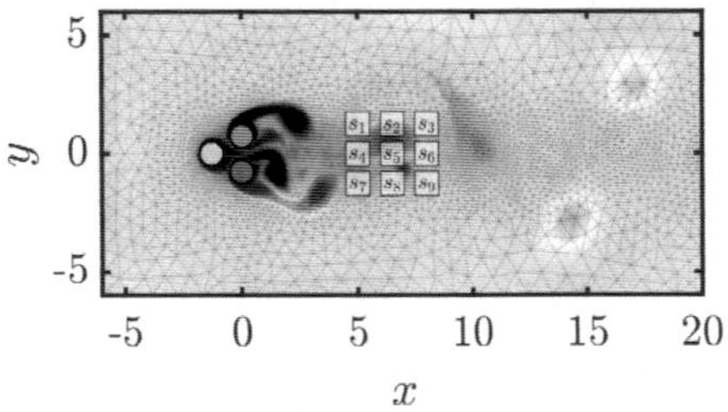

Figure 1. Visualization of the fluidic pinball plant. The flow domain is rendered by the instantaneous vorticity field, colors on the three cylinders indicate the magnitude of the corresponding rotation speeds, and s_1 to s_9 represent the locations of the in-flow sensors.

Table 1. Virtual Sensor locations and the measured velocity components.

Sensors	x/D	y/D	Velocity
s_1, s_2, s_3	5, 6.5, 8	1.25	streamwise
s_4, s_5, s_6	5, 6.5, 8	0	spanwise
s_7, s_8, s_9	5, 6.5, 8	-1.25	streamwise

the middle of the network to force a compressed representation of the input data ([9]). Figure 2 shows the architecture of the AE-based, control-oriented estimation method which will be used to estimate the flow state. The input of AE is the combined input vector $\boldsymbol{q}$. The encoder contains three fully-connected layers (64, 64, 5 nodes for each layer), and compresses the input data onto a 5-dimensional subspace (i.e. bottleneck). The decoder then upscale the compressed information onto high dimensional spaces with two 256-node hidden layers. The desired output is the instantaneous estimation of the flow field. However, given the high dimensional nature of the flow it is computationally expensive to estimate the velocity components on each grid point. Hence a lossless POD is performed, from which the velocity fluctuations ($\boldsymbol{u}_{\boldsymbol{b}}'$) can be expressed as the linear combination of the leading POD modes, such that:

$$\boldsymbol{u}_{\boldsymbol{b}}'(\boldsymbol{x}, t) = \sum_{k=1}^{N} a_k(\boldsymbol{b}, t)\boldsymbol{u}_k(\boldsymbol{x}) \tag{1}$$

By doing so the estimation of the flow field is equivalent to the prediction of the leading POD modal coefficients $\boldsymbol{a} = [a_1, a_2, ...a_N]^{\mathrm{T}}$. Since the first 78 POD modes contains 99% of the fluctuating energy N is set to 78 in this work, and these leading POD modal coefficients will act as the expected output from the output layer.

To investigate the effect of the data encoding, an estimation architecture based on the Deep Neural Network (DNN) is also adopted in this work. The schematic of the DNN estimation method is presented in Figure 3 and the detailed composition of DNN can be found in Table 2. To train both networks, the collected input-output data pairs are partitioned into a training set, a validation set, and a test set based on a random split of the control commands $\mathcal{B}$. The partition of the control commands follows a 80%-10%-10% ratio and each set is composed of all 100 time instances generated from the corresponding control commands. Other important parameters related to the DNN and AE architectures are also concluded in Table 2.

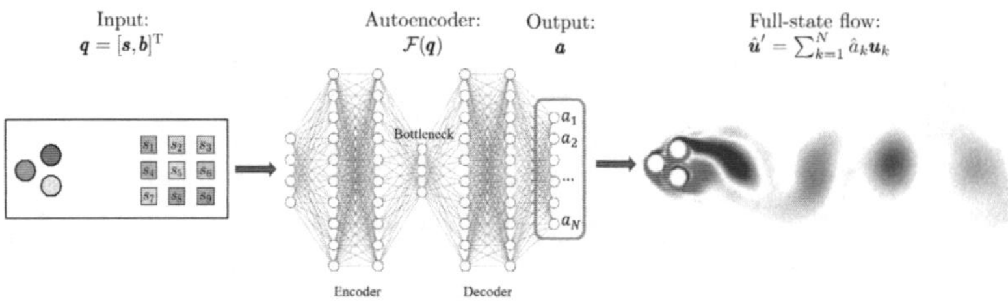

Figure 2. A schematic of AE estimator.

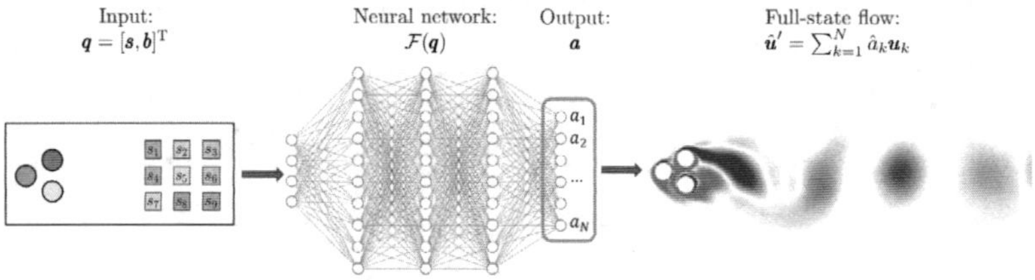

Figure 3. A schematic of DNN estimator.

Table 2. The architecture of DNN and AE.

Parameters	DNN	AE
Number of inputs	12 ($s \in \mathbb{R}^9$, $b \in \mathbb{R}^3$,)	12 ($s \in \mathbb{R}^9$, $b \in \mathbb{R}^3$)
Number of outputs	78 ($a \in \mathbb{R}^{78}$)	78 ($a \in \mathbb{R}^{78}$)
Hidden layer sizes	128,128,128	64,64,5,64,256,256
Activation function	ReLU	ReLU
Optimizer	Adam	Adam
Loss	MSE	MSE
Number of epochs	500	500

3. Results and Discussions

To start with we examine the performance of AE in comparison to DNN with two representative cases from the test set. The first case represents periodic vortex shedding under control, and the corresponding control command is $b = [0.5615, 0.7948, 1.5355]^T$. Figure 4 displays the comparison between the estimated POD modal coefficients and the target outputs from DNS data over all time instances under this control command. Here the estimates of three representative POD modes 1, 5, and 20 are presented in an attempt to evaluate estimation performance at different energy levels. Both AE and DNN are observed to be highly consistent to the DNS data, and both estimates can accurately describe the periodic behavior of the POD modal coefficients. Figure 5 visualizes a comparison between the estimated flow fields and the target DNS snapshot at a random time instance. The vorticity fields from both AE and DNN can accurately replicate the vortex shedding features in the original DNS snapshot.

The second case is a chaotic case in which no periodic vortex shedding takes place. In this case the control command is $b = [-0.2382, -1.2296, 1.1534]^T$. In general, a reduction of the estimation accuracy from both methods can be observed from Figure 6 in comparison to the previous case. Temporal evolution of a_1 from both AE and DNN estimates can basically follow the trend of the ground truth from DNS. However, clear discrepancies between DNS and sensor-

Journal of Physics: Conference Series **2367** (2022) 012024 doi:10.1088/1742-6596/2367/1/012024

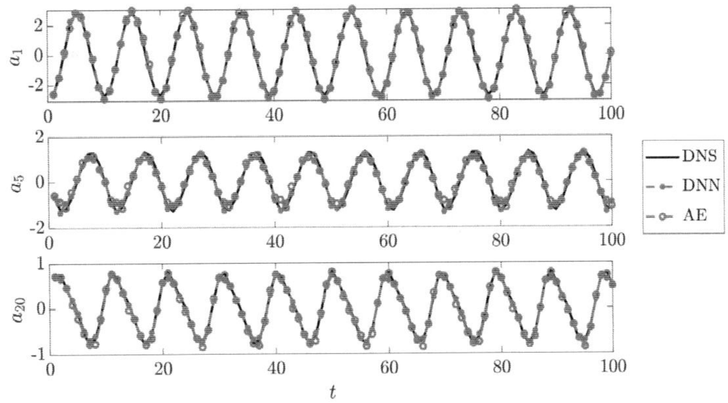

Figure 4. A comparison between estimated POD modal coefficients and the ground truth from DNS data in a periodic case ($\boldsymbol{b} = [0.5615, 0.7948, 1.5355]^{\mathrm{T}}$).

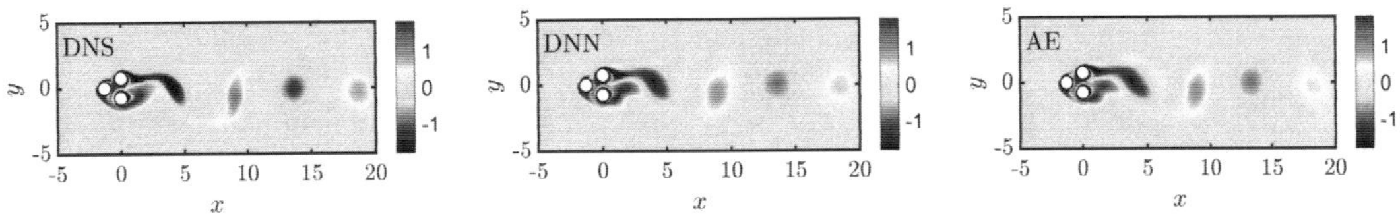

Figure 5. A comparison between the estimated vorticity fields and the DNS data in a random time instance of the exemplary periodic case.

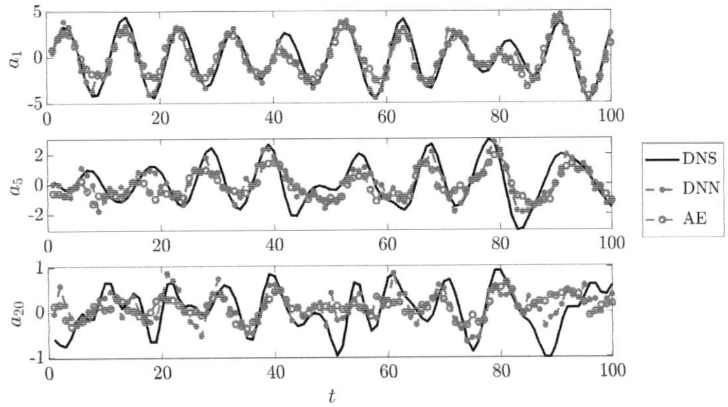

Figure 6. A comparison between estimated POD modal coefficients and the ground truth from DNS data in a chaotic case ($\boldsymbol{b} = [-0.2382, -1.2296, 1.1534]^{\mathrm{T}}$).

based estimates can be observed in higher order POD modes. Figure 7 provides a comparison between a DNS flow snapshot and the corresponding estimates from DNN and AE. Both DNN and AE estimates result in clear discrepancies comparing to the ground truth. Comparatively, the DNN estimate still outperforms AE regarding to the shapes, locations and intensities of the downstream vortical structures.

Besides the case studies, we perform an overall evaluation of the estimation methods under all 100 control commands in the test dataset. Each test case contains all time instances under

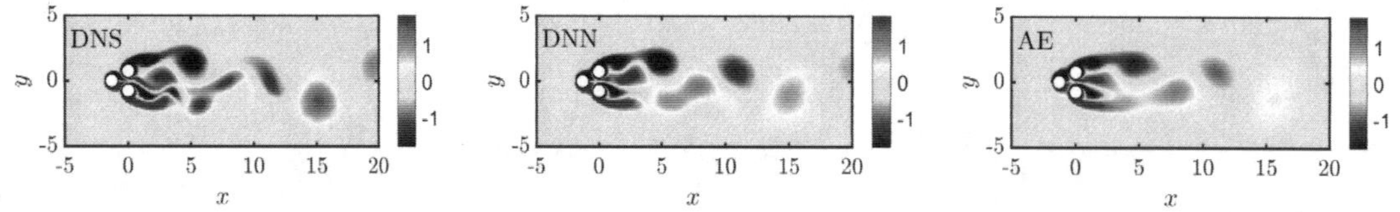

Figure 7. A comparison between the estimated vorticity fields and the DNS data in a random time instance of the exemplary chaotic case.

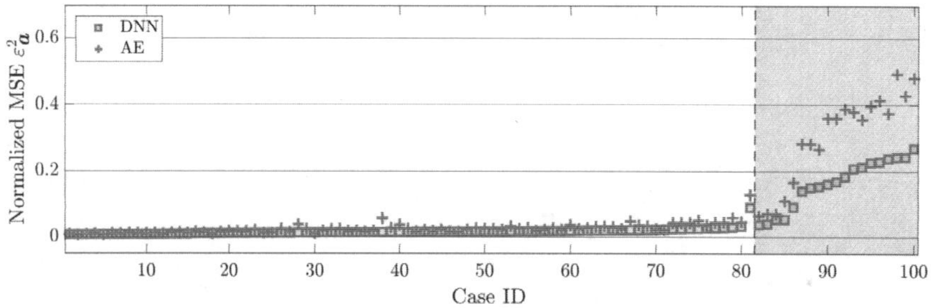

Figure 8. Distribution of normalized mean-squared-error for the prediction of POD expansion coefficients in all 100 test cases. The cases are classified into periodic (white background) and chaotic (gray background) categories. In each category, case IDs are sorted based on the error magnitude of DNN estimator.

a single control command. To evaluate the estimation performance from each control command in the test set, we define the estimation error of POD modal coefficients as:

$$\varepsilon_{\boldsymbol{a}}(\boldsymbol{b}) = \sqrt{\frac{\langle \|\boldsymbol{a} - \hat{\boldsymbol{a}}\|_2^2 \rangle_t}{\langle \|\boldsymbol{a}\|_2^2 \rangle_t}}. \tag{2}$$

Figure 8 presents the distribution of $\varepsilon_{\boldsymbol{a}}^2$ over all test cases. Here the test cases are classified into periodic and chaotic categories based on the characteristics of the downstream vortex shedding (see [10] for more details). For better comparison all cases in each category are sorted based on the error magnitude of DNN. When the flow is periodic, both AE and DNN are capable to generate accurate estimates of the POD modal coefficients and the errors are always at a similar level. This observation indicates the 5-dimensional compression from the encoder preserve the most important flow state information in the low-dimensional subspace. When the flow becomes chaotic, both kNN and DNN results in higher estimation errors than those in the periodic cases. In addition, the estimation errors from kNN become significantly higher than DNN. Hence we can conclude the encoding process results in the loss of partial information from sensor signals when the forced flow is chaotic.

Table 3 compares the squared normalized velocity estimation error $\varepsilon_{\boldsymbol{u}}^2$ as well as the squared normalized overall estimation error E^2, where

$$\varepsilon_{\boldsymbol{u}}(\boldsymbol{x}) = \sqrt{\frac{\langle \|\boldsymbol{u}_b'(\boldsymbol{x}) - \hat{\boldsymbol{u}}_b'(\boldsymbol{x})\|_2^2 \rangle_{\boldsymbol{b},t}}{\langle \|\boldsymbol{u}_b'(\boldsymbol{x})\|_2^2 \rangle_{\boldsymbol{b},t}}} \tag{3}$$

and

$$E = \sqrt{\frac{\langle \int_\Omega \|\boldsymbol{u}(\boldsymbol{x}) - \hat{\boldsymbol{u}}(\boldsymbol{x})\|_2^2 d\boldsymbol{x} \rangle_{\boldsymbol{b},t}}{\langle \int_\Omega \|\boldsymbol{u}(\boldsymbol{x})\|_2^2 d\boldsymbol{x} \rangle_{\boldsymbol{b},t}}} \tag{4}$$

In both equations $\langle\cdot\rangle_{\boldsymbol{b},t}$ represents the ensemble average over both control commands and time instances. From the comparison of $\varepsilon_{\boldsymbol{u}}(\boldsymbol{x})^2$, AE is seen to produce higher estimation errors than DNN not only in the region just behind the pinball but also in the further downstream locations. The comparison of E^2 includes errors from all test cases, as well as their periodic and chaotic components. In the periodic regime, both AE and DNN generate an overall estimation error at a similar level. However, when the forced flow becomes chaotic, the AE estimates will produce larger estimation error than DNN. This observation confirms that the encoding process discards partial information from the input vector which becomes essential in chaotic cases. The comparison of other relevant parameters, such as the number of parameters and online/offline time consumption, are also included in this table.

4. Conclusions

In this study we propose an autoencoder (AE) based neural network model to explore the least-order representation of the sensor-based estimation problem for multi-input, multi-output

Table 3. A parametric comparison between DNN and AE estimators.

Item	DNN	AE
Schematic	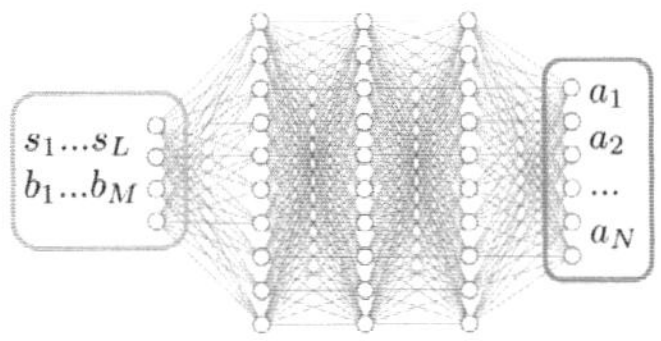	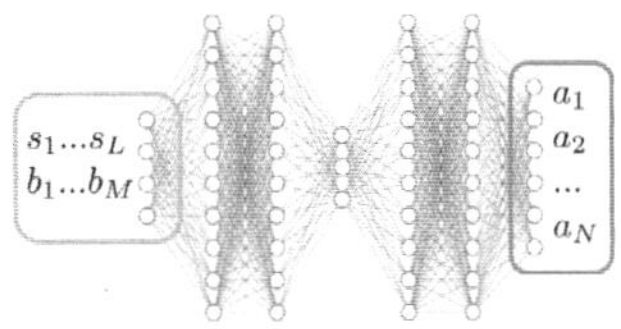
Number of parameters	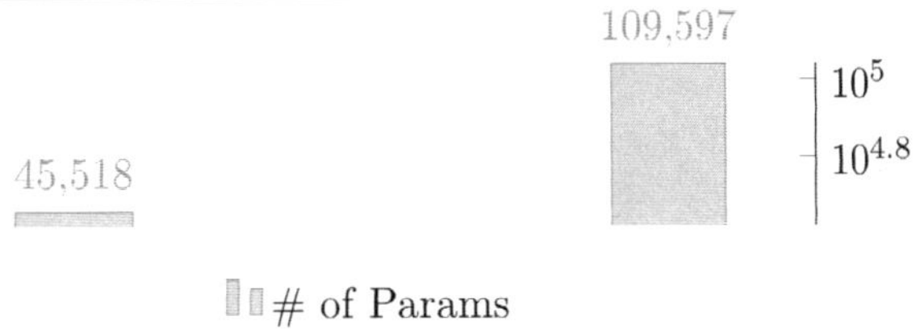	
Time cost (s)	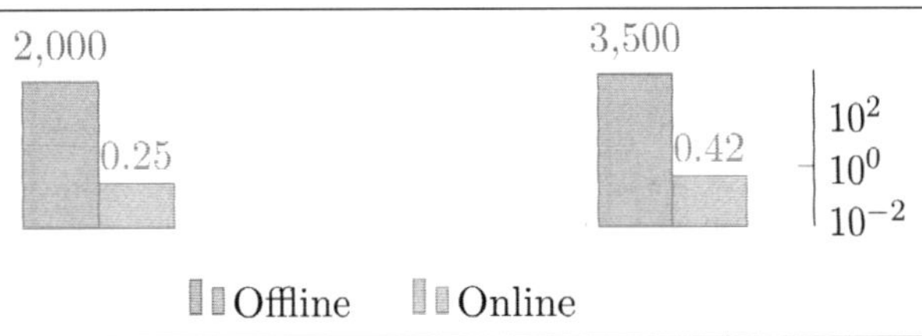	
Squared overall estimation error E^2	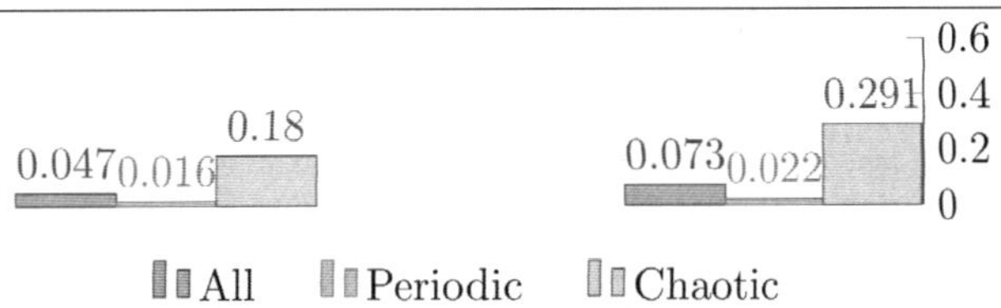	
Error distribution $\varepsilon_{\boldsymbol{u}}^2$	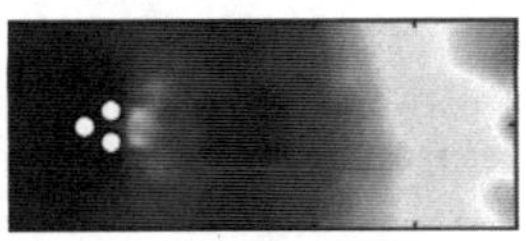	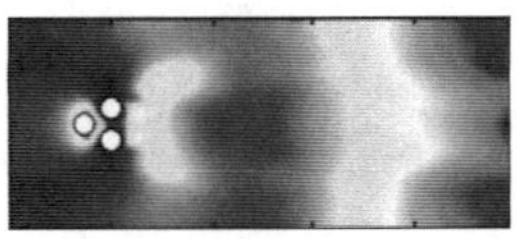

plants. The input vector are the MIMO signals, i.e. comprise the sensor signals and actuation commands. and output is the instantaneous flow field. This model is examined for the control-oriented flow estimation problem in the fluidic pinball plant. In this practical problem the AE architecture encodes the MIMO input vector into a five-dimensional subspace. To evaluate the effect of encoding-decoding, the estimation results are compared to the ones from a Deep Neural Network (DNN) architecture. In DNN no artificial compression is enforced to the input data. We find that when the controlled flow is periodic, the MIMO input vector can be effectively encoded into the five-dimensional subspace without a significant loss of accuracy. However, the low-dimensional features can not effectively extract all information from the input vector when the forced flow becomes chaotic. Consequently the estimation errors are notably increased in comparison to DNN. These results represent the low dimensional nature of the control-oriented estimation problem in MIMO plants and future study may focus on the efficient and effective extraction of the low dimensional subspace.

Acknowledgements
This work is supported by the National Science Foundation of China (NSFC) through grants 12172109 and 12172111 and by the Natural Science and Engineering grant of Guangdong province, China, through grant 2022A1515011492. We acknowledge valuable discussions with Guy Cornejo Maceda, Nan Deng, Marek Morzyński and Luc Pastur.

References

[1] S L Brunton, B R Noack, and P Koumoutsakos. “Machine learning for fluid mechanics”. In: *Annual Review of Fluid Mechanics* 52.1 (Jan. 2020), pp. 477–508.

[2] S L Brunton and B R Noack. “Closed-loop turbulence control: progress and challenges”. In: *Applied Mechanics Reviews* 67.5 (Aug. 2015).

[3] Z Deng, Y Chen, Y Liu, and K Kim. “Time-resolved turbulent velocity field reconstruction using a long short-term memory (LSTM)-based artificial intelligence framework”. In: *Physics of Fluids* 31.7 (July 2019), p. 075108.

[4] K Fukami, T Nakamura, and K Fukagata. “Convolutional neural network based hierarchical autoencoder for nonlinear mode decomposition of fluid field data”. In: *Physics of Fluids* 32.9 (Sept. 2020), p. 095110.

[5] L Guastoni, A Güemes, A Ianiro, S Discetti, P Schlatter, H Azizpour, and R Vinuesa. “Convolutional-network models to predict wall-bounded turbulence from wall quantities”. In: *Journal of Fluid Mechanics* 928 (2021), A27.

[6] R Ishar, E Kaiser, M Morzyński, D Fernex, R Semaan, M Albers, P S Meysonnat, W Schröder, and B R Noack. “Metric for attractor overlap”. In: *Journal of Fluid Mechanics* 874 (July 2019), pp. 720–755.

[7] G Y Cornejo Maceda, Y Li, F Lusseyran, M Morzyński, and B R Noack. “Stabilization of the fluidic pinball with gradient-enriched machine learning control”. In: *Journal of Fluid Mechanics* 917 (2021), A42.

[8] N Deng, B R Noack, M Morzyński, and L R Pastur. “Low-order model for successive bifurcations of the fluidic pinball”. In: *Journal of Fluid Mechanics* 884 (2020), A37.

[9] C M Bishop and N M Nasrabadi. *Pattern recognition and machine learning*. Vol. 4. 4. New York: Springer, 2006.

[10] Q Lin. “Fully Automated Control-Oriented Reduced-Order Modeling Examplified for the Fluidic Pinball”. Master Thesis. Harbin Institute of Technology, Dec. 2021.

Large Eddy Simulations of excited flames issuing from rectangular nozzles

Jakub Stempka[1] **and Artur Tyliszczak**[1]

[1] Faculty of Mechanical Engineering and Computer Science, Czestochowa University of Technology, Al. Armii Krajowej 21, 42-201 Czestochowa, Poland

E-mail: `jakub.stempka@pcz.pl`

Abstract. In the current study we assess an impact of passive and active flow control methods applied to the flames emanating from rectangular nozzles. The passive control is realised by varying a jet aspect ratio using differently shaped nozzles, while the active control relies on applying a forcing to the axial component of velocity. The analysis is performed through a series of high fidelity Large Eddy Simulations (LES) using a high order in-house solver. Combustion process is modelled using the detailed chemistry model of a hydrogen oxidation with a corresponding set of species transport equations. It is shown that the passive flow control affects the jet flames significantly only up to a moderate forcing frequency. For a high frequency excitation the effect of the aspect ratio is weak. A combination of both control methods is found as very efficient in altering important practical flame features like a lift-off height and flame shape.

1. Introduction

There is considerable interest in applying flow control methods to practical devices, e.g., burners, and combustion chambers. This is motivated by the potential for improvement of their operation, by obtaining the desired flame dynamic response [1, 2, 3]. The flow control methods can be classified into two main groups, i.e., passive and active methods. The first one relies on a geometrical shaping of flow domains (injector shapes, nozzles contraction, etc.) [1, 4] or addition of elements to the flow domain. During the operation of a given device, the passive approach does not require any additional energy input and is very efficient. For example, in our previous studies focused on non-reacting jets [4] and flames [5] it has been demonstrated that by shaping the jet/fuel nozzle one can generate significantly different flow dynamics. The second group - active methods, needs an input of energy for driving actuators with a required amplitude and frequency. The passive methods are simpler to realise in practice while the active methods require sophisticated design, however, they allow to adjust the excitation more precisely to varying flow conditions. In the present work we focus on the control of a hydrogen diffusion flame emanating from a rectangular nozzle. The main goal of this study is to combine passive and active control method to assess whether and to what extent, and in which regimes their joint impact improves the potential to alter the flame characteristics. In an early study of Gutmark and Grinstein [1] and a more recent work of Tyliszczak and Geurts [6] devoted to jets, it has been shown that by modifying an aspect ratio (AR) of rectangular nozzle one can substantially change the jet spreading rate, amplify the vortices growth and even cause the bifurcation phenomenon. Here, we follow this control strategy and additionally we combine it with an axial excitation of the main flow stream at a given amplitude and frequency. The research is performed by applying the LES method [7]. We use a high-order in-house numerical code and also a commercial ANSYS Fluent software. The obtained results confirm the ability to enhance the mixing process of the fuel and oxidizer and thus the combustion process, by appropriately modulating the sizes and time scales of emerging turbulent structures. This in turn allows obtaining the desired flame characteristics like a lift-off height or length of the flame.

2. Numerical set-up and methodology

2.1. Jet parameters

The conditions and composition of the investigated jet flames are similar to those studied by Cabra et al. [8]. The fuel jet is composed of hydrogen ($X_{H_2} = 0.25$) and nitrogen ($X_{N_2} = 0.75$) while its temperature is equal to $T_j = 305\ K$. It mixes with a hot co-flowing oxidizer ($T_{cf} = 1045\ K$, $X_{O_2} = 0.15$, $X_{H_2O} = 0.1$, $X_{N_2} = 0.75$) and auto-ignites at some distance downstream the nozzle. In the current study we consider jets emanating from rectangular nozzles with the aspect ratios $AR = L/h = 1 - 5$, where L and h are the dimensions of the nozzle exit highlighted in figure 1a. In all the cases the cross-sectional areas S are the same and correspond to a circle with the equivalent diameter $D_e = 2\sqrt{S/\pi} = 4.57\ mm$. The jets are characterized by the Reynolds number $Re = U_b D_e/\nu = 23\,600$ where U_b is the bulk velocity and ν is the kinematic viscosity of the fuel mixture. The aspect ratio plays the role of the passive control method. The active approach is employed by a modulation of the mean axial velocity by the sinusoidal signal with an amplitude equal to $10\%U_b$ and frequencies f defined in terms of the Strouhal number $St = fD_e/U_b$ in the range $St = 0.3 - 0.7$.

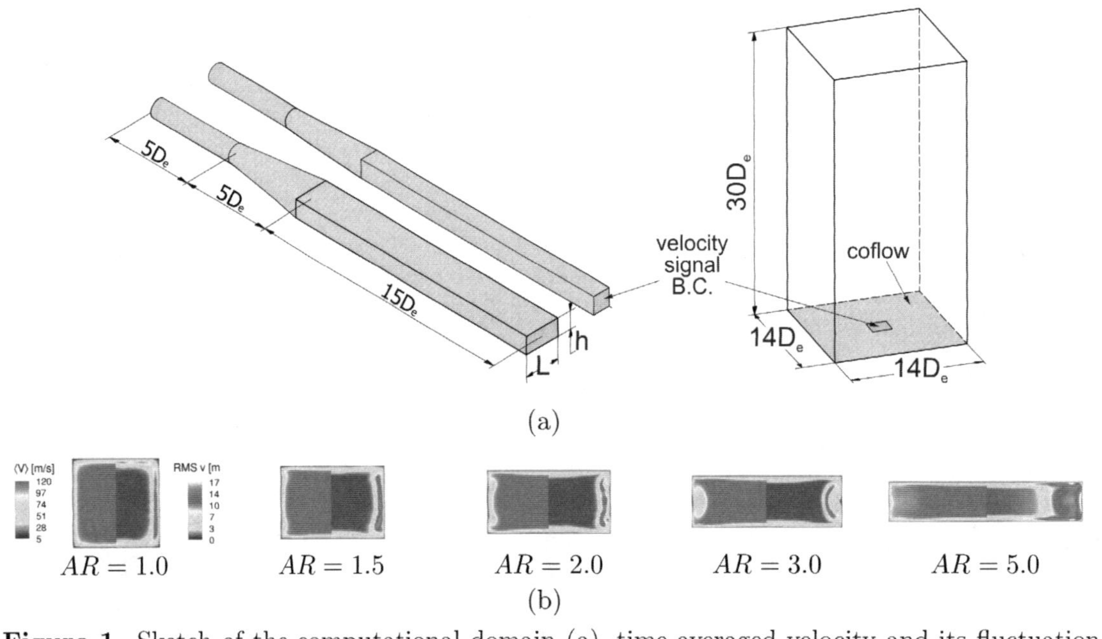

Figure 1. Sketch of the computational domain (a), time-averaged velocity and its fluctuation RMS distributions at the nozzle exit planes from the non-excited case (b).

2.2. Numerical procedure

The study was performed in two stages, both conducting the LES [7]. In the first stage of computations, we modelled the flow developing inside the nozzles of variable cross-sections. The end section of each nozzle was rectangular with different ARs (see figure 1a). At this stage, the simulations were carried out using ANSYS Fluent software [9], applying WALE subgrid-scale viscosity model. At the same time, we have been acquiring the unsteady velocity values from the outlets of the nozzles. Examples of the time-averaged axial velocity and its fluctuations RMS distributions at the outlets from non-excited cases are presented in figure 1b showing clearly the impact of the ARs. These values have been then used as the inlet boundary conditions in the second stage of simulations in the domain shown in figure 1a (right-hand side). At this stage, we carried out the computations using an in-house LES solver SAILOR [10] based on the high-order compact finite difference method with the Vreman subgrid-scale viscosity closure. The combustion process was modelled by applying a detailed chemistry model of H_2 oxidation developed by Mueller et al. [11] using a no-model approach for the calculation of the reaction source terms [12]. In the simulations carried out in Fluent, the domains of the nozzles were discretised using block-structured grids, with elements adjacent to the wall having a height of 15 μm, growing toward the axis and reaching the maximum size of 70 μm around it. The length of the elements in the axial direction was set to 1 mm. In computations carried out in SAILOR code a Cartesian grid was adopted, discretised using 192 $\times$ 192 $\times$ 384 points, which was compacted toward the axis and the exit plane. The minimum element sizes were equal to 110 μm in the radial directions and 190 μm in the axial direction. The dimension of D_e corresponded to approximately 26 cells at the exit plane. The accuracy of the modelling approach was verified in the simulations of non-reacting jets emanating from square nozzle, against the available experimental results, in the previous work [4]. Verification of the reacting cases against the Cabra experiment results for a circular nozzle can be found in [5].

3. Results

3.1. Flame topology

First, we focus on the overall structures of the flames. These are highlighted in figure 2, showing the iso-surfaces of the instantaneous temperature (T = 1200 K) interacting strongly with the vorticity field, visualized by a representative instantaneous vorticity iso-surface (ω = 18 1/s). It can be seen that the effect of the velocity excitation (active control) with a high frequency $St \geq 0.5$ leads to the emergence of the small-scale turbulent structures in the vicinity of the shear layers. Interestingly, a similar effect is obtained by increasing the nozzle aspect ratio above $AR \geq 3$ (passive control). The effect of applying either the nozzle AR or a high-frequency excitation was a significant distortion of the shear layers. In such cases, the flames were anchored further downstream of the nozzle. On the other hand, it can be seen that for the cases with low aspect ratios ($AR \leq 1.5$) excited with a low frequency ($St \leq 0.3$) the flames create two characteristic reaction zones. The first one is associated with the less disturbed shear layer close to the nozzle exit and the second one appears in a fully turbulent region, i.e., downstream of the end of the potential core. As it was already mentioned, more irregular and fine turbulent structures lead to a longer flame lift-off height, however, it can be noted that for higher excitations frequencies ($St \geq 0.5$) the change in AR of the nozzle has less influence on the flame lift-off height. Finally, by the visual inspection of the results, one can see that the two highest frequencies lead to similar flame topologies, on the other hand for $St = 0.3$ the impact of the excitation on the flame is significant only for the cases with $AR \geq 2.0$.

3.2. Time-averaged results

The time-averaged results showing the temperature distributions in the main cross-section plane parallel to the long edge of the nozzles are presented in figure 3. The qualitative behaviour of the flame shape and lift-off height reflects the previously described instantaneous solutions. Taking into account the time-averaged results one can notice that for the cases with $AR = 3.0$ to $AR = 5.0$ the application of the excitation with $St = 0.3$ causes only small changes in the flame shape. In the remaining configurations, the impact of the excitation is much more pronounced,

Journal of Physics: Conference Series **2367** (2022) 012025 doi:10.1088/1742-6596/2367/1/012025

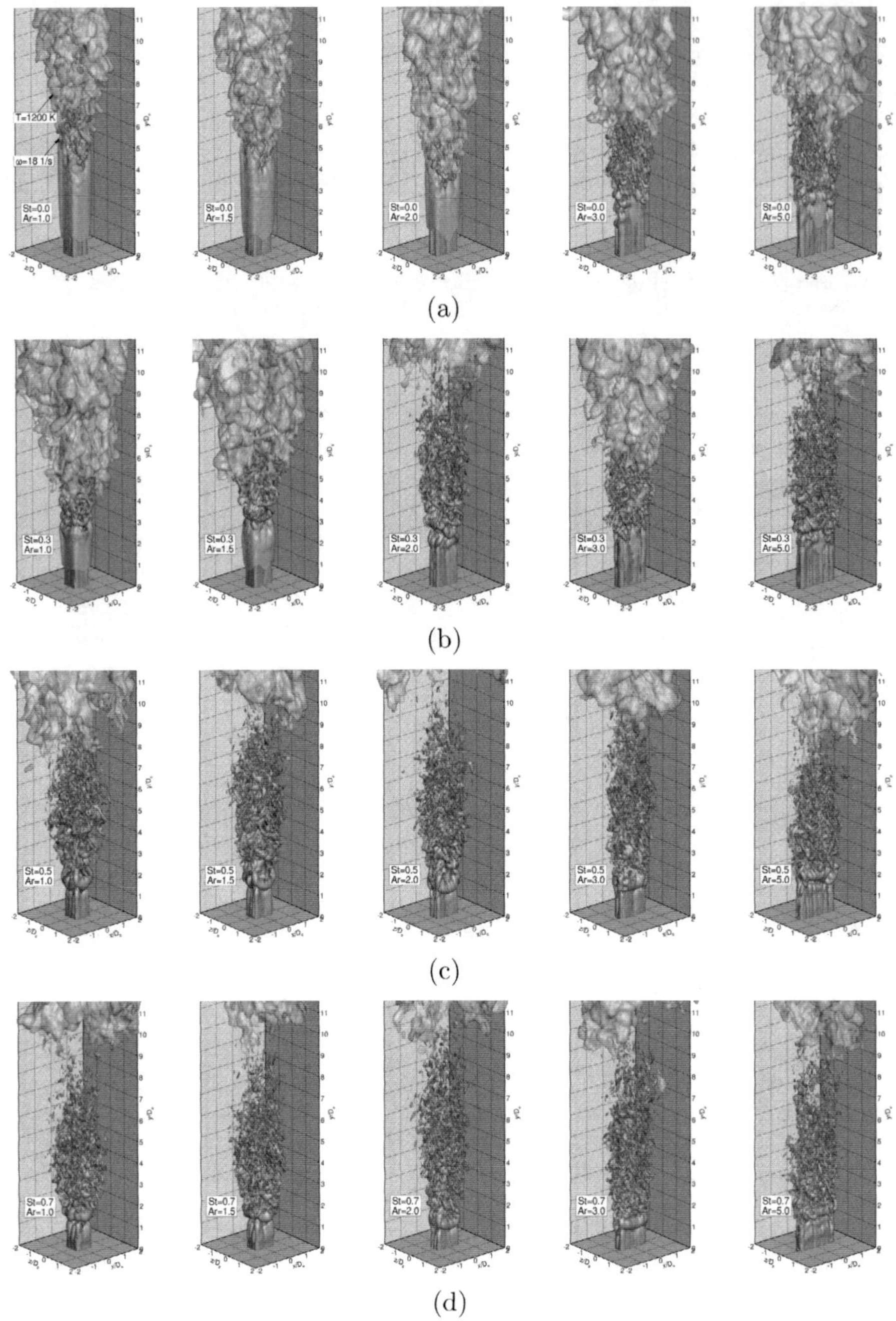

Figure 2. Iso-surfaces of the instantaneous vorticity ($\omega = 18$ 1/s) and temperature ($T = 1200$ K). The presented results are characterized by the following forcing frequencies: (a) $St = 0.0$, (b) $St = 0.3$, (c) $St = 0.5$ and (d) $St = 0.7$.

particularly for higher St. In the presented figures, the iso-contour of the stoichiometric mixture fraction value ($Z_{st} = 0.479$) is indicated by the green line. Regarding the localization of Z_{st} as a measure of mixedness one can see clear differences in the mixing process. For the non-excited cases with a low aspect ratio ($AR \leq 2.0$) the flame is attached to the stoichiometric surface. On the other hand, when the excitation is applied and the flame is lifted it can be seen that the high temperature regions are not aligned with Z_{st} iso-contour. This is due to the excessively entrained co-flowing gas, which flows towards the axis region and leads to the gap between the Z_{st} iso-contour and the flame base.

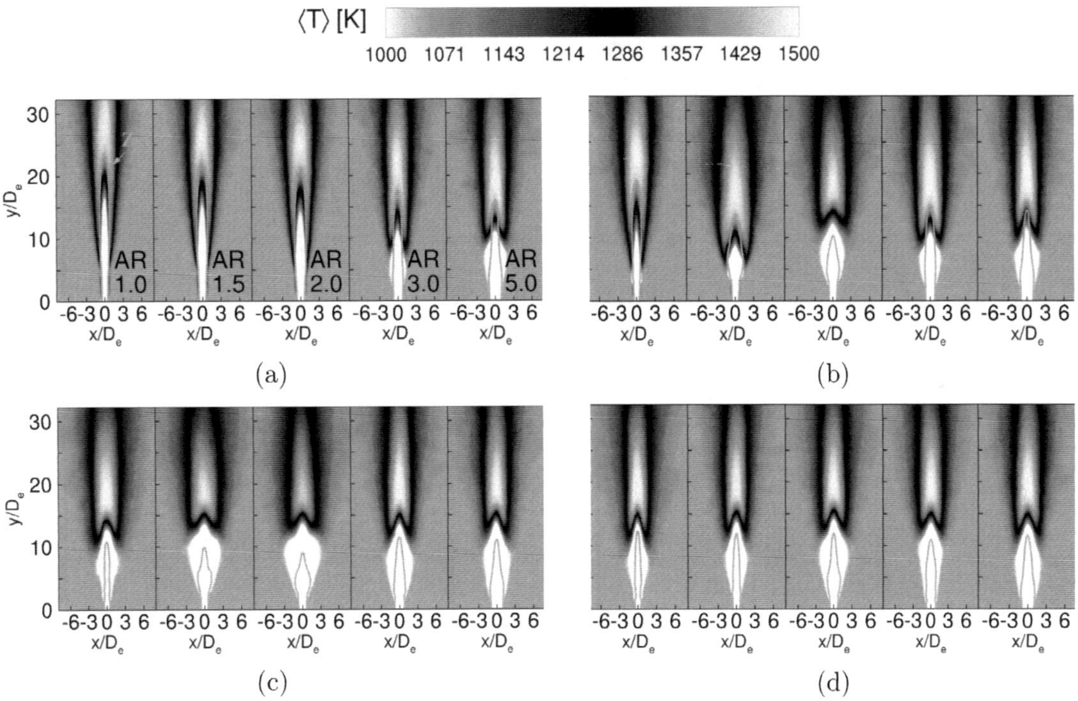

Figure 3. Time-averaged results showing temperature distributions on the middle plane. The presented results are characterized by the following forcing frequencies: (a) $St = 0.0$, (b) $St = 0.3$, (c) $St = 0.5$ and (d) $St = 0.7$.

3.3. Axial profiles

The corresponding axial profiles of the selected time-averaged quantities, i.e., $\langle T \rangle$, $\langle Y_{H_2} \rangle$, $\langle Y_{H_2O} \rangle$, $\langle Y_{O_2} \rangle$, are presented in figure 4.

The first set of the profiles ($St = 0.0$) highlights the sole influence of the passive control method on the flame and its structure in terms of the mass fractions of the major species. Prior to the combustion, the fuel and oxidizer need to be mixed. This process is strongly affected by the dynamics of the turbulent structures, which differ depending on the nozzle AR. For example, in the case of the nozzles with $AR \leq 2.0$ there is a deficit of the oxygen at the jet axis up to the location of about $15D_e$. Correspondingly, the decrease of the fuel mass fraction is shifted downstream compared to the cases with the higher ARs. Intensified mixing along the jet axis for the cases with $AR = 3$ and $AR = 5$ leads to faster growth of the axial temperature values and $\langle Y_{H_2O} \rangle$ mass fraction, which translates to the location of the flame tip closer to the nozzles. The peaks in $\langle Y_{O_2} \rangle$ in the near-filed of the jets with $AR = 3.0$ and 5.0 confirm also the higher entrainment level.

Excitation of the axial velocity with $St = 0.3$ clearly leads to the higher peaks in $\langle Y_{O_2} \rangle$ even in the cases that initially did not exhibit any maxima closer to the nozzle. The higher entrainment is most likely related with enhanced near-filed mixing along the jet axis. This partially explains

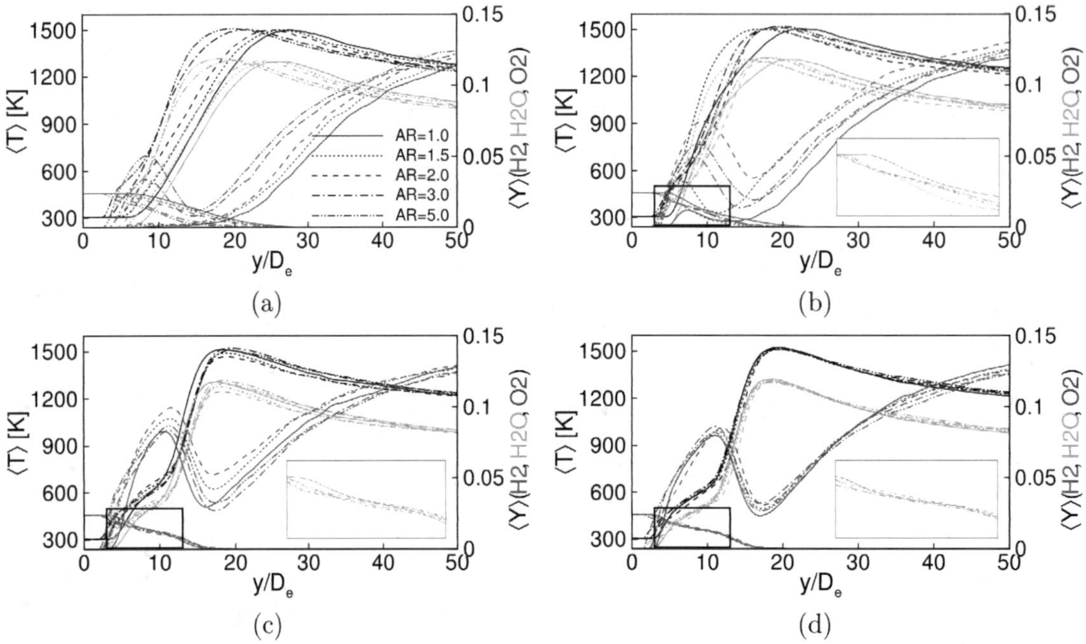

Figure 4. Axial profiles of the time-averaged quantities ($\langle T\rangle$, $\langle Y_{H_2}\rangle$, $\langle Y_{H_2O}\rangle$, $\langle Y_{O_2}\rangle$). The presented results are characterized by the following forcing frequencies: (a) $St = 0.0$, (b) $St = 0.3$, (c) $St = 0.5$ and (d) $St = 0.7$.

the fact that the flames positions are shifted toward the nozzle compared to the non-excited cases, as can be deduced analysing the localizations of the temperature maxima. Worth noticing is that the impact of the AR seems to be weakened at $St = 0.3$ since the profiles of $\langle T\rangle$ are closer to each other. In these cases, a significantly different O_2 entrainment from the co-flowing gas has practically no influence on the axial flame position. Apparently, its small amount is sufficient to initiate the combustion process. The fact that the $\langle T\rangle$ profiles are similar can be related to the $\langle Y_{H_2}\rangle$ values, which exhibit almost overlapping slopes. The most remarkable difference between the jets excited with frequency $St = 0.3$ and the remaining cases is that only for this particular value the influence of the nozzle AR is strong. As noted previously, the excitation with higher frequencies ($St = 0.5$ and 0.7) impacts the flame structure along the jet axis in a very similar way. The effect of the excitation is larger compared to the alteration of the flame dynamics caused by changing AR. More intense mixing in the near-field leads to larger entrainment of O_2 from the co-flow. However, unlike in other cases, the level of excitation prevents the flame to develop close to the nozzle, it can be seen that its base shifts downstream. This is evidenced by the inflexion point appearing in the profiles of $\langle T\rangle$ at $y/D_e \approx 12$ followed by a rapid increase of temperature due to the combustion initiation.

3.4. Radial profiles

Figures 5 and 6 show the radial profiles of the selected variables, i.e., $\langle T\rangle$, $\langle T'T'\rangle$, $\langle Y_{OH}\rangle$, from different axial locations, extracted from the nozzles symmetry planes parallel either to the long or short nozzle edges, respectively. The results are presented for the non-excited cases only. Analysing the radial profiles, starting from the ones placed nearest the nozzle at $y/D_e = 2$, one can readily notice the impact of the width of the nozzle on $\langle T\rangle$ while changing AR. In this location, the temperature maxima occur only for the cases $AR = 1.0$, 1.5 and 2.0. They are accompanied by the peaks in $\langle Y_{OH}\rangle$ that indicates the initialisation of the chemical reactions.

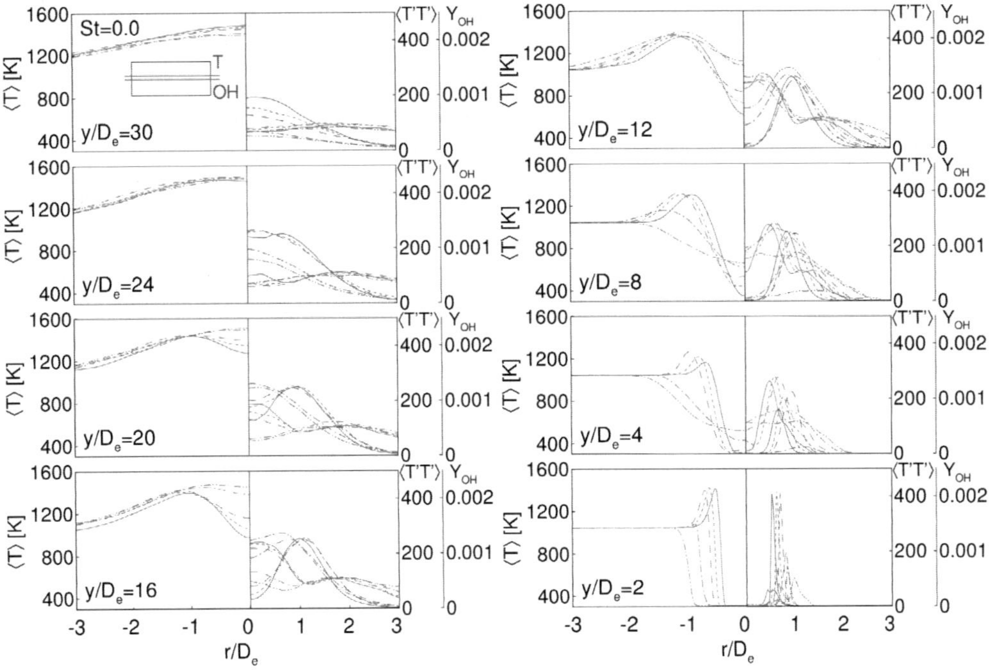

Figure 5. Radial profiles of $\langle T \rangle$, $\langle T'T' \rangle$, $\langle Y_{OH} \rangle$ at different axial locations, taken from the nozzle symmetry planes parallel to the long nozzle edge (linetypes as in figure 4).

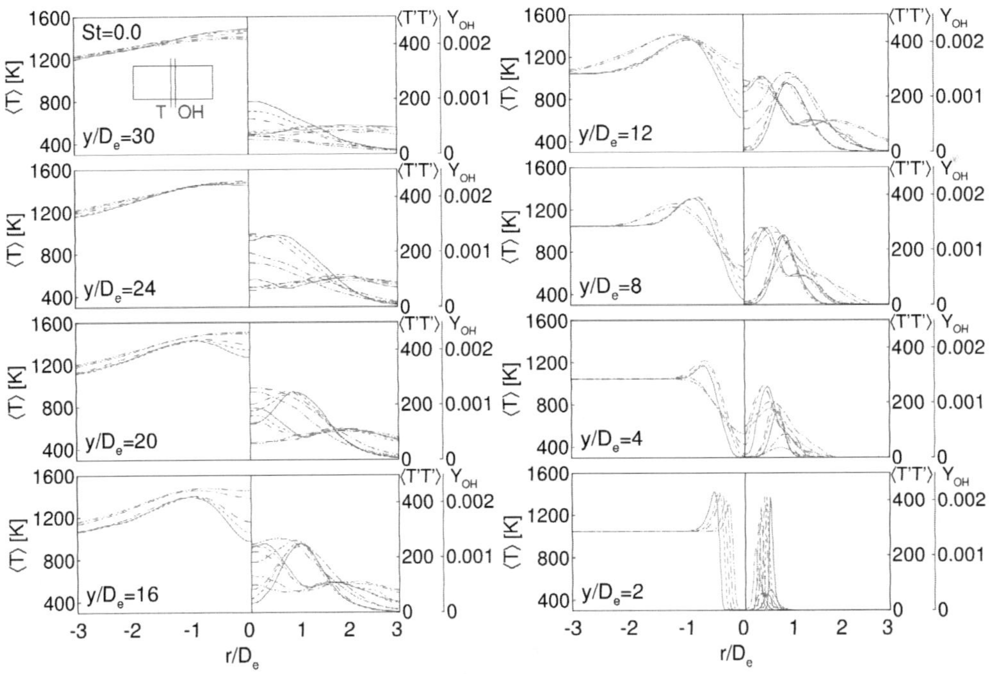

Figure 6. Radial profiles of $\langle T \rangle$, $\langle T'T' \rangle$, $\langle Y_{OH} \rangle$ at different axial locations, taken from the nozzle symmetry planes parallel to the short nozzle edge (linetypes as in figure 4).

These are attributed to the emergence of the first type of the flame as described in section 3.1. It extinguishes at about $y/D_e = 4$, which is evidenced by the lower values of $\langle T \rangle$. At $y/D_e = 8$ the reaction is initiated again due to the high amount of accumulated OH radical. At this location the combustion at the peripheries of the jet is not taking place in cases $AR = 3.0$ and 5.0. At further distances $y/D_e > 20$ the $\langle T \rangle$ profiles attain a self-similar state irrespectively of the nozzle AR. This is accompanied by a low level of temperature fluctuations RMS that can be interpreted as an indicator of a stable combustion regime.

Figure 6 shows the analogous results extracted from the symmetry plane of the nozzle parallel to the short edge. It can be seen that the values of presented variables are slightly different than in the corresponding perpendicular cross-section. This suggests that the character of the combustion process is related to the particular side of the jet. For example, in the location $y/D_e = 8$ the profiles indicate that the combustion already takes place. It is in opposition to what was observed in figure 5 where for the cases $AR = 3.0$ and 5.0 the combustion process has not started yet. The profiles taken from either of the two planes do not show any significant differences only starting from the distance $y/D_e = 16$. From this point the flames appear to be homogeneous along the azimuthal direction.

4. Conclusions

In the current study, we assessed the impact of the passive and active flow control methods on flames emanating from rectangular nozzles. The passive control was realised by varying the nozzle aspect ratio, while the active control relied on applying the excitation of the axial velocity component. It has been shown that both flow control approaches allow for the modification of the flame characteristics and can be used as a tool to influence the mixing process in the non-reacting regions, and thus to change the flame lift-off height, its length or spread. It has been found that the impact of the nozzle shape is strong only when the excitation frequency is low, in the analysed cases equal to $St = 0.3$. Above that value ($St = 0.5$ and 0.7) the flame topology and its dynamics are almost solely dependent on the excitation.

Acknowledgements

This work was supported by National Science Center in Poland (Grant 2018/31/B/ST8/00762) and the statutory funds of the Department of Thermal Machinery (Czestochowa University of Technology). The computations were carried out using PL-Grid infrastructure.

References

[1] Gutmark E J and Grinstein F F 1999 *Annu. Rev. Fluid Mech.* **31** 239–272
[2] Tyliszczak A and Geurts B J 2019 *Int. J. Heat Fluid Fl.* **80** 108488
[3] da Silva C B and Métais O 2002 *Phys. Fluids* **14** 3798–3819
[4] Tyliszczak A, Kuban Ł and Stempka J 2022 *Int. J. Heat Fluid Fl.* **94** 108944
[5] Kuban Ł, Stempka J and Tyliszczak A 2021 *Energies* **14** 554
[6] Tyliszczak A and Geurts B J 2015 *J. Turbul.* **16** 742–771
[7] Geurts B J 2003 *Elements of Direct and Large Eddy Simulation* (R.T. Edwards, Inc)
[8] Cabra R, Chen J, Dibble R, Karpetis A and Barlow R 2005 *Combust. Flame* **143** 491–506
[9] *Ansys Fluent User's Guide, Release 19.5, ANSYS, Inc.*
[10] Tyliszczak A 2014 *J. Comput. Phys.* **276** 438–467
[11] Mueller M A, Kim T J, Yetter R A and Dryer F L 1999 *Int. J. Chem. Kinet.* **31** 113–125
[12] Duwig C, Nogenmyr K J, Chan C and Dunn M J 2011 *Combust. Theor. Model.* **15** 537–568

Journal of Physics: Conference Series **2367** (2022) 012026 doi:10.1088/1742-6596/2367/1/012026

On the generalization of Calogero-Ahmed summation formulas

K Urbanowicz[1], **M Stosiak**[2] **and A Bergant**[3]

[1] West Pomeranian University of Technology, Faculty of Mechanical Engineering and Mechatronics, al. Piastów 17, 70-310 Szczecin, Poland

[2] Wrocław University of Science and Technology, Faculty of Mechanical Engineering, 7/9 Łukasiewicza St, 50-371 Wrocław, Poland

[3] Litostroj Power d.o.o. & University of Ljubljana, Faculty of Mechanical Engineering, Litostrojska Cesta 50, 1000 Ljubljana, Slovenia
E-mail: kamil.urbanowicz@zut.edu.pl

Abstract. The use of the Laplace transform gives the solution of water hammer equations in the frequency domain. The inverse transform of this solution over the years seemed impossible to derive, due to the significant complexity and the fact that the square root of the Bessel function was embedded in the argument of the resulting hyperbolic functions. In this work, we consider some generalizations that enable the determination of the modified Calogero-Ahmed infinite series. These generalizations will allow us in the near future (using the machine learning and artificial intelligence algorithms) a return to the time domain in a very wide range of the dynamic viscosity function, which plays the most important role in this complex fluid dynamic problem.

1. Introduction

The analysis of unsteady fluid flows in pressurized pipes plays a profound role at the stage of designing and modernizing the hydraulic piping systems. The durability of the installation is influenced not only by the design assumptions but also by those related to the operation of the installation. In systems, the most difficult situation occurs when there are dynamic changes of the flow rate. For example, when valves are suddenly closed, the waves of the increased pressure form and start to propagate and sometimes may have a destructive effect (pipe rupture). Another common way of blocking the flow is the stepwise shutdown (intentionally or not - power failure) of hydraulic devices, e.g. pumps. In both cases, when the flow is stopped, one experiences a phenomenon of water hammer. Its mathematical description is expressed by a system of partial differential equations of the hyperbolic type [1, 2]:

$$\begin{cases} \frac{\partial p}{\partial t} + \rho c^2 \frac{\partial v}{\partial x} = 0 \\ \rho \frac{\partial v}{\partial t} + \frac{\partial p}{\partial x} + \frac{2}{R}\tau_w = 0 \\ \tau_w = \tau_{q-s} + \tau_u = \frac{f\rho}{8} v|v| + \frac{2\mu}{R}\int_0^t w(t-u)\cdot\frac{\partial v(u)}{\partial t}du \end{cases}, \quad (1)$$

where: $p = p(x,t)$ – pressure at the pipe cross section, x – axial coordinate along the pipe, t – time, ρ – density of the liquid, c – pressure wave speed, $v = v(x,t)$ – mean flow velocity of the liquid at the pipe cross section, R – inner diameter of the pipe, τ_w – wall shear stress, f – Darcy-Weisbach friction factor, μ – dynamic viscosity of the liquid, $w(t-u)$ – weighting function.

To date, no complete analytical solution of the above system of equations in the time domain has been found. Holmboe [1] managed to find an analytical solution of pressure but in the Laplace domain $\tilde{p}(x,s)$ only. Recently [3-5], using Holmboe's solution, it was also possible to derive the missing solutions in the Laplace domain describing the transforms of velocity $\tilde{v}(x,s)$ and wall shear stress $\tilde{\tau}(x,s)$ in transient laminar pipe flow:

$$\begin{cases} \tilde{p}(x,s) = \mathbb{Z}\frac{\mathbb{M}}{s}\left[e^{-\frac{x}{c}\mathbb{M}\cdot s} - e^{-\frac{(2L-x)}{c}\mathbb{M}\cdot s} - e^{-\frac{(2L+x)}{c}\mathbb{M}\cdot s} + e^{-\frac{(4L-x)}{c}\mathbb{M}\cdot s} + e^{-\frac{(4L+x)}{c}\mathbb{M}\cdot s} - \cdots\right] + \frac{8\mu v_0}{sR^2}x \\ \tilde{v}(x,s) = -\frac{v_0}{s}\left[-e^{-\frac{x}{c}\mathbb{M}\cdot s} - e^{-\frac{(2L-x)}{c}\mathbb{M}\cdot s} + e^{-\frac{(2L+x)}{c}\mathbb{M}\cdot s} + e^{-\frac{(4L-x)}{c}\mathbb{M}\cdot s} - e^{-\frac{(4L+x)}{c}\mathbb{M}\cdot s} - \cdots\right] - \frac{v_0}{s} \\ \tilde{\tau}(x,s) = \frac{Rv_0\rho}{2}\left(-e^{-\frac{x}{c}\mathbb{M}\cdot s} - e^{-\frac{(2L-x)}{c}\mathbb{M}\cdot s} + e^{-\frac{(2L+x)}{c}\mathbb{M}\cdot s} + \cdots\right)(1-\mathbb{M}^2) - \frac{4\mu v_0}{sR} \end{cases}, \quad (2)$$

where: $\mathbb{Z} = \rho c v_0$ is the Joukowsky pressure rise formula, v_0 – initial pre-transient flow velocity; L – pipe length and $\mathbb{M}$ – dynamic viscosity function which is a square root of the ratio of Bessel functions:

$$\mathbb{M} = \sqrt{\frac{\mathrm{I}_0\left(\sqrt{s\frac{R^2}{\nu}}\right)}{\mathrm{I}_2\left(\sqrt{s\frac{R^2}{\nu}}\right)}} \;. \quad (3)$$

Attempts to inverse the transform and to return it back to the original, have been made recently. However, the research began long before the formulation of the currently used differential form of this model. Joukowsky and Allievi dealt with this phenomenon over a hundred years ago, ignoring the hydraulic resistance i.e. $\tau_w = 0$. Rich [6], developed Wood's solutions and derived (using inverse Laplace transform) a time-domain analytical solution for water hammer in an axisymmetric conduit assuming a quasi-steady friction term $\tau_w = \tau_{q-s}$. Holmboe [7] proposed an analytical pressure solution valid at the valve cross-section, but only for the first phase of the water hammer event. Brown [8] analyzed this flow situation and noticed an interesting asymptotic development of the dynamic viscosity function. Jayasinghe and Leutheusser [9], Karam and Leonard [10], Yamaguchi and Ichikawa [11] proposed analytical solutions that took into account the frequency-dependent resistance, but as based on additional assumptions, their practical scope of applicability is limited. A very interesting solution, taking into account the asymptotic expansion of Brown, was presented by Muto and Takahashi [12]. This solution is in fact a significant extension of the Holmboe's solution, and it is based on unsteady friction term $\tau_w = \tau_{q-s} + \tau_u$ resulting in an accurate solution, close to numerically predicted normalized pressure $\hat{p}$ and velocity $\hat{v}$ histories (NUM) – see figure 1 ($\hat{t}$ – is the dimensionless time). These simulations were performed for a specific value of an important dimensionless number called water hammer number $Wh = (\nu L)/(cR^2)$ [13]. In the 1990s, these issues were also dealt with by Wang et al. [14]. This group of authors presented an interesting integral-based solution, which was further investigated by Xuan et al. [15]. Solutions analogous to Rich's solution were developed by Sobey [16] and Jović [17]. Mei and Jing [18], analyzing this issue from a different perspective (for blood flow), also derived an interesting approximate solution, which may be particularly useful for the analysis of water systems.

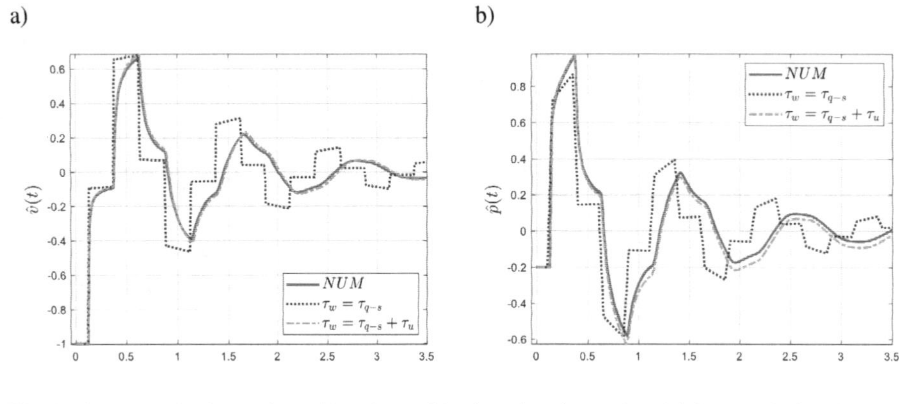

Figure 1. Dynamical waveform histories (midpoint of a pipe, $Wh = 0.05$): a) velocity, b) pressure

Due to the fact that there is still no complete solution to this interesting problem for transient laminar pipe flow (even for the simplest system made of reservoir-pipe-valve (RPV)), the authors of this paper have been investigating this issue for some time. In our first work [3] most promising models were compared and few novel analytical solutions useful for predicting velocity and pressure time histories were presented. In more recent papers the novel analytical solution for wall shear stress was derived and compared [4, 5]. If one carefully examine transforms in the Laplace domain (2) it is clear that the time domain solutions (for pressure, velocity and wall shear stress) will be based on inverse Laplace transforms of the following functions: $\mathbb{M}^2$, $\mathbb{M}$ and $e^{-as\mathbb{M}}$. Hence, steps were taken to learn about these solutions. The inverse transformation of the first function, i.e. $\mathbb{M}^2$, can be found in Brereton and Jiang paper [19]. As for the second function $\mathbb{M}$, a new approach was presented in [20], which can be used to define the final simplified form of this function in the time domain. However, knowledge of the Calogero-Ahmed series is necessary to obtain a complete simplified solution. The analysis of papers [21-23] indicated that the basic solutions of these series for low exponent values are known, but there is a lack of a universal function (similar to recursive formulas defining Rayleigh series - which are also based on zeros of Bessel functions) that define these functions for any exponent. The inverse transform of the exponent function $e^{-as\mathbb{M}}$ will be the subject of our further research.

This work describes the current stage of research aimed at a complete description of the analytical water hammer phenomenon. The complete solution when found, will help us analyzing all systems in which not only laminar but also turbulent unsteady pipe flows occur (taking into account the approach proposed and developed by García and Alvariño [24, 25]).

2. Calogero-Ahmed summation formulas

Recent research on the inverse of the function describing the dynamic viscosity function $\mathbb{M}$ to the time domain [20] revealed that the solution in the time domain can be presented as an infinite series of exponential functions based on zeros of the Bessel function:

$$\mathcal{L}^{-1}\{\mathbb{M}-1\} = 2\frac{\nu}{R^2}\sum_{i=1}^{\infty} e^{-\mu_i^2\hat{t}} + 6\frac{\nu}{R^2}\left[\hat{t}\sum_{i=1}^{\infty} e^{-\mu_i^2\hat{t}} + \sum_{i=1}^{\infty}\frac{e^{-\mu_i^2\hat{t}}}{\mu_i^2}\right] + 20\frac{\nu}{R^2}\left[\frac{\hat{t}^2}{2}\sum_{i=1}^{\infty} e^{-\mu_i^2\hat{t}} + \hat{t}\sum_{i=1}^{\infty}\frac{e^{-\mu_i^2\hat{t}}}{\mu_i^2}\left(\frac{6}{4}\right) + \sum_{i=1}^{\infty}\frac{e^{-\mu_i^2\hat{t}}}{\mu_i^4}\left(\frac{8-\mu_i^2}{4}\right)\right] + 70\frac{\nu}{R^2}\left[\frac{\hat{t}^3}{6}\sum_{i=1}^{\infty} e^{-\mu_i^2\hat{t}} + \frac{\hat{t}^2}{2}\sum_{i=1}^{\infty}\frac{e^{-\mu_i^2\hat{t}}}{\mu_i^2}\left(\frac{24}{12}\right) + \hat{t}\sum_{i=1}^{\infty}\frac{e^{-\mu_i^2\hat{t}}}{\mu_i^4}\left(\frac{38-4\mu_i^2}{12}\right) + \sum_{i=1}^{\infty}\frac{e^{-\mu_i^2\hat{t}}}{\mu_i^6}\left(\frac{54-9\mu_i^2}{12}\right)\right] + 252\frac{\nu}{R^2}\left[\frac{\hat{t}^4}{24}\sum_{i=1}^{\infty} e^{-\mu_i^2\hat{t}} + \frac{\hat{t}^3}{6}\sum_{i=1}^{\infty}\frac{e^{-\mu_i^2\hat{t}}}{\mu_i^2}\left(\frac{120}{48}\right) + \frac{\hat{t}^2}{2}\sum_{i=1}^{\infty}\frac{e^{-\mu_i^2\hat{t}}}{\mu_i^4}\left(\frac{220-20\mu_i^2}{48}\right) + \hat{t}\sum_{i=1}^{\infty}\frac{e^{-\mu_i^2\hat{t}}}{\mu_i^6}\left(\frac{350-55\mu_i^2}{48}\right) + \sum_{i=1}^{\infty}\frac{e^{-\mu_i^2\hat{t}}}{\mu_i^8}\left(\frac{512-106\mu_i^2+3\mu_i^4}{48}\right)\right] + \cdots, \tag{4}$$

where: I_0 and I_2 – modified Bessel functions of the first kind of zeroth and second order respectively, ν – kinematic viscosity of liquid, $\hat{t} = \frac{\nu}{R^2}t$ – dimensionless time, μ_i – following zeros of Bessel function $J_0(\mu_i) = 0$.

Dynamical viscosity function $\mathbb{M}$ plays the most important role in the Laplace domain water hammer solution (2) as it affects the impedance and wave propagation operator. The resultant polynomials appearing in the above solution (marked in red and blue) are the result of applying Calogero-Ahmed formulas, which are described by the infinite rational series based on Bessel's zeros. For example, **Appendix A** shows which Calogero-Ahmed sums made up the last polynomial (marked with blue colour), which in the numerator contains the term $3\mu_i^4$. These functions appear in derived solution (4) as the result of multiple convolution calculations. Readers interested in the details are advised to consult the work in [20]. In the works of Calogero [21, 22] and Ahmed and Calogero [23], the following solutions can be found for the modified Calogero-Ahmed (5) and the standard Calogero-Ahmed (6) summation formulas, respectively:

$$\sum_{\substack{k=1\\k\neq n}}^{\infty}\left(y_{n,p}-y_{k,p}\right)^{-b} y_{k,p}^{-a} = f(y_{n,p},p) \quad for\ a = 1 \ldots 3\ and\ b = 1 \ldots 3\ , \tag{5}$$

$$\sum_{\substack{k=1\\k\neq n}}^{\infty}\left(y_{n,p}-y_{k,p}\right)^{-c} = g(y_{n,p},p) \quad for\ c = 1 \ldots 7\ , \tag{6}$$

where: $f(y_n, p)$ and $g(y_n, p)$ – simplified solutions being rational functions, p – order of Bessel functions of the first kind and $y_{n,p} = \mu_{n,p}^2$ – the square of the selected zero of Bessel function of the first kind and p order. From this moment on, in order to simplify, we abandon the subscripts p giving the order of the Bessel function ($y_{n,p} \to y_n$ and $y_{k,p} \to y_k$).

As can be seen from the above equations (5) and (6), these formulas were derived by Calogero and Ahmed for limited exponent values. However, in the obtained analytical solution (4) of the inverse Laplace transform, in order to define successive terms (extend the application range), it is necessary to know Calogero-Ahmed series solutions (both standard and modified) for higher values of exponents. Hence, it seems important to designate general solutions that will be useful for higher values, i.e. $a > 3$; $b > 3$ and $c > 7$ so that it would be possible to determine successive coefficients of polynomials appearing in the derived solution (4) with the use of machine learning. Such a general solution can also help significantly in the learning purpose of the artificial intelligence, that can as well be useful to determine the successive polynomials.

3. Generalization of modified Calogero-Ahmed summation formulas

Finding a generalization of the modified Calogero-Ahmed series began with the need to find the two solutions in the original works of Calogero and Ahmed [21-23] that are missing there. These solutions were necessary to develop the formula (4) in the paper [20]. During our research it has been noticed that for the modified Calogero-Ahmed series the following recursive formula has been derived:

$$\sum_{\substack{k=1\\k\neq n}}^{\infty} \frac{y_n}{y_k^a (y_n - y_k)^b} = \sum_{\substack{k=1\\k\neq n}}^{\infty} \frac{1}{y_k^a (y_n - y_k)^{b-1}} + \sum_{\substack{k=1\\k\neq n}}^{\infty} \frac{1}{y_k^{a-1} (y_n - y_k)^b} . \tag{7}$$

The above formula (7) has been verified by us only for the zeros of the Bessel function of the same order. With its help, successive elements of the following matrix of possible solutions (figure 2) can be determined in a follow-up manner. On the edges (boundary) of this matrix there are functions based on Rayleigh series in the zero column (red fields), while in the zero row there are standard Calogero-Ahmed series (blue fields).

a	$\sum_{\substack{k=1\\k\neq n}}^{\infty} \frac{1}{y_k^a}$	$\sum_{\substack{k=1\\k\neq n}}^{\infty} \frac{1}{y_k^a (y_n - y_k)}$	$\sum_{\substack{k=1\\k\neq n}}^{\infty} \frac{1}{y_k^a (y_n - y_k)^2}$	$\sum_{\substack{k=1\\k\neq n}}^{\infty} \frac{1}{y_k^a (y_n - y_k)^3}$	→	$\sum_{\substack{k=1\\k\neq n}}^{\infty} \frac{1}{y_k^a (y_n - y_k)^b}$
↑	↑	↑	↑	↑	↑↗→	↑
3	$\sum_{\substack{k=1\\k\neq n}}^{\infty} \frac{1}{y_k^3}$	$\sum_{\substack{k=1\\k\neq n}}^{\infty} \frac{1}{y_k^3 (y_n - y_k)}$	$\sum_{\substack{k=1\\k\neq n}}^{\infty} \frac{1}{y_k^3 (y_n - y_k)^2}$	$\sum_{\substack{k=1\\k\neq n}}^{\infty} \frac{1}{y_k^3 (y_n - y_k)^3}$	→	$\sum_{\substack{k=1\\k\neq n}}^{\infty} \frac{1}{y_k^3 (y_n - y_k)^b}$
2	$\sum_{\substack{k=1\\k\neq n}}^{\infty} \frac{1}{y_k^2}$	$\sum_{\substack{k=1\\k\neq n}}^{\infty} \frac{1}{y_k^2 (y_n - y_k)}$	$\sum_{\substack{k=1\\k\neq n}}^{\infty} \frac{1}{y_k^2 (y_n - y_k)^2}$	$\sum_{\substack{k=1\\k\neq n}}^{\infty} \frac{1}{y_k^2 (y_n - y_k)^3}$	→	$\sum_{\substack{k=1\\k\neq n}}^{\infty} \frac{1}{y_k^2 (y_n - y_k)^b}$
1	$\sum_{\substack{k=1\\k\neq n}}^{\infty} \frac{1}{y_k}$	$\sum_{\substack{k=1\\k\neq n}}^{\infty} \frac{1}{y_k (y_n - y_k)}$	$\sum_{\substack{k=1\\k\neq n}}^{\infty} \frac{1}{y_k (y_n - y_k)^2}$	$\sum_{\substack{k=1\\k\neq n}}^{\infty} \frac{1}{y_k (y_n - y_k)^3}$	→	$\sum_{\substack{k=1\\k\neq n}}^{\infty} \frac{1}{y_k (y_n - y_k)^b}$
0	╳	$\sum_{\substack{k=1\\k\neq n}}^{\infty} \frac{1}{(y_n - y_k)}$	$\sum_{\substack{k=1\\k\neq n}}^{\infty} \frac{1}{(y_n - y_k)^2}$	$\sum_{\substack{k=1\\k\neq n}}^{\infty} \frac{1}{(y_n - y_k)^3}$	→	$\sum_{\substack{k=1\\k\neq n}}^{\infty} \frac{1}{(y_n - y_k)^b}$
╳	**0**	**1**	**2**	**3**	→	**b**

Figure 2. A solution matrix of modified Calogero-Ahmed series

A simple example of the usefulness and correctness of the formula (7) is presented in **Appendix B**. When analyzing some of the solutions presented in the Ahmed and Calogero paper (exactly the formulas (16) and (18) from [23]), which are characterized by the value $b = 1$ of the exponent from the difference $(y_n - y_k)^b$, a similarity between the solutions described in equation (5), and solutions that are Rayleigh series $\sigma_p^{(2a)} = \sum_{\substack{k=1\\k\neq n}}^{\infty} y_k^{-a} + y_n^{-a}$ was noticed. At this moment it is worth recalling that the Rayleigh series can be determined analytically using an interesting recursive solution developed by Meiman [26], Sneddon [27], Kishore [28] and relatively recently by Elizalde et al. [29]:

$$\sigma_p^{(2l)} = \sum_{m=1}^{\infty} \frac{1}{\mu_{m,p}^{2l}} = \left(\frac{1}{p+l}\right) \sum_{k=1}^{l-1} \sigma_p^{(2k)} \sigma_p^{(2l-2k)} \quad \text{for } l \geq 2 \ . \tag{8}$$

A formula for the case when $b = 1$ can be deduced by decomposing the solution for any a with the help of the formula (7) to Rayleigh series and a single Calogero-Ahmed series (field 1–0). Let us present it in a form analogous to that presented in the work by Ahmed and Calogero [23]:

$$y_n^{a+1} \sum_{\substack{k=1 \\ k\neq n}}^{\infty} (y_n - y_k)^{-1} y_k^{-a} = \sum_{l=1}^{a} \sigma_p^{(l)} y_n^l - \frac{1}{2}(p + 2a + 1) \ . \tag{9}$$

Interestingly, the solution formulated above (9), gives the correct value even in the case of $a = 0$. It is then the sum depending on the Reyleigh series disappears and the final formula coincides with the basic main Calogero formula:

$$y_n \sum_{\substack{k=1 \\ k\neq n}}^{\infty} (y_n - y_k)^{-1} = -\frac{1}{2}(p + 1) \ . \tag{10}$$

Solution (9) is for all modified Calogero-Ahmed series from column 1 of the solutions matrix (figure 2). This finding was an impulse for further search for other solutions. Deducing similarly, it was possible to define a general solution for the functions listed in the first line of the analyzed matrix ($a = 1$ and $b \geq 1$):

$$\sum_{\substack{k=1 \\ k\neq n}}^{\infty} (y_n - y_k)^{-b} y_k^{-1} = \sum_{l=1}^{b} y_n^{-l} \sum_{\substack{k=1 \\ k\neq n}}^{\infty} (y_n - y_k)^{-(b+1-l)} + y_n^{-b} \sum_{\substack{k=1 \\ k\neq n}}^{\infty} y_k^{-1} \ . \tag{11}$$

Decomposition of the selected functions from the main diagonal (fields: 1–1, 2–2, 3–3 etc. of the matrix figure 2) with a thorough investigation of formulas defined by equation (9) and equation (11) resulted in deriving the formula for the functions located on the main diagonal (case when $a = b$) of the solution matrix:

$$\sum_{\substack{k=1 \\ k\neq n}}^{\infty} (y_n - y_k)^{-a} y_k^{-a} = \sum_{l=1}^{a} \binom{a-2+l}{l-1} y_n^{-(a-1+l)} \left[\sum_{\substack{k=1 \\ k\neq n}}^{\infty} y_k^{-(a+1-l)} + \sum_{\substack{k=1 \\ k\neq n}}^{\infty} (y_n - y_k)^{-(a+1-l)} \right] \tag{12}$$

where $\binom{x}{y} = \frac{x!}{y!(x-y)!}$ are binomial coefficients that can be written with the help of the factorial function. One can see from the presented solution (12), that the decomposition of the modified Calogero-Ahmed function located on the main diagonal of the matrix ($a = b$) using the recursive formula (7) leads to a solution based on the basic Rayleigh functions and the standard Calogero-Ahmed series. Successful derivation of equations (9, 11 and 12) stimulated us to further investigate the decomposition of other functions (those for the general case when $a \neq b$ – for exemplary decomposition see **Appendix C**) from the solution matrix defined in figure 2. This analysis allowed us to find a general final solution for all possible cases of the modified summation formulas of Calogero and Ahmed:

$$\sum_{\substack{k=1 \\ k\neq n}}^{\infty} (y_n - y_k)^{-b} y_k^{-a} = \sum_{l=1}^{a} \binom{a+b-1-l}{b-1} y_n^{-(a+b-l)} \sum_{\substack{k=1 \\ k\neq n}}^{\infty} y_k^{-l} + \sum_{l=1}^{b} \binom{a+b-1-l}{a-1} y_n^{-(a+b-l)} \sum_{\substack{k=1 \\ k\neq n}}^{\infty} (y_n - y_k)^{-l} \ , \tag{13}$$

where: $\sum_{\substack{k=1 \\ k\neq n}}^{\infty} y_k^{-l} = \sum_{m=1}^{\infty} \frac{1}{\mu_{m,p}^{2l}} - \frac{1}{\mu_{n,p}^{2l}} = \sigma_p^{(2l)} - \frac{1}{\mu_{n,p}^{2l}}$.

To complete the theory about the inverse Laplace transform of the dynamic viscosity function, one more problem need to be solved. Namely, as results from the final general solution (13), it is always based on the standard Calogero-Ahmed series (6), and already Calogero [22] noted that: "... *an interesting question is whether it can yield a closed-form expression for the sum for all (positive integral) values of c*". So the last step look to be the derivation of a recursive solution (similar to Meimann's for the Rayleigh sums) for the standard Calogero-Ahmed functions:

$$\chi_p^{(l)} = \sum_{\substack{k=1 \\ k\neq n}}^{\infty} (y_n - y_k)^{-l} = \sum_{\substack{k=1 \\ k\neq n}}^{\infty} \frac{1}{\left(\mu_{n,p}^2 - \mu_{k,p}^2\right)^l} \ , \tag{14}$$

where: $\mu_{p,k}$ are the consecutive zeros of the first kind Bessel function of p order $J_p(\mu_{p,k}) = 0$. When such a general recursive formula is defined then a final result of every modified Calogero-Ahmed sum will be dependent only on the solutions of the main summation functions: Rayleigh and Calogero-Ahmed ones (known for the lowest powers), i.e.:

$$\sum_{m=1}^{\infty} \frac{1}{\mu_{m,p}^2} = \frac{1}{4(1+p)} \quad and \quad \sum_{\substack{k=1 \\ k\neq n}}^{\infty} \frac{1}{\mu_{n,p}^2 - \mu_{k,p}^2} = -\frac{(1+p)}{2\mu_{n,p}^2} \;. \tag{15}$$

4. Conclusions

The analysis of the water hammer solution obtained in the Laplace domain (2) showed that the most important role plays a function $\mathbb{M}$, which is defined as the dynamic viscosity function. It is this function that is responsible for the damping and dispersion of pressure waves. This function plays an important role not only in the solution for pressure $\tilde{p}$, in which it is multiplied by an infinite series of exponential functions (having $\mathbb{M}$ function in the exponents), but also in the solution for flow velocity $\tilde{v}$ and wall shear stress $\tilde{\tau}$. Knowing the inverse Laplace transforms of this function $\mathbb{M}$ and the expressions $e^{-as\mathbb{M}}$ it will be possible to define a complete analytical solution of water hammer type flow. Such a solution will help in finding more effective methods of securing pressure systems against negative effects, e.g. in more accurate detection of leaks as well as in an in-depth comparative analysis of existing numerical methods (e.g. method of characteristics).

This paper presents author's recent developments in general solutions for decomposing modified Calogero-Ahmed series into series built of basic functions, which are Rayleigh series and standard Calogero-Ahmed series. The derived novel formula (7) turned out to be very useful for obtaining the final general solution (13). The presented solutions are very important in order to calculate the coefficients of polynomials that appear in the inverse Laplace transformation of the dynamic viscosity function [20]. The last necessary step would be to automate the entire process of searching for the coefficients of these polynomials, either with the help of machine learning or artificial intelligence, with the goal to find a recursive formula similar to the Meiman-Sneddon-Kishore formula (describing the $\sigma_p^{(2a)}$ Rayleigh series) for standard Calogero-Ahmed series $\chi_p^{(l)}$ (14).

5. Appendix A

Final derived polynomial of the inverse Laplace transform of dynamic viscosity function (4) is a sum of twelve standard and modified Calogero-Ahmed summation formulas:

$$\left(\frac{512-106\mu_i^2+3\mu_i^4}{48\mu_i^8}\right) = \Bigg(-\sum_{\substack{j=1 \\ j\neq i}}^{\infty} \frac{1}{\left(\mu_i^2-\mu_j^2\right)^4} + \sum_{\substack{j=1 \\ j\neq i}}^{\infty} \frac{1}{\left(\mu_j^2-\mu_i^2\right)^4} - \frac{2}{\mu_i^2}\sum_{\substack{j=1 \\ j\neq i}}^{\infty} \frac{1}{\left(\mu_i^2-\mu_j^2\right)^3} + 2\sum_{\substack{j=1 \\ j\neq i}}^{\infty} \frac{1}{\mu_j^2\left(\mu_j^2-\mu_i^2\right)^3} - \frac{19}{6\mu_i^4}\sum_{\substack{j=1 \\ j\neq i}}^{\infty} \frac{1}{\left(\mu_i^2-\mu_j^2\right)^2} + \frac{19}{6}\sum_{\substack{j=1 \\ j\neq i}}^{\infty} \frac{1}{\mu_j^4\left(\mu_i^2-\mu_j^2\right)^2} + \frac{1}{3\mu_i^2}\sum_{\substack{j=1 \\ j\neq i}}^{\infty} \frac{1}{\left(\mu_i^2-\mu_j^2\right)^2} - \frac{1}{3}\sum_{\substack{j=1 \\ j\neq i}}^{\infty} \frac{1}{\mu_j^2\left(\mu_j^2-\mu_i^2\right)^2} + \frac{9}{2}\sum_{\substack{j=1 \\ j\neq i}}^{\infty} \frac{1}{\mu_j^6\left(\mu_j^2-\mu_i^2\right)} - \frac{9}{2\mu_i^6}\sum_{\substack{j=1 \\ j\neq i}}^{\infty} \frac{1}{\left(\mu_i^2-\mu_j^2\right)} + \frac{3}{4\mu_i^4}\sum_{\substack{j=1 \\ j\neq i}}^{\infty} \frac{1}{\left(\mu_i^2-\mu_j^2\right)} - \frac{3}{4}\sum_{\substack{j=1 \\ j\neq i}}^{\infty} \frac{1}{\mu_j^4\left(\mu_j^2-\mu_i^2\right)} \Bigg) \tag{A1}$$

6. Appendix B

The sum of the matrix field 0–1 + 1–0 (where the first value is the column number and the second one is the row number) gives the following result:

$$\sum_{\substack{k=1 \\ k\neq n}}^{\infty} \frac{1}{y_k} + \sum_{\substack{k=1 \\ k\neq n}}^{\infty} \frac{1}{(y_n - y_k)} = y_n \sum_{\substack{k=1 \\ k\neq n}}^{\infty} \frac{1}{y_k(y_n - y_k)} \tag{B1}$$

One should note that the sum of y_k^{-1} is not a Rayleigh sum, as the term y_n is omitted during this summation. However, this sum can easily be written as a Rayleigh sum:

$$\sum_{\substack{k=1 \\ k\neq n}}^{\infty} \frac{1}{y_k} = \sum_{k=1}^{\infty} \frac{1}{y_k} - \frac{1}{y_n} = \sigma_p^{(2)} - \frac{1}{y_n} \tag{B2}$$

In general, for any exponent a, the following formula is valid:

$$\sum_{\substack{k=1 \\ k\neq n}}^{\infty} \frac{1}{y_k^a} = \sum_{k=1}^{\infty} \frac{1}{y_k^a} - \frac{1}{y_n^a} = \sigma_p^{(2a)} - \frac{1}{y_n^a} \tag{B3}$$

In a similar way, for example, a function from the matrix field 1−2 can be determined as the sum of the elements 0−2 + 1−1 multiplied by $1/y_n$, etc.

7. Appendix C

A typical decompositions of the modified Calogero-Ahmed series into basic series, i.e. Rayleigh and standard Calogero-Ahmed series, are presented below:

$$\sum_{\substack{k=1\\k\neq n}}^{\infty}\frac{1}{y_k^4(y_n-y_k)^3}=\frac{1}{y_n{}^3}\sum_{\substack{k=1\\k\neq n}}^{\infty}\frac{1}{y_k^4}+\frac{3}{y_n{}^4}\sum_{\substack{k=1\\k\neq n}}^{\infty}\frac{1}{y_k^3}+\frac{6}{y_n{}^5}\sum_{\substack{k=1\\k\neq n}}^{\infty}\frac{1}{y_k^2}+\frac{10}{y_n{}^6}\sum_{\substack{k=1\\k\neq n}}^{\infty}\frac{1}{y_k}+\frac{1}{y_n{}^4}\sum_{\substack{k=1\\k\neq n}}^{\infty}\frac{1}{(y_n-y_k)^3}+ \frac{4}{y_n{}^5}\sum_{\substack{k=1\\k\neq n}}^{\infty}\frac{1}{(y_n-y_k)^2}+\frac{10}{y_n{}^6}\sum_{\substack{k=1\\k\neq n}}^{\infty}\frac{1}{(y_n-y_k)} \tag{C1}$$

With the help of the above solution, it is possible to determine the final simplified form of these series. It is a case not provided by Ahmed and Calogero in their works [21-23]:

$$\sum_{\substack{k=1\\k\neq n}}^{\infty}\frac{1}{y_k^4(y_n-y_k)^3}=\frac{1}{y_n{}^3}\left(\sigma_p^{(8)}-\frac{1}{y_n{}^4}\right)+\frac{3}{y_n{}^4}\left(\sigma_p^{(6)}-\frac{1}{y_n{}^3}\right)+\frac{6}{y_n{}^5}\left(\sigma_p^{(4)}-\frac{1}{y_n{}^2}\right)+\frac{10}{y_n{}^6}\left(\sigma_p^{(2)}-\frac{1}{y_n}\right)+ \frac{1}{y_n{}^4}\sum_{\substack{k=1\\k\neq n}}^{\infty}\frac{1}{(y_n-y_k)^3}+\frac{4}{y_n{}^5}\sum_{\substack{k=1\\k\neq n}}^{\infty}\frac{1}{(y_n-y_k)^2}+\frac{10}{y_n{}^6}\sum_{\substack{k=1\\k\neq n}}^{\infty}\frac{1}{(y_n-y_k)} \tag{C2}$$

where: $\sigma_p^{(8)}=\frac{5p+11}{2^8(p+1)^4(p+2)^2(p+3)(p+4)}$; $\sigma_p^{(6)}=\frac{2}{2^6(p+1)^3(p+2)(p+3)}$; $\sigma_p^{(4)}=\frac{1}{2^4(p+1)^2(p+2)}$; $\sigma_p^{(2)}=\frac{1}{2^2(p+1)}$ and

$$\sum_{\substack{k=1\\k\neq n}}^{\infty}\frac{1}{(y_n-y_k)}=-\frac{p+1}{2y_n};\ \sum_{\substack{k=1\\k\neq n}}^{\infty}\frac{1}{(y_n-y_k)^2}=\frac{y_n-(p+1)(p+5)}{12y_n{}^2};\ \sum_{\substack{k=1\\k\neq n}}^{\infty}\frac{1}{(y_n-y_k)^3}=\frac{y_n-2(p+1)(p+3)}{16y_n{}^3}. \tag{C3}$$

For $p>0$, it can be seen that determining the result with this method will be a challenge, so let's assume that we are looking for the result for $p=0$ then:

$$\sigma_0^{(8)}=\frac{11}{12288};\ \sigma_0^{(6)}=\frac{2}{384};\ \sigma_0^{(4)}=\frac{1}{32};\ \sigma_p^{(2)}=\frac{1}{4}\ \text{and}\ \sum_{\substack{k=1\\k\neq n}}^{\infty}\frac{1}{(y_n-y_k)}=-\frac{1}{2y_n};\ \sum_{\substack{k=1\\k\neq n}}^{\infty}\frac{1}{(y_n-y_k)^2}=\frac{y_n-5}{12y_n{}^2};\ \sum_{\substack{k=1\\k\neq n}}^{\infty}\frac{1}{(y_n-y_k)^3}= \frac{y_n-6}{16y_n{}^3} \tag{C4}$$

This gives:

$$F(y_n)=\sum_{\substack{k=1\\k\neq n}}^{\infty}\frac{1}{y_k^4(y_n-y_k)^3}=\frac{1}{y_n{}^3}\left(\frac{11}{12288}-\frac{1}{y_n{}^4}\right)+\frac{3}{y_n{}^4}\left(\frac{2}{384}-\frac{1}{y_n{}^3}\right)+\frac{6}{y_n{}^5}\left(\frac{1}{32}-\frac{1}{y_n{}^2}\right)+\frac{10}{y_n{}^6}\left(\frac{1}{4}-\frac{1}{y_n}\right)+\frac{1}{y_n{}^4}\frac{(y_n-6)}{16y_n{}^3}+ \frac{4}{y_n{}^5}\frac{(y_n-5)}{12y_n{}^2}-\frac{10}{y_n{}^6}\frac{1}{2y_n}=\frac{11y_n{}^4+192y_n{}^3+2304y_n{}^2+35584y_n-332288}{12288y_n{}^7} \tag{C5}$$

Numerical check showed that this is a correct result (figure 3), as the waveform calculated from the infinite sum ("Inf Sum" – left hand of eq. C5) is consistent with the polynomial waveform ("Poly sol" – right hand of eq. C5).

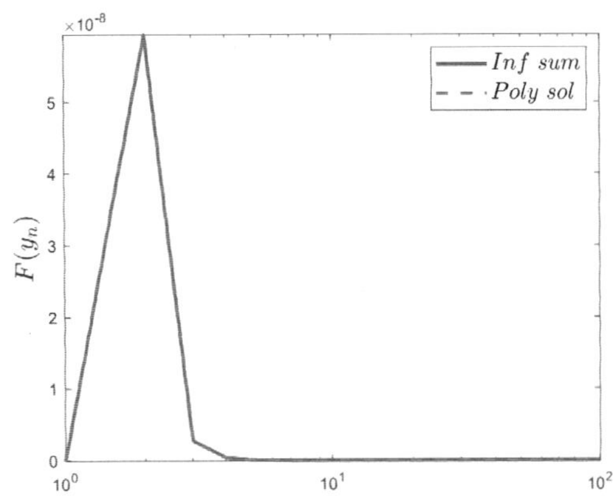

Figure 3. Exemplary comparison of infinite summation case

In the above figure 3 the $F(y_n)$ are the values of analyzed modified Calogero-Ahmed series (C5), and k is the index of summation.

8. Acknowledgements

A. Bergant gratefully acknowledges the support of the Slovenian Research Agency (ARRS) conducted through the research project L2-1825 and the programme P2-0162.

9. References

[1] Holmboe E L 1964 *Viscous distortion in wave propagation as applied to waterhammer and short pulses* (Doctoral Thesis, Carnegie Institute of Technology)
[2] Zielke W 1966 *Frequency-dependent friction in transient pipe flow* (Doctoral Thesis, University of Michigan)
[3] Urbanowicz K, Jing H, Bergant A, Stosiak M and Lubecki M 2021 *Proc. of Fluids Engineering Division Summer Meeting FEDSM2021*, FEDSM2021-65920
[4] Urbanowicz K, Bergant A, Stosiak M and Lubecki M 2022 *Structural Integrity* **24** 57
[5] Urbanowicz K, Bergant A, Stosiak M and Towarnicki K 2022 *Structural Integrity* **24** 69
[6] Rich G R 1945 *Trans. of ASME* **67(5)** 361
[7] Holmboe E L and Rouleau W T 1967 *Journal of Basic Engineering* **89** 174
[8] Brown F T and Nelson S E 1965 *Journal of Fluids Eng*ineering **87(2)** 504
[9] Jayasinghe D A P and Leutheusser H J 1970 *Proc of the Symposium Stockholm, International Association for Hydraulic Research* paper E2
[10] Karam J T and Leonard R G 1973 *ASME J. Fluids Eng.* **95 (4)** 498
[11] Yamaguchi K and Ichikawa T 1972 *Bulletin of the JSME* **15(88)** 1197
[12] Muto T and Takahashi K 1985 *Bulletin of the JSME* **28(244)** 2325
[13] Urbanowicz K *et al* 2021 *J. Phys.: Conf. Ser.* **1736** 012040
[14] Wang X Q, Sun J G and Sha W T 1995 *Journal of Hydrodynamics* Ser. B **2** 51
[15] Xuan L J, Mao F and Wu J Z 2012 *Acta Mech Sin* **28(2)** 266
[16] Sobey R J 2004 *Journal of Hydroinformatics* **6(3)** 187
[17] Jović V 2013 *Analysis and modelling of non-steady flow in pipe and channel networks* (Wiley)
[18] Mei C C and Jing H 2016 *Mathematical Biosciences* **280** 62
[19] Brereton G J and Jiang Y 2005 *Phys. Fluids* **17** 118104
[20] Urbanowicz K, Bergant A, Grzejda R and Stosiak M 2022 *Materials* **15** 4364
[21] Calogero F 1977 *Lett. Nuovo Cimento* **20(7)** 254
[22] Calogero F 1977 *Lett. Nuovo Cimento* **20(13)** 476
[23] Ahmed S and Calogero F 1978 *Lett. Nuovo Cimento* **21(15)** 531
[24] García-García F J and Alvariño P F 2019 *European Journal of Mechanics / B Fluids* **74** 200
[25] García-García F J and Alvariño P F 2019 *Journal of Fluid Mechanics* **881** 420
[26] Meiman N N 1956 *Dokl. Akad. Nauk SSSR* **108** 190
[27] Sneddon I N 1960 *Glasgow Mathematical Journal* **4(3)** 144
[28] Kishore N 1963 *Proc. Amer. Math. Soc.* **14** 527
[29] Elizalde E, Leseduarte S and Romeo A 1993 *Journal of Physics A: Mathematical and General* **26(10)** 2409

Journal of Physics: Conference Series **2367** (2022) 012027 doi:10.1088/1742-6596/2367/1/012027

Density diffusion in low Mach number flows

J Pozorski[1] **and A Kajzer**[1]

[1]Institute of Fluid-Flow Machinery, Polish Academy of Sciences, Fiszera 14, 80-231 Gdańsk, Poland

E-mail: jp@imp.gda.pl, akajzer@imp.gda.pl

Abstract. In the realm of compressible viscous flow modelling, we briefly revisit the debate on a possible inconsistency of the Navier-Stokes (NS) equations. Then, we recall a recent proposal from the literature, put forward by M. Svärd. One of its features is the mass diffusive term in the continuity equation. The presence of density diffusion in the Svärd model reduces dispersive numerical errors when simple centred 2nd order, numerical diffusion free, spatial schemes are used, as confirmed in the simulations of a doubly-periodic shear layer at Ma = 0.05 and Re = 10^4. Further reduction of the dispersive errors at the spatial discretisation level is possible by more sophisticated approximation techniques.

1. Introduction

The equations governing viscous fluid flows were formulated nearly two centuries ago by Navier (1823) and Stokes (1845). They are considered accurate within the assumption of fluid continuum at a local thermodynamic equilibrium, as extensively validated experimentally and commonly accepted for prediction of various flow cases and a plethora of practical purposes, including turbulence, strongly compressible flows featuring shocks, etc. In the liteature, there have been several attempts to revisit the formulation of governing equations of a compressible Newtonian flow, see [1, 2, 3] and references therein. Recently, Svärd proposed a quite different approach [4] to remedy some physical inconsistencies of the compressible Navier-Stokes equations, which simultaneously seems to properly address some critics [3] towards the model of Brenner [1]. A common feature of those attempts is the appearance of an extra diffusive term in the continuity equation as well as some modifications to the momentum and energy equations. However, the Svärd concept is rather radical: it goes beyond the Newton and Fourier constitutive laws. The issue seems to be controversial, as it touches some fundamental notions of thermomechanics of continuum and its underlying mesoscopic formulation on the level of the Boltzmann equation. We will not enter the very debate here (except for some general thoughts presented in Section 2). Rather, in the present work we intend to focus on: (i) the relation, in chosen aspects, of the novel flow model involving mass diffusion proposed by Svärd to well-established Navier-Stokes equations (Section 3); (ii) the numerical properties of the two respective approaches (Section 4). As an illustrative flow example, the doubly-periodic shear layer is chosen (Section 5).

2. On the mass diffusive formulations

The governing equations of fluid mechanics on the macroscopic, or continuum, level involve the fields of relevant variables; of particular interest here are the fields of density $\rho(\mathbf{x}, t)$ and velocity $\mathbf{u}(\mathbf{x}, t)$. These variables are assumed here to be continuous and smooth, i.e. differentiable w.r.t. the spatial coordinates and time. For the present reasoning, we assume the fluid to be homogeneous (not a mixture of components or phases) and admit no sources of mass which may otherwise stem from chemical reactions or phase changes.

The well known form of the mass conservation equation in a material continuum is readily derived in the Lagrangian approach using the notion of fluid element as the crucial concept. Denoting the mass of the fluid element by $\delta m = \rho\delta V$, the usual procedure assumes that $d(\delta m)/dt = 0$ which expresses the fact that δV is a material volume. Then, using the kinematic condition $d(\delta V)/dt = \delta V \nabla \cdot \mathbf{u}$, one gets the continuity equation $d\rho/dt = -\rho\nabla \cdot \mathbf{u}$. [NB: The material volume δV is large enough for the associated macroscopic quantities to be well defined, but small enough for them be constant within it.] The same result is obtained in the Eulerian description when one considers the balance of fluid mass within a fixed small volume (the control volume): the mass may change due to the *convective* flux through the boundary of that volume. The above form of continuity equation holds independently of the constitutive relationship for the medium considered, be it fluid or solid. Actually, the Lagrangian formulation is readily grasped for deformable solids, being the exact description at both micro- and macroscopic levels.

The key assumption about the material volume δV in the reasoning above is that it contains, over time, the same quantity of matter. On the microscopic description level, this would mean: the same number of molecules, which is evident in solids (where even the molecules remain the same), but less so in fluids. If one considers now a small control volume (the Eulerian description again) and a non-homogeneous fluid consisting of a (generally compressible) carrier phase and a passive admixture or a tracer, then its concentration $C = C(\mathbf{x}, t)$ will evolve also due the *diffusive* mechanism, satisfying $dC/dt + C\nabla \cdot \mathbf{u} = \nabla \cdot D\nabla C$ where D is a diffusivity coefficient.

One may now argue (following the arguments raised by Svärd [4]) that a homogeneous fluid of a non-uniform initial density distribution in 1D (such as a slightly perturbed, sine-wave pattern of a length scale L) will undergo the diffusive mechanism as well with the diffusive mass flux proportional to $D\nabla\rho$. In our perception, this way of reasoning might serve as an unrigorous justification of the appearance of the density diffusion term in the continuity equation, as proposed by Svärd, in the form $d\rho/dt = -\rho\nabla \cdot \mathbf{u} + \nabla \cdot D\nabla\rho$. Moreover, his derivation was done for gases where the mass and momentum (and heat) transport are governed by the same underlying molecular mechanism; hence the Schmidt number is of the order of one here (same for the Prandtl number) or, equivalently, $D \sim \nu$ with ν (the momentum transport coefficient) being the kinematic viscosity. Given that $\nabla\rho \sim \rho_0/L$ (where ρ_0 is a reference density) and the diffusive and convective time scales can be estimated as $\tau_d = L^2/D$ and $\tau_c = L/U$ respectively, this indicates that the significance of the mass diffusion term may be limited to flow regions of strong density gradients and low Reynolds numbers ($\tau_d \leq \tau_c$ means $Re \leq 1$). If this estimation is correct, then the physical relevance of diffusive mass term may be qualified as an issue of curiosity rather than controversy.

It is readily observed that the diffusive term in the continuity equation (if tenable) may only be of significance in the variable-density regions of viscous flow. It will disappear in flows of constant density and in inviscid compressible flows. Historically, the former include the realm of classical hydrodynamics (governed by the incompressible N-S eqs.) and the latter refer to many situations in aerodynamics (governed by the Euler eqs.). Actually, the effect of viscosity in compressible flows is basically limited to the shocks, contact waves and near-wall regions.

There have been some earlier attempts to introduce the mass diffusion term in the mass conservation equation and also critics against it (L.D. Landau). Apparently, Brenner [1, 2] was the first to propose a comprehensive derivation. He introduced a bi-velocity model with the mass velocity (referring to small, macroscopic solid particles) and the volume velocity (referring to molecular-type tracer) [5, 6]. In compressible flows the two quantities may differ by the diffusive flux of volume. This is somewhat unusual for a classical view on homogeneous continuous media where a single quantity is introduced at a point. The idea has been criticised in [3] on the grounds of physical inconsistency: the non-respect of the Galilean invariance and the lack of angular momentum conservation. Brenner [2] formulated a rebuttal and stated that his model represented a "hydrodynamic-level manifestation of fluid's biased Brownian motion, a fundamental contribution which was previously overlooked." There have also been attempts to derive the mass diffusion terms starting from a finer-level, mesoscopic description in terms of the Boltzmann equation [7, 8]. The outcome is sometimes termed the extended Navier-Stokes equations; however, this name refers to the formulation accounting for the flows of finite Knudsen numbers.

Recently, Svärd [4] formulated a novel mathematical model of viscous gas dynamics. His main motivation was to remedy some inconsistencies of the Navier-Stokes equations: (i) the lack of built-in mechanism to guarantee the positivity of density (particularly important from the numerical standpoint); (ii) relaxation to thermodynamic equilibrium has to induce macroscopic flow; (iii) issues with modelling multi-component fluids; (iv) inconsistency of chosen types of boundary conditions. These issues are claimed in [4] to be solved by the novel model. A crucial aspect of its derivation is strictly keeping the Eulerian frame of reference, which eliminates the necessity of introducing the concept of fluid element (which is ill-defined, according to Svärd's reasoning). Also, in his derivation Svärd consequently takes into account the underlying microscopic (molecular) structure of the fluid. A controversial yet interesting (and rather convincing) outcome of his reasoning is that a macroscopic flow model does not have to satisfy the conservation of the angular momentum which is lost due to the "coarse-graining" towards the macroscopic formulation. Using the same argument Svärd claims that macroscopic models can not assure the uniform motion of the fluid centre-of-mass. Nevertheless, the model of Svärd is rotationally invariant and conserves the angular momentum if the relevant diffusion coefficient is constant. Since this coefficient is related to the viscosity, it is independent of the velocity and therefore the model is Galilean invariant. This features address the critics in [3].

The novel mass difussive model was subsequently validated on a number of compressible flow cases [9]. To date, there seem to have been no fundamental critics to it in the literature.

Interestingly, the explicit (i.e. *à la* artificial viscosity) mass diffusion term in the continuity equation has been proposed on purely pragmatic grounds in computational fluid dynamics when solving the discrete system of flow equations [10]. Another example is the so-called δ– SPH approach in the Lagrangian particle method of Smoothed Particle Hydrodynamics [11].

3. Mathematical models

The compressible Navier-Stokes (CNS) equations which model the thermodynamically perfect, viscous and heat conducting gas flows are:

$$\partial_t \rho = -\nabla \cdot (\rho \mathbf{u}), \tag{1}$$

$$\partial_t (\rho \mathbf{u}) = -\nabla \cdot (\rho \mathbf{u}\mathbf{u} + p\mathbf{I}) + \nabla \cdot \boldsymbol{\tau}, \tag{2}$$

$$\partial_t E = -\nabla \cdot ((E + p)\mathbf{u}) + \nabla \cdot (\boldsymbol{\tau} \cdot \mathbf{u} + \kappa \nabla T), \tag{3}$$

where E is the total energy, p is the pressure, T is the temperature and κ is the heat conductivity coefficient; $\mathbf{I}$ is the unit tensor. The above equations express the laws of conservation of mass, momentum and energy. The viscous stress tensor in the Newtonian fluid is defined by:

$$\boldsymbol{\tau} = \mu \left(\nabla \mathbf{u} + \nabla \mathbf{u}^T - \frac{2}{3} (\nabla \cdot \mathbf{u}) \mathbf{I} \right) \tag{4}$$

where μ is the dynamic viscosity of the fluid. The total energy is the sum of internal and kinetic energy of the gas:

$$E = \frac{p}{\gamma - 1} + \frac{\rho |\mathbf{u}|^2}{2} \tag{5}$$

where $\gamma = c_p/c_v$ is the ratio of specific heats (or heat capacities) and we set $\gamma = 7/5$ in this work. The system is closed with the equation of state $p = \rho RT$ which defines the perfect gas, where R is the gas constant.

For brevity, without getting into physical arguments and details of the derivation, we recall below the system of equations for viscous gas [9] that according to Svärd should supersede Equations (1)–(3):

$$\partial_t \rho = -\nabla \cdot (\rho \mathbf{u}) + \nabla \cdot \left(\frac{\mu}{\rho} \nabla \rho \right), \tag{6}$$

$$\partial_t (\rho \mathbf{u}) = -\nabla \cdot (\rho \mathbf{u}\mathbf{u} + p\mathbf{I}) + \nabla \cdot \left(\frac{\mu}{\rho} \nabla (\rho \mathbf{u}) \right), \tag{7}$$

$$\partial_t E = -\nabla \cdot ((E + p)\mathbf{u}) + \nabla \cdot \left(\frac{\mu}{\rho} \nabla E \right). \tag{8}$$

The system will be referred to as the Eulerian Svärd Flow Model (ESFM). We emphasize that μ is the dynamic viscosity coefficient as indicated in [9]; primarily Svärd admitted other choices [4].

Please note that for the incompressible (constant density) case, the diffusive term in Equation (6) vanishes. For inviscid fluid, the flow model of Svärd correctly reduces to the Euler equations of gas dynamics.

Moreover, it is interesting to compare (w.r.t. the diffusive terms) the Svärd model to chosen analytical regularisations of the Euler equations [12] and modified equations stemming from the Lax-Friedrichs type schemes.

We also point here that in [13] we have proposed a simplified model for nearly incompressible flows, established by Equations (6) and (7) (neglecting the energy equation) and the isothermal equation of state.

4. Numerical method

To solve the ESFM governing equations we use the so-called method of lines (MOL): the spatial discretisation is done first, leading to a system of ordinary differential equations (ODEs) which are then solved using a suitable ODE solver. Since we will consider spatial schemes which do not introduce numerical diffusion, the equations have to be integrated in time using a method whose stability region contains a subset of the imaginary axis, e.g. the Runge-Kutta methods of third (or higher) order. Then the stability is independent of the grid spacing for arbitrary value of the physical diffusion coefficients.

Below we present the spatial discretisation in two-dimensional case (the three-dimensional extension is straightforward). We restrict ourselves to uniform cartesian grids with cell size Δx.

The CNS and ESFM equations presented in the previous section are in fact already written in the flux form which in two dimensions reads

$$\partial_t \mathbf{Q} = -\partial_x \mathbf{F} - \partial_y \mathbf{G} \tag{9}$$

where $\mathbf{u} = [u, v]$ and

$$\mathbf{Q} = [\rho, \rho u, \rho v, E]^T. \tag{10}$$

For the CNS:

$$\mathbf{F} = \left[\rho u, \quad \rho u^2 + p - \tau_{xx}, \quad \rho uv - \tau_{xy}, \quad (E + p)u - u\tau_{xx} - v\tau_{xy} - \kappa \partial_x T\right]^T, \tag{11}$$

while for the ESFM:

$$\mathbf{F} = \left[\rho u - \frac{\mu}{\rho}\partial_x \rho,\ \rho u^2 + p - \frac{\mu}{\rho}\partial_x(\rho u),\ \rho uv - \frac{\mu}{\rho}\partial_x(\rho v),\ (E + p)u - \frac{\mu}{\rho}\partial_x E\right]^T. \tag{12}$$

The fluxes **G** are defined by analogy. Now, the MOL consists in solving the following set of ODEs:

$$\frac{d\mathbf{Q}_{i,j}}{dt} = -\frac{1}{\Delta x}\left(\mathbf{F}_{i+\frac{1}{2},j} - \mathbf{F}_{i-\frac{1}{2},j} + \mathbf{G}_{i,j+\frac{1}{2}} - \mathbf{G}_{i,j-\frac{1}{2}}\right), \tag{13}$$

where $\mathbf{Q}_{i,j} = \mathbf{Q}(x_i, y_j)$, $\mathbf{F}_{i+\frac{1}{2},j} = \mathbf{F}(x_{i+\frac{1}{2}}, y_j)$, etc., for each grid cell indexed by (i,j).

The discrete fluxes are computed as follows. The derivatives of the parabolic parts are approximated by the usual two-point schemes and for the stress tensor τ the required components are averaged cell-centred quantities (resulting in a six-point stencil). For example we have:

$$(\partial_x T)_{i+\frac{1}{2},j} = (T_{i+1.j} - T_{i,j})/\Delta x \tag{14}$$

and

$$(\partial_y u)_{i+\frac{1}{2},j} = (u_{i+1.j+1} - u_{i+1,j-1} + u_{i.j+1} - u_{i,j-1})/(4\Delta x). \tag{15}$$

For the other terms of the parabolic parts the averaging on the cell faces is used, e.g.:

$$(\mu/\rho)_{i+\frac{1}{2},j} = \frac{2\mu}{\rho_{i,j} + \rho_{i+1,j}} \tag{16}$$

and

$$(u\tau_{xx})_{i+\frac{1}{2},j} = \frac{u_{i,j} + u_{i+1,j}}{2}(\tau_{xx})_{i+\frac{1}{2},j}. \tag{17}$$

As it is well known, the discretisation of the hyperbolic part is a non-trivial task and many different choices may be considered. First, let us point that we use the interpolation/extrapolation of the primitive variables $[\rho, u, v, p]$ in all the schemes. The choice of primitive variables is less straightforward than operating with the conserved variables (or cell-centred fluxes) but it brings advantage in the context of numerical oscillations in the solution or even its stability. For example, for the usual centred schemes the use of primitive variables automatically enforces the so-called split or skew-symmetric form (when 2nd-order accurate interpolation is applied), which is stable, contrary to the more traditional divergence form.

As the centred scheme for the hyperbolic flux $\mathbf{F}^{h,cen}_{i+\frac{1}{2},j}$ we use the split form, see e.g. [14]:

$$\mathbf{F}^{h,cen}_{i+\frac{1}{2},j} = \begin{bmatrix} \rho_{i+\frac{1}{2},j} u_{i+\frac{1}{2},j} \\ \rho_{i+\frac{1}{2},j} u_{i+\frac{1}{2},j} u_{i+\frac{1}{2},j} + p_{i+\frac{1}{2},j} \\ \rho_{i+\frac{1}{2},j} u_{i+\frac{1}{2},j} v_{i+\frac{1}{2},j} \\ \left(\frac{p_{i+\frac{1}{2},j}}{\gamma-1} + \frac{1}{2}\rho_{i+\frac{1}{2},j}\left(u^2_{i+\frac{1}{2},j} + v^2_{i+\frac{1}{2},j}\right) + p_{i+\frac{1}{2},j}\right) u_{i+\frac{1}{2},j} \end{bmatrix}, \tag{18}$$

where $f_{i+\frac{1}{2},j} = (f_{i+1,j} + f_{i,j})/2$ for $f = \rho, u, v, p$. The divergence form $\mathbf{F}_{i+\frac{1}{2},j} = (\mathbf{F}_{i+1,j} + \mathbf{F}_{i,j})/2$ based on the conserved variables leads to stability loss, for explanations and details see [14]. The above spatial discretisation has also been used in [13].

Although the ESFM equations are purely parabolic we have found that the above discretisation still causes visible dispersive errors. To reduce them we also investigated a central scheme based on extrapolation (it enables the use of the divergence form) instead of interpolation

$$\mathbf{F}^{h,cen-Fromm}_{i+\frac{1}{2},j} = \frac{1}{2}(\mathbf{F}^L_{i+\frac{1}{2},j} + \mathbf{F}^R_{i+\frac{1}{2},j}), \quad \mathbf{F}^L_{i+\frac{1}{2},j} = \mathbf{F}^h(\rho^L_{i+\frac{1}{2},j}, u^L_{i+\frac{1}{2},j}, v^L_{i+\frac{1}{2},j}, p^L_{i+\frac{1}{2},j}) \tag{19}$$

where $f^L_{i+\frac{1}{2},j} = f_{i,j} + \frac{1}{4}(f_{i+1,j} - f_{i-1,j})$ for $f = \rho, u, v, p$. This extrapolation scheme was introduced by Fromm [15]; however, we point that we do not apply any upwinding. Another possiblity to reduce the dispersive errors is the use of higher order centred schemes.

For the time integration we use the 3rd-order, three-stage, strong stability preserving Runge-Kutta method [16] setting the Courant number to 0.475. The constraint on the time step stemming from the diffusive terms is much less restrictive than the one stemming from the hyperbolic terms in the flow case considered below so it does not have to be taken into account.

5. Results

As a test case we have chosen the doubly periodic shear layer [17] which involves high shear and vorticity. The computational domain is the unit square with side lengths $L_x = L_y = L = 1$ m. The flow is periodic in both x and y directions. The reference velocity scale U_{ref} is chosen such that the Mach number does not exceed 0.05 during the whole simulation. The simulation is performed up to dimensionless time $t^+ = tL/U_{ref}$ equal 1. The initial conditions define two fluid layers sliding on each other in the opposite directions in the x-plane and a small velocity perturbation in the y-plane:

$$u(x,y) = \begin{cases} U_{ref}\tanh(b(y - L/4)) & \text{if} \quad y \leq L/2, \\ U_{ref}\tanh(b(3L/4 - y)) & \text{otherwise,} \end{cases} \tag{20}$$

where $b = 80$ and

$$v(x,y) = 0.05U_{ref}\sin(2\pi(x + L/4)). \tag{21}$$

This leads (in a well resolved simulation) to the creation of two vortices, see Figure 2. The fluid properties are set as the air parameters at normal conditions with the exception of viscosity which is artificialy set so that the resulting Reynolds number is $\mathrm{Re} = \rho_0 U_{ref} L/\mu = 10^4$. As the reference data we take the results of the truly incompressible simulation [17].

In Figure 1 we show the instantaneous density profiles obtained with the CNS and ESFM models using the two considered spatial discretisation schemes. Clearly, in the ESFM simulation the spurious oscillations are weaker, however, still visible. This unwanted effect is significantly reduced by the application of the centred-Fromm scheme. Also, the extrapolation-based scheme brings more improvement for the ESFM than for the CNS.

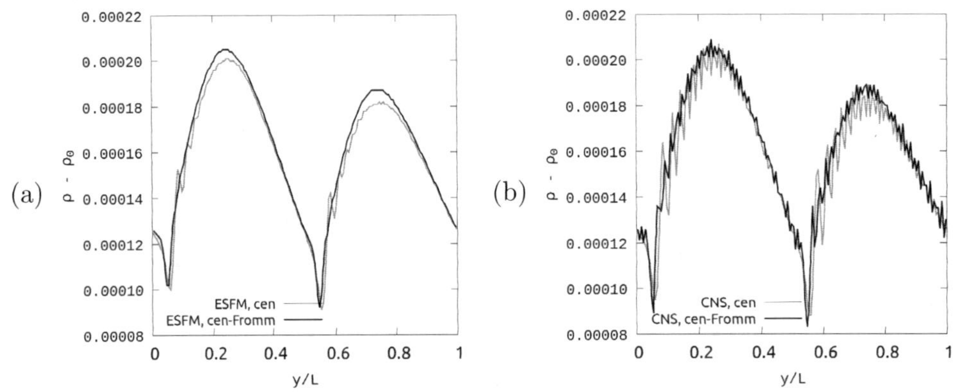

Figure 1. The density profiles along the line $y = L/2$ at $t^+ = 1$ from the ESFM (a) and the CNS model (b) using different spatial discretisation schemes on 160^2 grid.

In Figure 2 we present the vorticity fields at $t^+ = 1$ obtained with ESFM using different spatial resolutions. At the resolution of 128^2 both spatial schemes give a flawed result: two spurious vortices are created (the effect is known for numerical diffusion-free schemes [17]). The centred-Fromm scheme allows to get rid of the spurious vortices at lower spatial resolution than the centred interpolation based scheme. Notice also the symptoms of convergence of the vorticity extremal values visible in the case of centred-Fromm scheme (lower variation w.r.t. the spatial resolution).

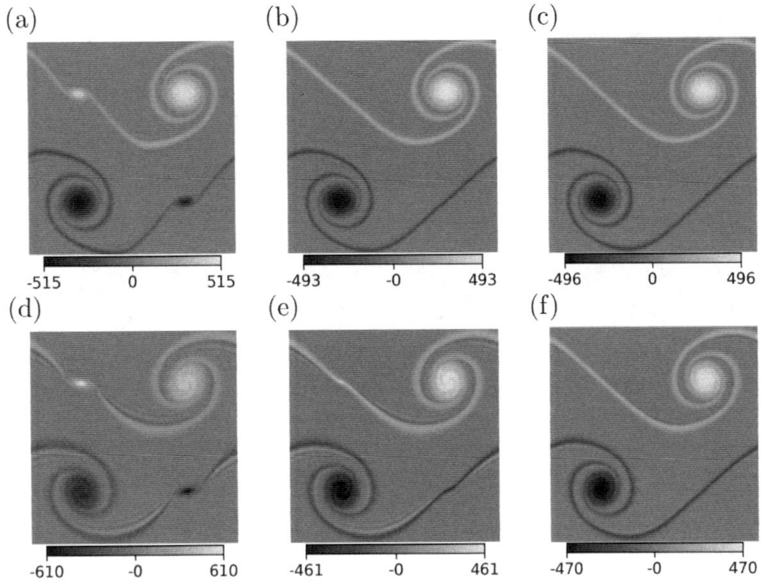

Figure 2. Comparison of the vorticity fields at $t^+ = 1$ obtained for resolutions $128^2, 160^2$, and 256^2 (from left to right). Schemes applied: centred Fromm (a-c) and centred (d-f).

In Figure 3 we present the quantitative analysis of the ESFM simulation; the histories of the total kinetic energy k and enstrophy Ω are shown. The agreement with the reference data is good in the case of both schemes. The dispersive errors introduced by the interpolation based scheme cause the overestimation of enstrophy for underresolved simulations. This discrepancy is smaller for the centred-Fromm scheme. The kinetic energy is slightly overestimated in the case of the centred-Fromm scheme and underestimated for the interpolation based scheme; in both cases the error decreases when the mesh is refined. The results obtained with CNS model are in general very similar to the ones obtained with ESFM and are therefore not shown.

6. Summary

We have briefly discussed the issue of mass diffusive term in the continuity equation which is one of the salient features of a novel mathematical model of viscous gas dynamics, recently formulated by Svärd [4] and subsequently validated on a number of compressible flow cases by Dolejši and Svärd [9].

Given some experience of the present authors regarding explicit discretisation schemes for weakly-compressible flow models [18], we have attempted to study a low Mach number flow both using the compressible Navier-Stokes equations and the new mass diffusive model. We have compared the CNS and ESFM models for the simulations of the nearly incompressible flow at relatively high Reynolds number, involving high vorticity (doubly periodic shear layer). The advantage of the ESFM model is the reduction of the dispersive errors due to the purely parabolic type of the governing equations. However, some numerical improvements are still required to increase the resolving power and reduce the (still) present spurious effects in the density (oscillations) and velocity (unphysical vortex structure) fields. As a more general remark we point that the ESFM equations have simpler structure of the parabolic part (in comparison to its CNS counter-part) which does not require the use of multidimensional stencils (easier algorithimcs; less memory access is required). As expected, since alike CNS the ESFM model reduces to incompressible N-S equations for low Mach numbers, we obtained good qualitative and quantitative agreement with the results of the truly incompressible simulations.

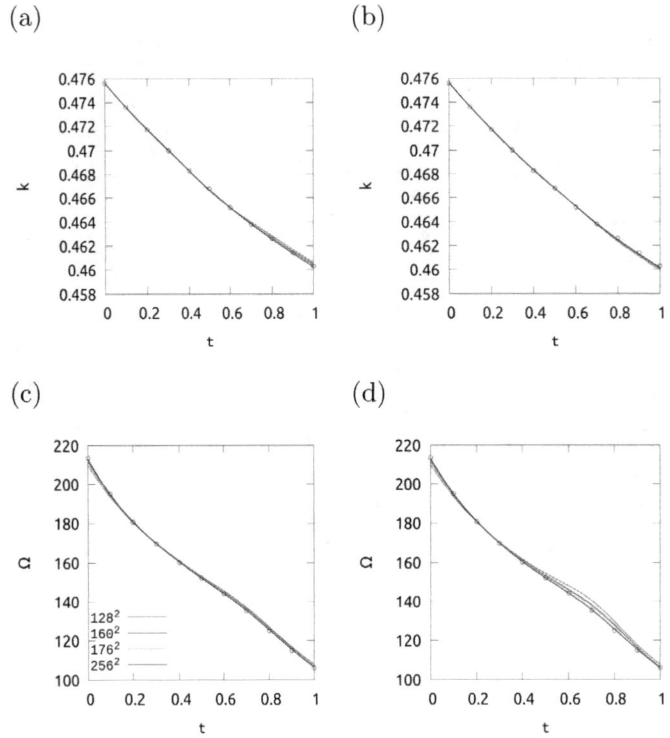

Figure 3. The history of the kinetic energy k (a and b) and the enstrophy (c and d) obtained with the ESFM using the centred Fromm scheme (a and c) and simple centred scheme (b and d) at different resolutions. Symbols (○) denote the reference data [17].

In our opinion, an interesting possibility for future work is to study numerical schemes (possibly improved and more efficient than presented above) to solve the ESFM equations in low Mach number regime, taking into account the purely parabolic character of the governing equations.

References

[1] Brenner H 2005 *Physica A* **349** 60–132
[2] Brenner H 2012 *Int. J. Eng. Sci.* **54** 67–98
[3] Öttinger H C, Struchtrup H and Liu M 2009 *Phys. Rev. E* **80** 056303
[4] Svärd M 2018 *Physica A* **506** 350–375
[5] Brenner H 2012 *Phys. Rev. E* **86** 016307
[6] Brenner H 2013 *Phys. A* **392** 558–566
[7] Chakraborty G and Durst F 2007 *Phys. Fluids* **19** 088104
[8] Han G, Liu X, Huang J, Nawnit K and Sun L 2020 *Phys. Fluids* **32** 022001
[9] Dolejši V and Svärd M 2021 *J. Comput. Phys.* **427** 110068
[10] Jameson A, Schmidt W and Turkel E 1981 *AIAA 5th Computational Fluid Dynamics Conf.* **81** 1259–1277
[11] Molteni D and Colagrossi A 2009 *Comput. Phys. Commun.* **180** 861–72
[12] Guermond J L and Popov B 2014 *SIAM J. Appl. Math.* **74** 284–305
[13] Kajzer A and Pozorski J 2022 *(submitted to Comput. Math. Appl.; revised after review)*
[14] Pirozzoli S 2010 *J. Comp. Phys.* **229** 7180–7190
[15] Fromm J E 1968 *J. Comp. Phys.* **3** 176–189
[16] Gottlieb R K and Shu C-W 1998 *Math. Comp.* **67** 73–85
[17] Minion M L and Brown D L 1997 *J. Comp. Phys.* **138** 734–765
[18] Kajzer A and Pozorski J 2018 *Comput. Math. Appl.* **76** 997–1013

Simulations of Johnson-Segalman viscoelastic fluids flows in shear-thinning regime

Tomáš Bodnár[1,2] **and Adélia Sequeira**[3]

[1] Faculty of Mechanical Engineering, Czech Technical University in Prague, Karlovo Náměstí 13, 121 35 Prague 2, Czech Republic.
[2] Institute of Mathematics, Czech Academy of Sciences, Žitná 25, 115 67 Prague 1, Czech Republic.
[3] Department of Mathematics and CEMAT, Instituto Superior Técnico, University of Lisbon, Av. Rovisco Pais, 1049–001 Lisbon, Portugal.

E-mail: Tomas.Bodnar@fs.cvut.cz

Abstract. This paper presents selected numerical simulations performed using Johnson–Segalman model with specific setting of the convected derivative resulting in shear-thinning behavior of the model. The results are compared with the generalized Oldroyd-B model, which is typically used to simulate shear-thinning viscoelastic flows. Both models are solved using an in-house finite-volume solver on structured grid simulating steady flow in three-dimensional axisymmetric channel with smooth contraction. The results presented here focus on the flow recirculation behind the contraction and the structure of the forces emanating from the viscoelastic stress tensor.

1. Introduction

The mathematical modeling and numerical simulations of viscoelastic fluids flows play an important role in many industrial, environmental and biomedical applications. Quite often the viscoelastic fluids also exhibit the so-called *shear-thinning* behavior, meaning that in the simple shear flow the fluid apparent viscosity decreases when the rate of deformation $\dot{\gamma}$ increases. Such behavior can typically be attributed to the microstructure of the fluid.

In mathematical models of viscoelastic fluids, the shear-thinning property is usually parametrized separately from the general viscoelastic behavior, by making the model coefficients shear rate dependent. This paper deals with the possibility to rather change the convected derivative in the model rather than generalizing (making shear rate dependent) its coefficients.

2. Mathematical model

The motion of incompressible homogeneous fluid can be described by the balance laws for mass and linear momentum, leading to the following system of partial differential equations:

$$\rho \frac{d\boldsymbol{u}}{dt} = \mathrm{div}\mathbf{T} - \boldsymbol{\nabla} p \quad (1)$$

$$\mathrm{div}\, \boldsymbol{u} = 0 \quad (2)$$

with $\boldsymbol{u}$ being the fluid velocity, ρ is the constant density and p the pressure. The material time derivative is defined by $d/dt = \partial/\partial t + \boldsymbol{u} \cdot \nabla$. Stress tensor $\mathbf{T}$ has to be defined by a suitable constitutive law to account for the rheological behavior of the fluid. Such constitutive relation links the stress in fluid to flow kinematics. For Newtonian fluids the stress tensor $\mathbf{T}$ is directly proportional to the rate of deformation tensor $\mathbf{D} = ((\nabla \boldsymbol{u}) + (\nabla \boldsymbol{u})^T)/2$, leading to constitutive relation in the form $\mathbf{T} = 2\mu\mathbf{D}$, where proportionality coefficient μ is the (constant) dynamic viscosity. With this Newtonian fluid assumption, the system (1)-(2) corresponds to the well know incompressible Navier-Stokes equations.

The Johnson–Segalman model [1],[2], is defined by the constitutive relation in the form:

$$\mathbf{T} + \lambda_1 \frac{\delta \mathbf{T}}{\delta t} = 2\mu \left(\mathbf{D} + \lambda_2 \frac{\delta \mathbf{D}}{\delta t} \right) \tag{3}$$

where the parameters λ_1 and λ_2 have the dimension of time and correspond to relaxation and retardation time scales of the model. The convected time derivative (denoted by $\frac{\delta \cdot}{\delta t}$) has to be used instead of the material time-derivative to guarantee the model frame indifference, i.e. assuring that the rheological behavior is independent of rigid body motion. Different choices of the convected derivative in (3) result into different rheological models, including among others the classical Oldroyd-A and -B models as special cases (see, e.g., [3], [2]) .

In the equation (3) the viscosity μ can be decomposed as $\mu = \mu_s + \mu_e$ with μ_s and μ_e being the solvent and the elastic viscosity coefficients respectively. The corresponding time scales λ_1 and λ_2 are:

$$\lambda_1 = \frac{\mu_e}{G} \quad , \qquad \lambda_2 = \lambda_1 \frac{\mu_s}{\mu_s + \mu_e} \tag{4}$$

where G is the fluid elastic modulus. In the same way the stress tensor $\mathbf{T} = \mathbf{T}_s + \mathbf{T}_e$ can be decomposed into its Newtonian (solvent) part $\mathbf{T}_s$ and elastic extra stress part $\mathbf{T}_e$ satisfying a constitutive relation of Maxwell type:

$$\mathbf{T}_s = 2\mu_s \mathbf{D} \quad \Longrightarrow \quad \mathbf{T}_s = 2\frac{\lambda_2}{\lambda_1}\mu\,\mathbf{D} \tag{5}$$

$$\mathbf{T}_e + \lambda_1 \frac{\delta \mathbf{T}_e}{\delta t} = 2\mu_e \mathbf{D} \quad \Longrightarrow \quad \mathbf{T}_e + \lambda_1 \frac{\delta \mathbf{T}_e}{\delta t} = 2\left(1 - \frac{\lambda_2}{\lambda_1}\right)\mu\,\mathbf{D} \tag{6}$$

Convected Derivatives

Most of the commonly used convected derivatives can be written in the following form:

$$\left(\frac{\delta \mathbf{M}}{\delta t}\right)_a = \frac{d\mathbf{M}}{dt} - \mathbf{WM} + \mathbf{MW} + a(\mathbf{DM} + \mathbf{MD}) \qquad a \in [-1;1] \tag{7}$$

Such general form is known under the name *Gordon–Schowalter derivative*, where $\mathbf{D}$ and $\mathbf{W}$ are the symmetric and skew-symmetric parts respectively of the velocity gradient tensor $\nabla \boldsymbol{u} = \mathbf{D} + \mathbf{W}$, the adjustable parameter $a = \xi - 1$, with ξ often denoted as *slip parameter*. Specific choices of parameter a lead to different convective derivatives. For example the Upper-convected derivative is obtained for $a = -1$, leading to Oldroyd-B model. The Co-rotational derivative arises by setting $a = 0$, while $a = 1$ leads to the Lower-Convected derivative in Oldroyd-A model.

3. Shear-thinning viscoelastic models

Generalized Oldroyd-B model

Models employing Upper- or Lower- convected derivatives (i.e. using $a = \pm 1$ in (7)) predict constant shear viscosity, so they do not capture the fluid shear-thinning behavior. In spite of this, the Oldroyd-B model (with $a = -1$) is often used to develop generalized viscoelastic fluids models where the shear-thinning effects are introduced by making the model coefficients variable, depending on the local shear-rate. In case of the model presented here, the (originally constant) viscosity coefficient is substituted by a function $\mu(\dot{\gamma})$ that decays with the increasing shear rate. From the many available formulas, the generalized *Cross* model is considered further:

$$\mu(\dot{\gamma}) = \mu_\infty + \frac{\mu_0 - \mu_\infty}{(1 + (\lambda\dot{\gamma})^b)^a} \tag{8}$$

Here μ_0 and μ_∞ are the asymptotic viscosity values for low and high shear rates. Model parameters are usually obtained from suitable experimental data. In ([4]) the following setup is used for blood flow modeling:

$$\mu_0 = 1.6 \cdot 10^{-1}\text{Pa·s} \qquad \mu_\infty = 3.6 \cdot 10^{-3}\text{Pa·s}$$
$$a = 1.23,\ b = 0.64 \qquad \lambda = 8.2\ \text{s}$$

Many other models can be used to simulate the shear-rate dependent viscosity of blood [5], [6].

Johnson–Segalman model
The apparent shear viscosity (naturally embedded in the model (3), without any generalization of its coefficients) has the form:

$$\mu(\dot{\gamma}) = \mu_\infty + \frac{\mu_0 - \mu_\infty}{1 + [1 - a^2]\lambda_1^2\dot{\gamma}^2} \tag{9}$$

The last term on the right-hand side of (9) can be associated with the apparent elastic viscosity μ_e. In case of $a^2 = 1$ the predicted viscosity is constant, but for all remaining values $a \in (-1; 1)$ the apparent viscosity $\mu(\dot{\gamma})$ decays (asymptotically to zero) with increasing shear rate $\dot{\gamma}$.

Based on [2], [7] it is possible to obtain closed form expressions for the first and second normal stress coefficients Ψ_1 and Ψ_2:

$$\Psi_1 = 2\lambda_1 \left[\mu(\dot{\gamma}) - \mu \cdot \left(\frac{\lambda_2}{\lambda_1}\right)\right] \tag{10}$$

$$\Psi_2 = -(1 + a)\lambda_1 \left[\mu(\dot{\gamma}) - \mu \cdot \left(\frac{\lambda_2}{\lambda_1}\right)\right] = -\frac{(1 + a)}{2}\Psi_1 \tag{11}$$

In all cases the first normal stress difference $\Psi_1 > 0$, but the second normal stress difference $\Psi_2 = 0$ just in case $a = -1$ (i.e. for Oldroyd-B model), while $\Psi_2 < 0$ for any $a > -1$. Real fluids have either zero or small in magnitude second normal stress difference, so the realistic choices of parameter a should remain close to $a = -1$ (here the values $a = -0.9, -0.8, -0.7$ were used).

4. Numerical simulations

The results presented in this paper were obtained using an in-house developed finite-volume code. The spatial discretization is performed using a simple central scheme on structured multi-block grid with hexahedral cells. The multistage Runge-Kutta method is used for time-stepping. More information about the numerical algorithms used for the presented simulations were published in [8, 9] with detailed description for the present cases in [7].

4.1. Computational domain

The presented computational test case was established in [4], based on smoothly stenosed blood vessel with nominal diameter $D = 6.2$ mm. The minimal diameter is $D/2$, meaning that the cross-section area reduces to 1/4 of the original value, which causes important flow speed-up in the constricted part of the vessel.

4.2. Model Parameters

Generalized Oldroyd-B model — The numerical values of parameters are adopted from [4] and were already used in number of our previous studies [8, 9]. The generalized apparent viscosity is defined by (8), and also the remaining viscoelastic model parameters are taken from [4]:

$$\mu_e = 4.0 \cdot 10^{-4}\,\text{Pa}\cdot\text{s} \qquad \mu_s = 3.6 \cdot 10^{-3}\,\text{Pa}\cdot\text{s}$$
$$\lambda_1 = 0.06\text{ s} \qquad \lambda_2 = 0.054\text{ s}$$
$$\mu_0 = \mu = \mu_s + \mu_e \qquad \rho = 1050\text{ kg}\cdot\text{m}^{-3}$$

The Johnson–Segalman model — The Johnson–Segalman model was adjusted for the shear thinning regime so that the viscosities μ_0 and μ_∞ together with the retardation time scale λ_2 were preserved the same as for the above Oldroyd-B model [7]. Using the viscosity relations (4) it is possible to get the expressions for solvent and elastic viscosity μ_s and μ_e, respectively

$$\mu_s = \mu\frac{\lambda_2}{\lambda_1} = \mu\frac{\mu_s}{\mu_s + \mu_e} = \mu_\infty\,, \qquad \mu_e = \mu - \mu_s = \mu_0 - \mu_\infty \tag{12}$$

From known viscosities μ_0, μ_∞ and retardation time λ_2 the relaxation time λ_1 can be obtained:

$$\mu_s = \mu\frac{\lambda_2}{\lambda_1} \Longrightarrow \frac{\mu_s}{\mu} = \frac{\mu_\infty}{\mu_0} = \frac{\lambda_2}{\lambda_1} \qquad \Longrightarrow \qquad \lambda_1 = \lambda_2\frac{\mu_0}{\mu_\infty} \tag{13}$$

In this way the asymptotic apparent shear viscosities of the Johnson–Segalman model are the same as for the generalized Oldroyd-B model. The remaining parameter $a \in (-1;1)$ in the convected derivative will be independently adjusted to assess its effect on the model predictions.

5. Numerical results

The numerical results presented in this section can be considered a complement to the set of results previously published in [7], where the velocity field obtained using different models was discussed. Here the focus is on some specific flow features and namely on the effects of the viscoelastic stress tensor in Johnson–Segalman model.

The flow in the contracted channel is characterized by a sudden flow acceleration when the fluid enters the part with reduced cross-section, which results in jet-like flow structure. Once the fluid passes the narrowest channel point, the flow expands and the fast core of the flow decelerates. For the smallest flow rate $Q = 0.5\text{ cm}^3/\text{s}$ there is no flow separation observed for any of the models, but for faster flows with $Q = 1.0$ and $Q = 2.0\text{ cm}^3/\text{s}$, the flow behind the minimal cross-section is detached, forming large recirculation zones in the near wall region. This is the typical flow feature found in generalized Oldroyd-B model results, that should be reproduced by the Johnson–Segalman model. The fields of axial velocity are shown together with streamlines in figures 1, 2 and 3 for the three flow rates solved always for the generalized Oldroyd-B model (as a reference) and by the Johnson–Segalman model with three different values of the parameter $a = -0.9$, $a = -0.8$ and $a = -0.7$.

The comparison of flow fields in the case $Q = 0.5\text{ cm}^3/\text{s}$ shown in figure 1 shows that the velocity field as well as its general character differs visibly between the Oldroyd-B and Johnson–Segalman model for all settings of the parameter a. The flow expands faster towards the wall when using Johnson–Segalman model, leaving the core of the flow slower than in the case of generalized Oldroyd-B model. No flow separation is observed.

For faster flows at flow rates $Q = 1.0\text{ cm}^3/\text{s}$ and $Q = 2.0\text{ cm}^3/\text{s}$ shown in figure 2 and 3 the situation is different. The flow separation occurs behind the contraction, leading to large recirculation regions with back flow (negative sign of axial velocity component), which is marked in the figures by the gray color. Depending on setting of the parameter a, similar recirculation is also observed in the simulations using Johnson–Segalman model. By comparing the flow fields obtained using Johnson–Segalman model for different settings of the parameter a, it seems to be

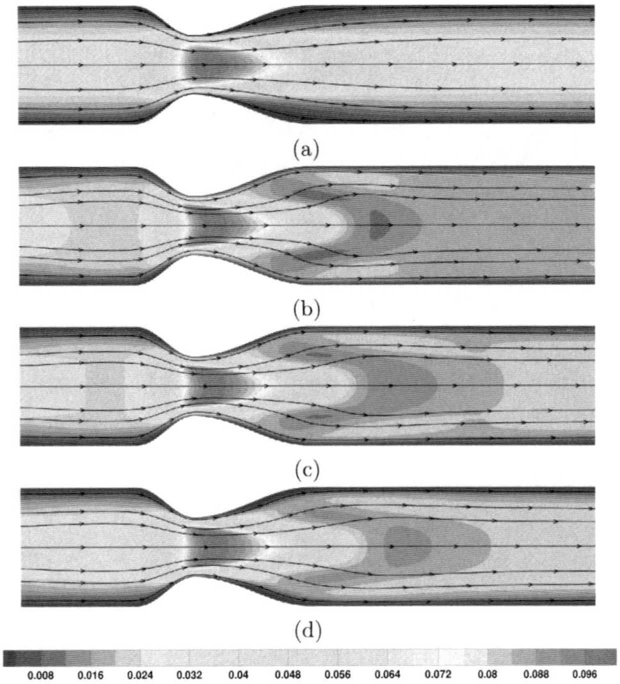

Figure 1: Axial velocity [m/s] and streamlines for $Q = 0.5$ cm^3/s. (a) Generalized Oldroyd-B model , (b) Johnson–Segalman model $a = -0.9$, (c) $a = -0.8$, (d) $a = -0.7$.

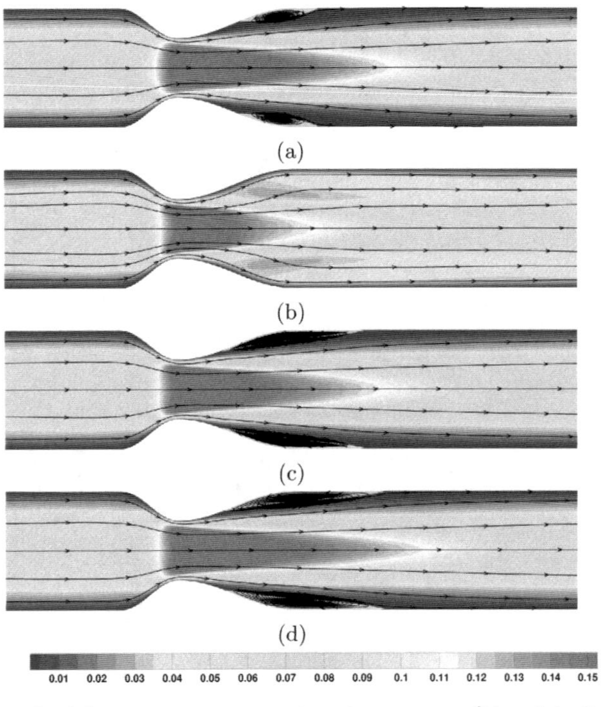

Figure 2: Axial velocity [m/s] and streamlines for $Q = 1.0$ cm^3/s. (a) Generalized Oldroyd-B model, (b) Johnson–Segalman model $a = -0.9$, (c) $a = -0.8$, (d) $a = -0.7$.

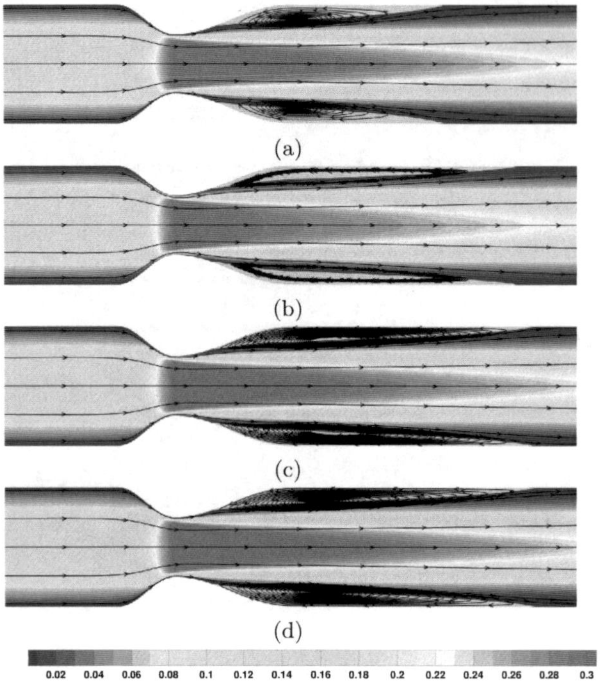

Figure 3: Axial velocity [m/s] and streamlines for $Q = 2.0$ cm^3/s. (a) Generalized Oldroyd-B model, (b) Johnson–Segalman model $a = -0.9$, (c) $a = -0.8$, (d) $a = -0.7$.

possible to obtain results (at least flow fields) very similar to the predictions of the generalized Oldroyd-B model (by simply adjusting the parameter a). This is however only possible at higher flow rates (velocities) shown in figures 2 and 3 for $Q = 1.0$ cm^3/s and $Q = 2.0$ cm^3/s.

Deeper insight into the forces acting on the fluid, can be obtained by comparison of the divergence of the elastic stress tensor, i.e. div$\mathbf{T}_e$, which represents the additional force due to elasticity for different settings of the parameters. Figures 4, 5, and 6 are showing the axial and radial components of the force emanating from the divergence of the elastic stress tensor $\mathbf{T}_e$. The axial component of the div$\mathbf{T}_e$ corresponds to force accelerating/decelerating the fluid flow

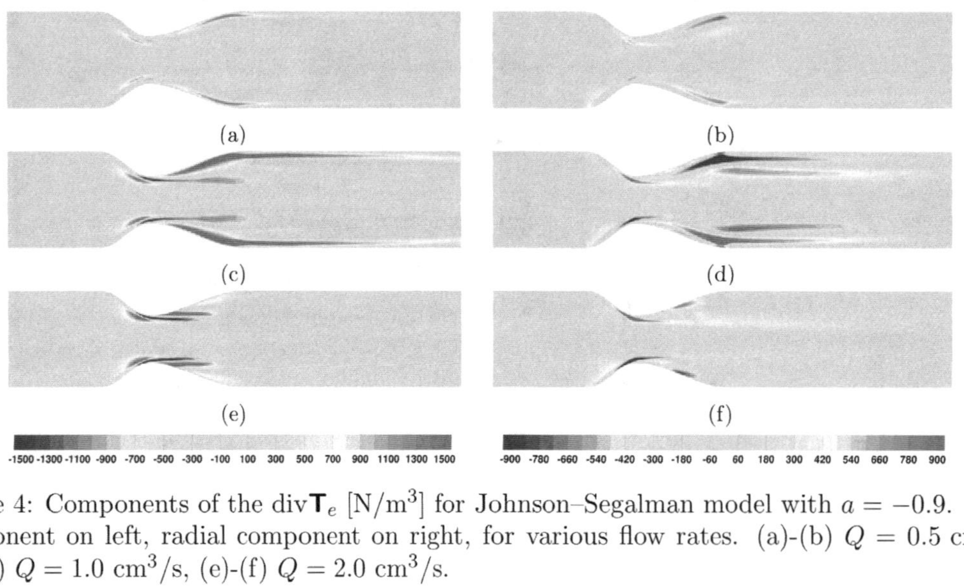

Figure 4: Components of the div$\mathbf{T}_e$ [N/m^3] for Johnson–Segalman model with $a = -0.9$. Axial component on left, radial component on right, for various flow rates. (a)-(b) $Q = 0.5$ cm^3/s, (c)-(d) $Q = 1.0$ cm^3/s, (e)-(f) $Q = 2.0$ cm^3/s.

Journal of Physics: Conference Series **2367** (2022) 012028 doi:10.1088/1742-6596/2367/1/012028

in the corresponding direction. Obviously the local character of this force changes dramatically with flow rate Q, while the magnitude of the force rather depends on the parameter a. Similar observation can be made for the radial component of the elastic force pushing the fluid from or towards the channel axis.

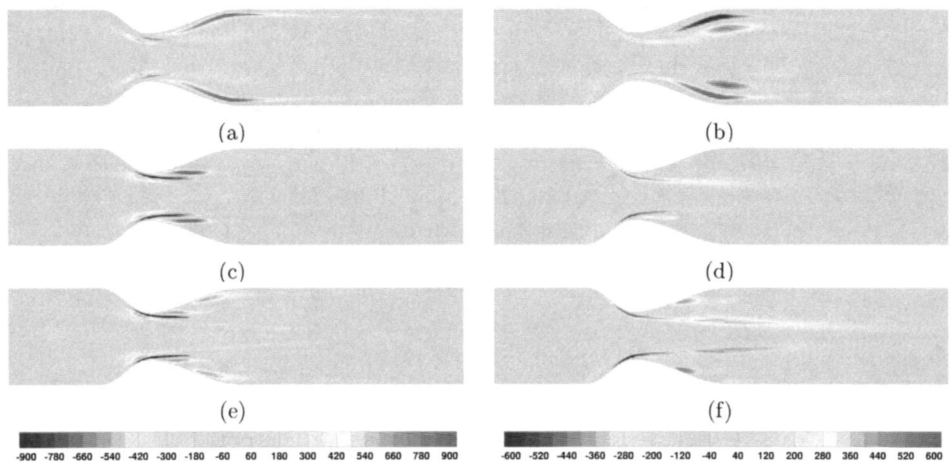

Figure 5: Components of the div$\mathbf{T}_e$ [N/m^3] for Johnson–Segalman model with $a = -0.8$. Axial component on left, radial component on right, for various flow rates. (a)-(b) $Q = 0.5$ cm^3/s, (c)-(d) $Q = 1.0$ cm^3/s, (e)-(f) $Q = 2.0$ cm^3/s.

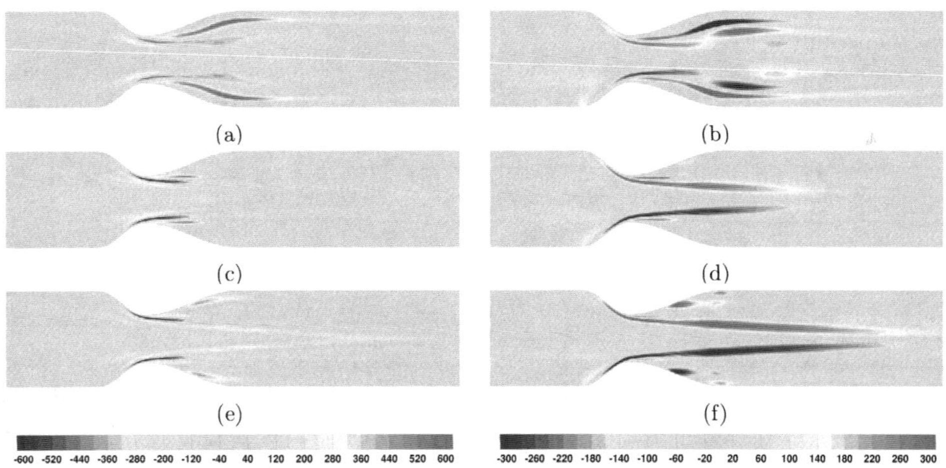

Figure 6: Components of the div$\mathbf{T}_e$ [N/m^3] for Johnson–Segalman model with $a = -0.7$. Axial component on left, radial component on right, for various flow rates. (a)-(b) $Q = 0.5$ cm^3/s, (c)-(d) $Q = 1.0$ cm^3/s, (e)-(f) $Q = 2.0$ cm^3/s.

It is interesting to note that in large parts of the domain the elastic forces are quite low, with higher values just in relatively small regions either close to the wall or in the constricted region of the channel. Important is that the forces are acting in relatively thin layers with steep gradients and changes in sign.

6. Conclusions and remarks

The simulations presented in this paper have brought some additional information on the viscoelastic stress effects in the Johnson–Segalman model, while used in the shear-thinning regime with the convected derivative parameter $a > -1$. In certain flow regimes, the results obtained using Johnson–Segalman model (with properly chosen parameter a) are very close to solutions obtained using the more common generalized Oldroyd-B model. This mutual agreement between the flow fields, including the recirculation zones, is only possible due to shear-thinning behavior of the Johnson–Segalman model with settings $a > -1$.

This study only compares the outputs of different mathematical models (using different rheological laws). Based on these results it is not possible to conclude whether any of the presented models is better than the others. The models are different, aimed for different fluids, although they share some of the most significant features, the viscoelasticity and the shear thinning. This paper has shown that when it comes to shear-thinning features in the flow, the Johnson–Segalman model (without any modification) is able to replicate the results of the shear-thinning generalization of the Oldroyd-B model. On the other hand, there are clear differences between the outputs of both models in certain flow regimes.

One of the possible explanations is the second normal stress difference Ψ_2 given by (11), which is zero for Oldroyd-B model (even generalized) with $a = -1$, but is always non-zero for Johnson–Segalman model with $a \in (-1; 1)$. Therefore one of the key questions in deciding whether to use the Johnson–Segalman or Oldroyd-B model is if the fluid of interest exhibits non-zero second normal stress difference Ψ_2. This data is however very seldom available for real fluids (including blood considered in presented simulations).

Future investigation should focus on more detailed comparison of results, including other fluids, flow regimes and geometries, with complete comparison of full stress tensor (including pressure).

Acknowledgments

The financial support for T.B. was partly provided by the European Regional Development Fund-Project "Center for Advanced Applied Science" No.CZ.02.1.01/0.0/0.0/16 019/0000778 and partly by the *Praemium Academiae* of Š. Nečasová. The support for A.S. from Fundação para a Ciência e a Tecnologia-FCT (Portugal) under the Projects UIDB/04621/2020 and UIDP/04621/2020 of CEMAT/IST-ID and the Project PTDC/MAT-APL/7076/2020, is also acknowledged.

References

[1] Johnson M W and Segalman D 1977 *Journal of Non-Newtonian Fluid Mechanics* **2** 255–270
[2] Bird R, Armstrong R and Hassager O 1987 *Dynamics of Polymeric Liquids* (John Wiley & Sons, New-York)
[3] Oldroyd J 1958 *Proceedings of Royal Society, London, Ser. A* **245** 278–297
[4] Leuprecht A and Perktold K 2001 *Computer Methods in Biomechanics and Biomechanical Engineering* **4** 149–163
[5] Galdi G, Rannacher R, Robertson A and Turek S (eds) 2008 *Hemodynamical Flows - Modeling, Analysis and Simulation* (*Oberwolfach Seminars* vol 37) (Birkäuser)
[6] Yeleswarapu K 1996 *Evaluation of continuum models for characterizing the constitutive behavior of blood* Ph.D. thesis University of Pittsburgh
[7] Bodnár T and Sequeira A 2022 *Fluids* **7**
[8] Bodnár T and Sequeira A 2010 *Advances in Mathematical Fluid Mechanics* ed Rannacher R and Sequeira A (Springer Verlag) pp 83–104
[9] Bodnár T, Rajagopal K and Sequeira A 2011 *Mathematical Modelling of Natural Phenomena* **6** 1–24

Kinematic simulations as a subgrid-scale model for particle motion: *a priori* LES of homogeneous isotropic turbulence

Michał Rajek[1] **and Jacek Pozorski**[1]

[1] Institute of Fluid-Flow Machinery, Polish Academy of Sciences, Fiszera 14, 80-231 Gdańsk, Poland

E-mail: mrajek@imp.gda.pl, jp@imp.gda.pl

Abstract. In this work we investigated the impact of filtering and subgrid-scale modeling on particle settling velocity and collision-related statistics in a turbulent flow. To reduce the complexity of this task we first studied the motion of inertial particles in the low-pass filtered homogeneous and isotropic turbulence, which was subsequently enriched with the subgrid-scale velocity components obtained from a frozen high-pass filtered velocity field. Particular emphasis has been put on the radial distribution function and the radial relative velocity of nearly touching particles both in the presence or absence of the gravitational settling. These statistics are the key input parameters to the kinematic collision kernel which is of crucial importance in determining the collision rate of inertial particles in a turbulent flow. Furthermore, kinematic simulations were selected as a means of enhancing the fluid velocity at particle locations. We analyzed a wide range of Stokes numbers, i.e. a measure of particle inertia, and, in contrast to scientific premises found in the literature, we observed no improvement of particle statistics when the low-pass filtered fluid velocity was enriched with both a synthetic or spectrally-filtered small-scale structures. We discuss the shortages of any frozen-velocity-based subgrid-scale model in predicting both single- and two-point particle statistics. We also indicate that in some cases, particularly concerning the collision rate of particles suspended in homogeneous and isotropic turbulence, subgrid-scale contribution in the particle equation of motion can be neglected.

1. Introduction

Turbulent transport and spatial distribution of a dispersed phase suspended in the form of small inertial particles, bubbles or droplets are of importance to many industrial and environmental applications. This includes for instance cyclone separation, fuel injection, sediment transport in rivers or seas, warm rain formation or more generally – cloud processes, still considered as the major cause of uncertainty in weather forecasting [1].

To account for the dispersed phase in a turbulent flow one can choose from different methods extensively reviewed by Kuerten [2] and Elghobashi [3]. For dilute suspensions of particles smaller than the Kolmogorov length scales, the dispersed entities are safely approximated as point-particles (mass-points) and tracked individually leading to the so-called point-particle direct numerical- or large eddy-simulations (DNS and LES, respectively). Due to prohibitively high computational cost, DNS, although being the most accurate approach, is still not feasible for most practical simulations, i.e. beyond relatively low Reynolds numbers and simple geometries. For this reason, LES have recently emerged as a promising tool for particle-laden flows. By means of spatial filtering only the large, energy containing scales are resolved, whereas the small-scale contribution is accounted for through an additional term in the momentum conservation equation, often expressed in terms of an artificial viscosity. Due to the lack of small-scale motions, particle statistics in DNS and LES diverge with decreasing particle inertia [4]. For this reason, point-particle LES typically require "two-stage" subgrid-scale (SGS) modeling. Ideally, the "first-stage" ensures the large-scale motions to be properly resolved, while the second precisely enhances particle motions. An ideal "single-stage" structural SGS model, in which the small-scale velocity field is generated using the available quantities, and can thereafter be used to both determine the SGS stress tensor in the momentum equation and enhance the fluid velocity at particle locations, remains unreachable.

A more profound understanding of the dispersed phase dynamics, particularly concerning preferential concentration, preferential sweeping, particle collisions and/or coalescence, is critical for further advancements in the aforementioned fields. Squires & Eaton [5] followed by Wang & Maxey [6] observed that inertial particles tend to agglomerate in the regions of high vorticity and low strain rate. Fessler et al. [7] experimentally investigated the instantaneous particle concentration in a turbulent channel flow and confirmed the aforementioned numerical findings. Sundaram & Collins [8] were the first to study the motion of finite-volume particles taking into account interparticle collisions. Ray & Collins [9] performed the DNS of isotropic turbulence and studied the impact of filtering on the radial distribution function (RDF), a measure of particle clustering, and the radial relative velocity (RRV). These two-point statistics are of importance to accurate prediction of particle collision kernel. The Authors reported that with increasing filter size the RDF decreases for Stokes numbers $St < 1.5$ and increases for $St > 1.5$. Due to filtering, the velocity field is deprived of small-scale fluctuations which essentially control the clustering of inertial particles at low St numbers. On the other hand, the lack of subgrid scales suppresses the dispersion of large particles so that clustering is favored. Ray & Collins [10] were the first to investigate the impact of kinematic simulations (KS) on particle statistics in a low-pass filtered DNS (FDNS) of isotropic turbulence, excluding the gravitational settling. The Authors reported accurate prediction of preferential concentration for particles of $St > 2$. However, particle clustering in the absence of gravitational settling is already well predicted for $St > 2$ by means of FDNS without any additional SGS model as discussed in Section 3. This work was followed by Zhou et al. [11] who combined the approximate deconvolution method (ADM) with KS to study the relative dispersion of flow tracers in LES of isotropic turbulence. The Authors used ADM to supplement the standard spectral eddy viscosity model of Chollet & Lesieur [12] and, consequently, to enhance scales near the cutoff. KS, on the other hand, was used to include explicit SGS motions. The Authors reported significant improvement in the prediction of Lagrangian dispersion of passive particles for LES 32^3 and 64^3 when compared to the reference DNS 512^3. Zhou et al. [13] investigated the application of the combined ADM and KS for particle laden isotropic turbulence. The Authors did not report any further improvement when compared to FDNS enriched with KS reported by Ray & Collins [10].

Modeling of dispersed turbulent two-phase flows still poses a serious challenge and many issues are to be resolved. This includes for instance: (i) the development of SGS closures for the Lagrangian equation of particle motion to enhance the predictions of both one- and two-point particle statistics; (ii) the development of specific source terms in the governing equations to account for particle-to-turbulence momentum transfer; (iii) accurate treatment of particle-particle aerodynamic interactions. In this work we take a necessary step towards the development of an accurate SGS model for particle motion in *a posteriori* LES of isotropic turbulence. To reduce the complexity of this task, comprising essentially "two-stages" of SGS modeling, we decided to perform a point-particle FDNS often referred to as *a priori* or ideal LES, i.e. deprived of any "first-stage" SGS modeling errors. In the case of FDNS, however, a greater caution in drawing conclusions is required as usually, due to inaccurate "first-stage" SGS modeling, particle statistics in *a posteriori* LES may considerably diverge from its *a priori* counterpart.

2. Numerical methods

The equations governing an incompressible fluid motion are given by:

$$\frac{\partial \mathbf{u}}{\partial t} = \mathbf{u} \times \boldsymbol{\omega} - \nabla\left(\frac{p}{\rho} + \frac{1}{2}\mathbf{u}^2\right) + \nu\nabla^2\mathbf{u} + \mathbf{f}(\mathbf{x}, t), \qquad \nabla \cdot \mathbf{u} = 0, \tag{1}$$

where $\mathbf{u}$ is the velocity field, $\boldsymbol{\omega} = \nabla \times \mathbf{u}$ is the vorticity field, p is the pressure field, ρ and ν are the density and kinematic viscosity of the fluid, respectively. The term $\mathbf{f}(\mathbf{x}, t)$ is an artificial forcing used to sustain turbulence. In this work we used the stochastic forcing proposed by Eswaran and Pope [14] which is non-zero for the first two shells in wavenumber space, i.e. $|\mathbf{k}| < \sqrt{8}$.

2.1. *Direct Numerical Simulations (DNS)*

We performed the DNS of homogeneous isotropic turbulence using the standard pseudo-spectral method [15] in a periodic cube with an edge length of 2π. In Fourier space, equations (1) become:

$$\left(\frac{\partial}{\partial t} + \nu k^2\right)\hat{\mathbf{u}}(\mathbf{k}, t) = \mathbf{P}(\mathbf{k})\mathcal{F}(\mathbf{u} \times \boldsymbol{\omega}) + \hat{\mathbf{f}}(\mathbf{k}, t) \qquad \text{for} \quad k \leq k_{\text{max}}, \tag{2}$$

where $\hat{\mathbf{u}}$ is the fluid velocity in Fourier space, $\mathcal{F}$ denotes the Fourier transformation and $\mathbf{P} = \delta_{ij} - k_i k_j/k^2$ $(i, j = 1, 2, 3)$ is the projection tensor. The computational domain was uniformly discretized with $N = 256^3$ points allowing to accurately resolve a turbulent flow with the Reynolds number based on the Taylor microscale $Re_\lambda \approx 107$. The spatial resolution was monitored by the value of $k_{\text{max}}\eta$ which was kept larger than 1.5. The two-way momentum coupling was safely neglected as the particle volume fraction was kept low, i.e. $\Phi_V < 2.25 \times 10^{-5}$.

2.2. *Filtered direct numerical simulation (FDNS)*

FDNS was performed by applying a low-pass sharp spectral filter which removes all the Fourier coefficients beyond a prescribed cutoff wavenumber k_c according to equation (3a):

$$(a)\ \hat{\tilde{\mathbf{u}}} = \begin{cases} \hat{\mathbf{u}}(\mathbf{k}, t) & \text{if } |\mathbf{k}| \leq k_c \\ 0 & \text{otherwise,} \end{cases} \qquad (b)\ \hat{\mathbf{u}}_{\text{sgs}} = \begin{cases} \hat{\mathbf{u}}(\mathbf{k}, t_0) & \text{if } |\mathbf{k}| > k_c \\ 0 & \text{otherwise,} \end{cases} \tag{3}$$

In this work we set $k_c = 30$ which corresponds to *a posteriori* LES with a 64^3 grid resolution. Such methodology was also employed to extract and explore the impact of a high-pass filtered frozen subgrid-scale velocity field (FFSGS) according to equation (3b) in which $t_0 \approx 15T_e$ with T_e being the large-eddy turnover time. Ergo, the FFSGS was extracted from a statistically steady, homogeneous and isotropic velocity field. Such velocity field is expected to be non-divergent and is meant to reconstruct, at least to some extent, the DNS energy spectrum.

2.3. *Discrete particle motion*

Inertial particles were tracked using the Lagrangian point-particle approach as the particle diameter d_p was set to be smaller than the Kolmogorov length scale η $(d_{p,\text{max}} \leq 0.2\eta)$. If the particle-fluid density ratio is assumed to be much larger than 1 (here $\rho_p/\rho = 1000$) then the particle equation of motion acquires the following form:

$$\frac{\mathrm{d}\mathbf{V}_p}{\mathrm{d}t} = -\frac{\mathbf{V}_p - (\tilde{\mathbf{u}} + \mathbf{u}_{\text{sgs}})}{\tau_p} + \mathbf{g}, \qquad \frac{\mathrm{d}\mathbf{x}_p}{\mathrm{d}t} = \mathbf{V}_p, \tag{4}$$

where $\mathbf{V}_p$ is the particle velocity, $\tilde{\mathbf{u}}$ and $\mathbf{u}_{\text{sgs}}$ are the fluid large-scale and small-scale velocities at the particle locations, respectively, $\mathbf{g}$ is the gravitational acceleration, $\mathbf{x}_p$ is the particle position and $\tau_p = \rho_p d_p^2/18\rho\nu = \tau_K St$ is the particle response time; τ_K is the Kolmogorov time scale.

2.4. Kinematic Simulations (KS)

In this work we studied the KS as an SGS model for particle motion with an emphasis on its impact on particle settling velocity and collision-related statistics. Following the idea of Fung et al. [16], a turbulent-like velocity field can be generated as follows:

$$\mathbf{u}_{\text{sgs}}(\mathbf{x},t) = \sum_{n=1}^{N}\sum_{m=1}^{M} \mathbf{a}_{nm} \times \hat{\mathbf{k}}_{nm} \cos(\mathbf{k}_{nm}\cdot\mathbf{x}+\omega_n t) + \mathbf{b}_{nm}\times\hat{\mathbf{k}}_{nm}\sin(\mathbf{k}_{nm}\cdot\mathbf{x}+\omega_n t), \quad (5)$$

where $\hat{\mathbf{k}}_{nm} = \mathbf{k}_{nm}/|\mathbf{k}_n|$. The energy is essentially distributed over N shells of radius $|\mathbf{k}_n|$, each containing M randomly oriented wavevectors. The velocity field is non-divergent, statistically stationary, homogeneous and isotropic. Coefficients $\mathbf{a}_{nm}$ and $\mathbf{b}_{nm}$ are randomly selected vectors with independent and normally distributed components, each with zero mean and variance $\sigma_n^2 = E(k_n)\Delta k_n$, where $E(k_n)$ is a prescribed energy spectrum and $\Delta k_n = (k_{n+1} - k_{n-1})/2$ for $2 \leq n \leq N-1$, $\Delta k_1 = (k_2 - k_1)/2$ and $\Delta k_N = (k_N - k_{N-1})/2$. For a comprehensive description of the aforementioned algorithm the reader is referred to [10]. In this work $N = k_{\max} - k_c$ to match all the missing wavenumbers, i.e. $k_n \in [k_c+1, k_{\max}]$. In contrast to other Authors [10, 11, 13] who usually select M large enough to obtain an acceptable statistical convergence of the reconstructed energy spectrum, we set $M = 1$ to reduce the computational cost of the KS model. The prescribed energy spectrum is retrieved through a proper resampling of $\mathbf{a}_{nm}$ and $\mathbf{b}_{nm}$ coefficients. The reference energy spectrum $E(k)$ was obtained by means of the DNS described in Section 2.1. The frequencies ω_n in equation (5) determine the unsteadiness associated with the wave mode n. For the purpose of the current study, we chose the frequency ω_n to be proportional to the eddy-turnover time of the wave mode n according to $\omega_n = \lambda[k_n^3 E(k_n)]^{1/2}$, with $\lambda = 0.4$. In this work the unsteady KS, equation (5), is referred to as UKS. We also explored a frozen KS, hereinafter described as FKS. In the latter, ω_n is essentially set equal to 0.

2.5. Particle statistics

We studied both the capability of KS to predict the settling velocity enhancement [6] of inertial particles subjected to gravity in the low-pass filtered homogeneous isotropic turbulence and its impact on RDF and RRV of nearly touching particles. The aforementioned statistics were computed according to [13] as:

$$RDF: g_r(r) = \frac{n_p/V_s}{[n(n-1)/2]/V}, \qquad RRV: w_r(r) = \left\langle \left| \left[\mathbf{V}_p^{[2]}(\mathbf{x}_p+\mathbf{r}) - \mathbf{V}_p^{[1]}(\mathbf{x}_p)\right] \cdot \frac{\mathbf{r}}{|\mathbf{r}|} \right| \right\rangle, \quad (6)$$

where n is the total number of particles, V is the fluid volume, n_p is the number of particles within a shell volume V_s at a distance r from a particle of interest, $\mathbf{V}_p^{[1]}$ and $\mathbf{V}_p^{[2]}$ are the velocities of particles located at $\mathbf{x}_p$ and $\mathbf{x}_p + \mathbf{r}$, respectively, $\langle\cdot\rangle$ denotes an ensemble average taken over all particle pairs.

3. Results

In this work we set particles as water droplets, $\rho_p = 1$ g/cm^3, moving in a turbulent atmosphere of $\rho = 0.001$ g/cm^3 and $\nu = 0.17$ cm^2/s. The magnitude of gravitational acceleration g was taken to be equal to 980 cm/s^2 and the average dissipation rate of the turbulent energy of the flow ϵ was considered equal to 1000 cm^2/s^3. As a trade-off between the computational cost and statistical uncertainty we simultaneously tracked a total number of $n = 64^3$ particles.

3.1. Settling velocity enhancement

As a result of the preferential sweeping [6], the settling velocity of small inertial particles subjected to gravity is enhanced when compared to their terminal velocity, $V_T = \tau_p g$, in quiescent fluid. Figure 1 shows the impact of both filtering and different types of SGS velocity fields on the particle average settling velocity $\langle V_s \rangle$. In FDNS, the average settling rate is accurately predicted for particles of the $St \gtrsim 0.6$ and $Fr \gtrsim 3$ as the dynamics of such entities is dominated by gravity rather than small-scale fluctuations. It is therefore justified to state that any additional SGS velocity field seen by such particles will have a minor impact on settling velocity statistics. As expected, if the fluid velocity at the particle locations, equation (4), is enriched with a model-based SGS velocity field, no impact is observed in the aforementioned range of Stokes numbers. The role of gravity decreases with decreasing St and therefore it is expected that the dynamics of smaller particles, say $St \lesssim 0.6$, is mostly driven by large, energy containing scales and, at least to some extent, by the small-scale velocity fluctuations. For smaller particles none of the SGS models being explored was capable of retrieving the settling velocity enhancement observed in DNS. This is at least partly explained by the essential character of a frozen velocity field, i.e. the lack of kinematical and dynamical interactions between the resolved and SGS scales. Figure 2(a) shows the PDF of the second invariant of the velocity gradient tensor Q normalized by its instantaneous standard deviation σ_Q. The Q-criterion is typically used to identify regions of large vorticity $Q > 0$ or strain $Q < 0$. It is readily observed that, as a result of filtering, such regions are less probable in the FDNS velocity field. In contrast to other Authors [13], the FDNS preserves the overall non-Gaussian nature of the PDF observed in DNS, particularly concerning large vorticity and, to a minor extent, large strain. If the filtered velocity field is enriched with both the FFSGS and FKS, only a moderate improvement is observed. Due to the lack of both the sweeping effect (i.e. the advection of small scales by larger ones) and non-linear interactions between the SGS scales, the FKS generates a Gaussian-like velocity field when compared to FFSGS, Figure 2(b). The *a priori* LES enriched with the aforementioned frozen SGS velocity fields seems to be incapable of retrieving the instantaneous flow structures observed in the DNS.

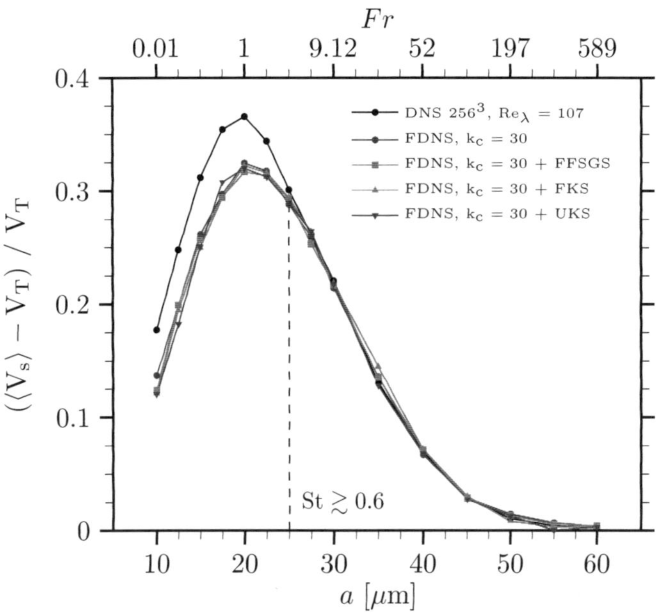

Figure 1: The impact of filtering and subsequent SGS enhancements on particle settling velocity; a is the particle radius.

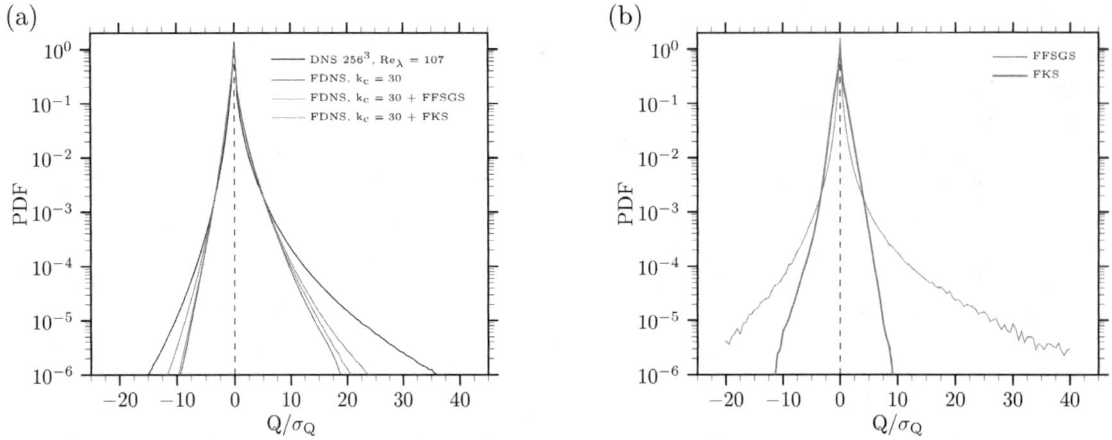

Figure 2: PDF of the normalized second invariant of the velocity gradient tensor: (a) full velocity field – the impact of filtering and subsequent SGS enhancements, (b) pure SGS velocity field.

3.2. Radial distribution function at contact

Figures 3(a) and 3(b) show the RDF at contact in the presence or absence of the gravitational settling, respectively. It is readily observed that *a priori* LES sufficiently well predicts RDF for $St \gtrsim 1.2$ when the particles are subjected to gravity and $St \gtrsim 1.6$ without any downward acceleration. In both cases the RDF is decreased when compared to FDNS, indicating enhanced mixing of particles. This can be at least partly explained by an infinite lifetime of the frozen velocity field structures resulting in an extended interaction between particles and subgrid-scales.

3.3. Radial relative velocity at contact

Figures 4(a) and 4(b) show the radial relative velocity at contact normalized with the Kolmogorov velocity scale v_K in the presence or absence of the gravitational settling, respectively. Without any downward acceleration *a priori* LES accurately predicts the RRV at contact for

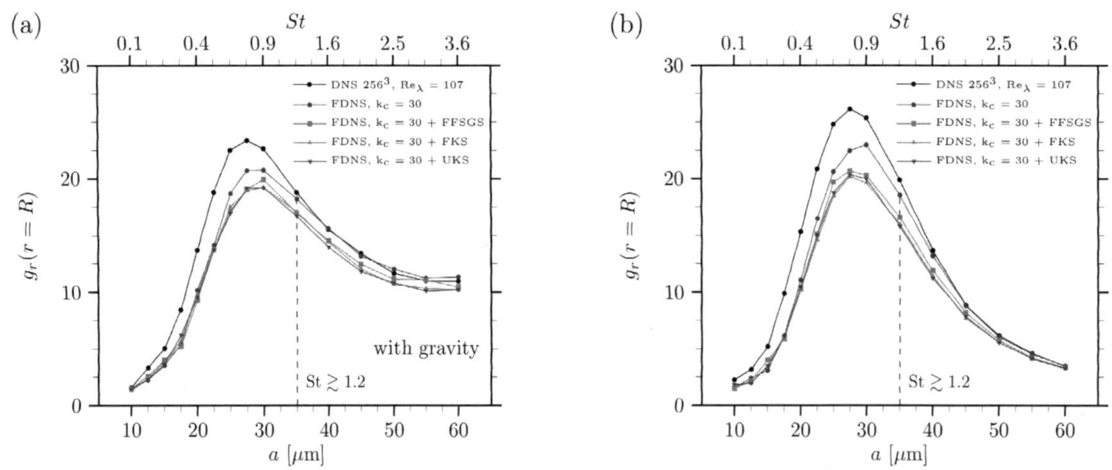

Figure 3: The impact of filtering and subsequent SGS enhancements on the radial distribution function at contact ($r = R = 2a$): (a) with gravity, (b) without gravity; a is the particle radius.

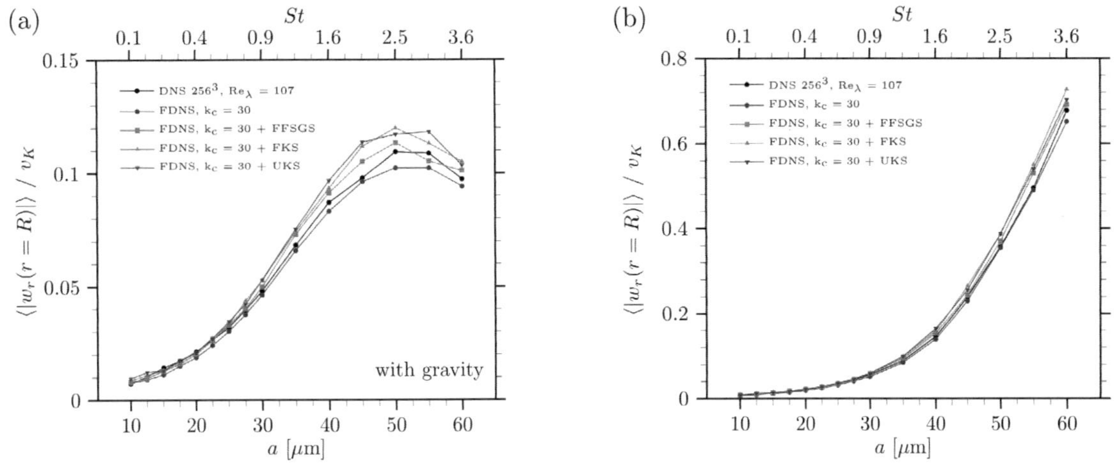

Figure 4: The impact of filtering and subsequent SGS enhancements on the radial relative velocity at contact ($r = R = 2a$): (a) with gravity, (b) without gravity; a is the particle radius.

all St being explored. On the other hand, however, the RRV is slightly underestimated when the particles are subjected to gravity. If the fluid velocity seen by particles was enriched with any model-based SGS fluctuations, the RRV at contact was, in contrast, slightly overestimated. These results are consistent with the findings discussed in the previous section.

3.4. Kinematic collision kernel

The collision rate of particles can be evaluated through the kinematic collision kernel defined as:

$$\Gamma^K = 2\pi R^2 w_r(r = R) g_r(r = R), \tag{7}$$

where $R = d_p = 2a$ is the collision radius in a monodispersed system. Figures 5(a) and 5(b) show the kinematic collision kernel in the presence or absence of gravitational settling, respectively. In *a priori* LES the errors in predicting RDF and RRV at contact partly cancel each other, resulting in sufficiently well predicted collision rates when compared to the reference DNS.

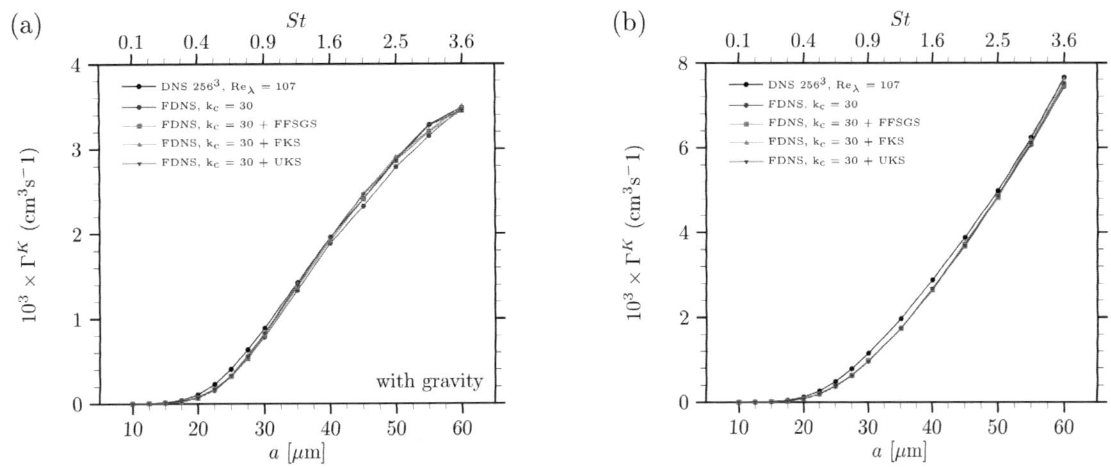

Figure 5: The impact of filtering and subsequent SGS enhancements on the collision frequency: (a) with gravity, (b) without gravity; a is the particle radius.

4. Conclusions

We studied the impact of kinematic simulations on particle settling velocity and collision-related statistics in *a priori* LES of homogeneous isotropic turbulence. We particularly considered the RDF and RRV at contact with both the presence or absence of the gravitational settling.

Due to the essential drawbacks of a frozen velocity field, i.e. the lack of the sweeping effect and SGS non-linear interactions, KS is incapable of retrieving the instantaneous flow structures observed in DNS, and therefore cannot properly enrich particle motions in *a priori* LES.

In the present case it seems that *a priori* LES without any explicit enhancement in the particle equation of motion sufficiently well predicts collision rate of particles for a wide range of Stokes numbers obviating the need for any "second-stage" SGS modeling. Settling velocity enhancement is accurately predicted only for $St \gtrsim 0.6$ as the dynamics of such particles is mostly affected by gravity and large-scale motions. In the case of $St \lesssim 0.6$, for which the role of gravity is progressively less important, the settling velocity is diminished due to the lack of SGS motions. However, a greater caution in drawing conclusions is required when analyzing *a priori* LES results as the "first-stage" SGS modeling is still insufficient in the case of *a posteriori* LES.

Acknowledgments

The work was supported by the National Science Centre (NCN, Poland), research project 2018/30/Q/ST8/00341, the Interdisciplinary Centre for Mathematical and Computational Modelling (ICM), grant G87-1145 and statutory activities of IMP PAN Gdańsk.

References

[1] Grabowski W W and Wang L P 2013 Growth of cloud droplets in a turbulent environment *Annu. Rev. Fluid Mech.* **45** 293–324
[2] Kuerten J G M 2016 Point-particle DNS and LES of particle-laden turbulent flow – a state-of-the-art review *Flow Turbul. Combust.* **97** 689–713
[3] Elghobashi S 2019 Direct numerical simulation of turbulent flows laden with droplets or bubbles *Annu. Rev. Fluid Mech.* **51** 217–44
[4] Marchioli C 2017 Large-eddy simulation of turbulent dispersed flows: a review of modelling approaches *Acta Mech.* **228** 741–71
[5] Squires K D and Eaton J K 1991 Preferential concentration of particles by turbulence *Phys. Fluids A* **3** 1169–78
[6] Wang L P and Maxey M R 1993 Settling velocity and concentration distribution of heavy particles in homogeneous isotropic turbulence *J. Fluid Mech.* **256** 27–68
[7] Fessler J R, Kulick J D and Eaton J K 1994 Preferential concentration of heavy particles in a turbulent channel flow *Phys. Fluids* **6** 3742–49
[8] Sundaram S and Collins L R 1997 Collision statistics in an isotropic particle-laden turbulent suspension. Part 1. Direct numerical simulations *J. Fluid Mech.* **335** 75–109
[9] Ray B and Collins L R 2011 Preferential concentration and relative velocity statistics of inertial particles in Navier-Stokes turbulence with and without filtering *J. Fluid Mech.* **680** 488–510
[10] Ray B and Collins L R 2014 A subgrid model for clustering of high-inertia particles in large-eddy simulations of turbulence *J. Turbul.* **15** 366–85
[11] Zhou Z, Wang S and Jin G 2018 A structural subgrid-scale model for relative dispersion in large-eddy simulation of isotropic turbulent flows by coupling kinematic simulation with approximate deconvolution method *Phys. Fluids* **30** 105110
[12] Chollet J P and Lesieur M 1981 Parameterization of small scales of three-dimensional isotropic turbulence utilizing spectral closure *J. Atmos. Sci.* **38** 2747–57
[13] Zhou Z, Wang S and Jin G 2020 A structural subgrid-scale model for the collision-related statistics of inertial particles in large-eddy simulations of isotropic turbulent flows *Phys. Fluids* **32** 095103
[14] Eswaran V and Pope S 1988 An examination of forcing in direct numerical simulations of turbulence *Comput. Fluids* **16** 257–78
[15] Orszag S A and Patterson G S 1972 Numerical simulation of three-dimensional homogeneous isotropic turbulence *Phys. Rev. Lett.* **28** 76–9
[16] Fung J C H, Hunt J C R, Malik N A and Perkins R J 1992 Kinematic simulation of homogeneous turbulence by unsteady random Fourier modes *J. Fluid Mech.* **236** 281–318

Journal of Physics: Conference Series **2367** (2022) 012030 doi:10.1088/1742-6596/2367/1/012030

Simulation of sloshing in the moon pool with Smoothed Particle Hydrodynamics

E Spricigo[1,2] and J Pozorski[1]

[1]Institute of Fluid-Flow Machinery, Polish Academy of Sciences
Fiszera 14, 80-231 Gdańsk, Poland
[2]ESTECO SpA, Padriciano 99, 34149 Trieste, Italy

E-mail: `espricigo@imp.gda.pl`, `jp@imp.gda.pl`

Abstract. In maritime heavy industry, the moon pool is an opening through the hull of some vessels. We replicate a generic design of the pool with the Lagrangian approach Smoothed Particle Hydrodynamics (SPH) whose mathematical model efficiently simulates the interphasial surface. In this study we present the capability of our in-house SPH code to handle the sloshing in the moon pool. We validate the SPH approach to flows for the sloshing tank benchmark and show the stability of this approach for the non-trivial case of the moon pool.

1. Introduction

Knowing that the multiphase flows characterize many industrial processes and everyday life, we are motivated to study these phenomena to achieve a better understanding and current technology improvements. In nowadays physics and engineering the numerical simulations facilitate this work to designers and technicians.

Among the many existing CFD approaches, we have chosen the Weakly Compressible SPH (WCSPH). It is a Lagrangian method where the discretization is particle-based: the particles feature the fluid properties and move inside the computational domain, as the fluid does. The implementation of each particle interaction with its neighbors is done using an interpolating function, called kernel. Therefore, the evolution of interphasial surfaces comes out naturally in SPH and no special treatment (such as interface tracking, capturing, etc.) is needed.

After having tested a number of benchmarks in our in-house code [1, 2], we moved to an industrial application: the moon pool. This case was suggested by ESTECO SpA, the optimization software provider located in Trieste (Italy), which is a partner of the present research. Maritime studies often focus on surface wave phenomena, interaction between solid body and fluid, rotational and translational motion of the ship. Experiments on ship dynamics are quite complicated and usually require the design of a scaled model. Especially for this reason, the CFD is widely used for research in marine engineering. The most frequent approach for the Eulerian moon pool simulations is the Volume of Fluid (VOF) [3, 4, 5, 6], a surface-capturing model. Since SPH is particularly efficient for violent impact flows, for macro-scale and multiphase problems, we think it might be a viable alternative to VOF in selected applications.

This work aims to prove the functionalities of the SPH in-house code reproducing the sloshing phenomenon. The next section provides SPH basics, in the third section the reader will find the description of the sloshing tank benchmark and the moon pool case, the results are shown in the fourth section. The last section is reserved to conclusions and future work.

2. The Smoothed Particle Hydrodynamics approach

In 1977 Monaghan and Gingold [7] and Lucy [8] formulated a new approach for astrophysics in which the discretization was Lagrangian: the mesh was replaced by particles. About twenty years ago SPH has been adopted for fluid dynamics including multiphase flows [9]; interestingly, the basics of this new CFD approach continue to be a matter of research [10, 11, 12, 13].

In SPH the motion of the particle i^{th} is based on the interaction with its neighbors (a generic particle j^{th}), which mathematically translates in the kernel function. One can think about the kernel as a bell with i^{th} in the center; the farther the neighbouring particles are from the center, the less is their influence. This concept is applicable to any particle property.

2.1. SPH approximation

Theoretically, the continuous form approximation of a scalar (or vector) field A at the position $\mathbf{r}$ exploits the definition of Dirac delta function δ

$$A(\mathbf{r}) = \int_V A(\mathbf{r}')\delta(\mathbf{r} - \mathbf{r}')d\mathbf{r}' \tag{1}$$

where V is the volume considered. In SPH the Dirac δ is replaced by the kernel function $W(\mathbf{r}, h)$

$$A(\mathbf{r}) \simeq \int_V A(\mathbf{r}')W(\mathbf{r} - \mathbf{r}', h)d\mathbf{r}' \tag{2}$$

that is dependent on the particles relative position $\mathbf{r} - \mathbf{r}'$ and its size is proportional to h. The parameter h is the *smoothing length* and determines the range where the interpolation is possible, meaning $W(\mathbf{r}, h) \equiv 0$ for $r \geq h$. The kernel should satisfy three major conditions: it should be an even function with compact support (i.e. convergent to $\delta(\mathbf{r})$ with $h \to 0$), and normalised

$$\int_\Omega W(\mathbf{r}, h)d\mathbf{r} = 1. \tag{3}$$

In this work the quintic Wendland kernel function is employed.

The discrete forms of equation (2) and the corresponding derivative are

$$A(\mathbf{r}) \simeq \sum_j A(\mathbf{r}_j)W(\mathbf{r}_i - \mathbf{r}_j, h)V_j \tag{4}$$

$$\nabla A(\mathbf{r}) \simeq \sum_j A(\mathbf{r}_j)\nabla W(\mathbf{r}_i - \mathbf{r}_j, h)V_j. \tag{5}$$

2.2. Governing equations

The governing equations for viscous fluid dynamics in the Lagrangian approach are

$$\frac{d\rho}{dt} = -\rho\nabla \cdot \mathbf{u} \tag{6}$$

$$\frac{d\mathbf{u}}{dt} = -\frac{1}{\rho}\nabla p + \frac{1}{\rho}\nabla \cdot (\mu\nabla\mathbf{u} + \mu\nabla\mathbf{u}^{\mathrm{T}}) + \frac{1}{\rho}\mathbf{f}_{st} + \mathbf{f}_b \tag{7}$$

$$\frac{d\mathbf{r}}{dt} = \mathbf{u}. \tag{8}$$

The continuity equation (6) is approximated in SPH as [14]

$$\langle\rho\rangle_i = m_i \sum_j W(\mathbf{r}_i - \mathbf{r}_j, h) = m_i\theta_i \tag{9}$$

where θ_i is the measure of particle number density, the inverse of particle volume.

In the momentum equation (7) the surface tension $\mathbf{f}_{st}$ is modelled with the Continuum Surface Force method (CSF), $\mathbf{f}_b$ refers to body forces and the pressure gradient [14] reads

$$\left\langle\frac{\nabla p}{\rho}\right\rangle_i = \frac{1}{m_i}\sum_j \left(\frac{p_i}{\theta_i^2} + \frac{p_j}{\theta_j^2}\right)\nabla W_{ij}. \tag{10}$$

The viscous force takes into account the variability of viscosity on the interface and it is based on the work by [11]

$$\left\langle\frac{\nabla\cdot\tau}{\varrho}\right\rangle_i = \frac{1}{\rho_i}\sum_j \left[V_j\frac{\overline{\mu}_{ij}}{r_{ij}}(n+2)(\mathbf{u}_{ij}\cdot\mathbf{e}_{ij})\mathbf{e}_{ij}\frac{\partial W}{\partial r_{ij}} + V_j\frac{\overline{\mu}_{ij}}{r_{ij}}\mathbf{u}_{ij}\frac{\partial W}{\partial r_{ij}}\right]; \tag{11}$$

precisely, $\mathbf{u}_{ij} = \mathbf{u}_i - \mathbf{u}_j$ is the relative velocity and $\mathbf{r}_{ij} = \mathbf{r}_i - \mathbf{r}_j$ the relative position, $\mathbf{e}_{ij} = \mathbf{r}_{ij}/|\mathbf{r}_{ij}|$ is the unit vector and $\overline{\mu}_{ab} = (2\mu_a\mu_b)/(\mu_a + \mu_b)$ is the inter-particle dynamic viscosity.

Since we are using the WCSPH approach, the pressure closure equation is the linear equation of state

$$p = s^2(\rho - \rho_0) \tag{12}$$

where s is the artificial speed of sound and ρ_0 is the reference density.

The mild and periodic motion of the cases presented in this work does not require a turbulence model.

3. Sloshing tank benchmark and moon pool case

In this section the models to validate the sloshing in the tank and the surface wave dynamics in the moon pool case are presented.

3.1. Sloshing tank benchmark

To validate our code for sloshing we decided to reproduce the case detailed in [15], but instead of running a free-surface simulation (liquid only) we arrange the case to be multiphase. Hence, a 1 m × 0.5 m closed tank is designed (see figure 1a), the water phase is 0.1 m deep and the rest is the air phase. On the fluid acts a rectilinear sinusoidal force

$$\mathrm{f}_{b,x}(t) = a_0(t) = A(2\pi f_w)^2\sin(2\pi f_w t) \tag{13}$$

where the displacement amplitude is $A = 0.5\,\mathrm{m}$, the excitation frequency is $f_w = 0.3 f_0$ with $f_{(m-1)} = \omega_m/(2\pi)$. Considering the first mode $m = 1$, the natural frequencies f_0 depend on

$$\omega_m = \sqrt{g\frac{\pi m}{L}\tanh\left(\frac{\pi m}{L}H\right)} \tag{14}$$

where g is the acceleration of gravity, L is the length of the tank and H is the water level.

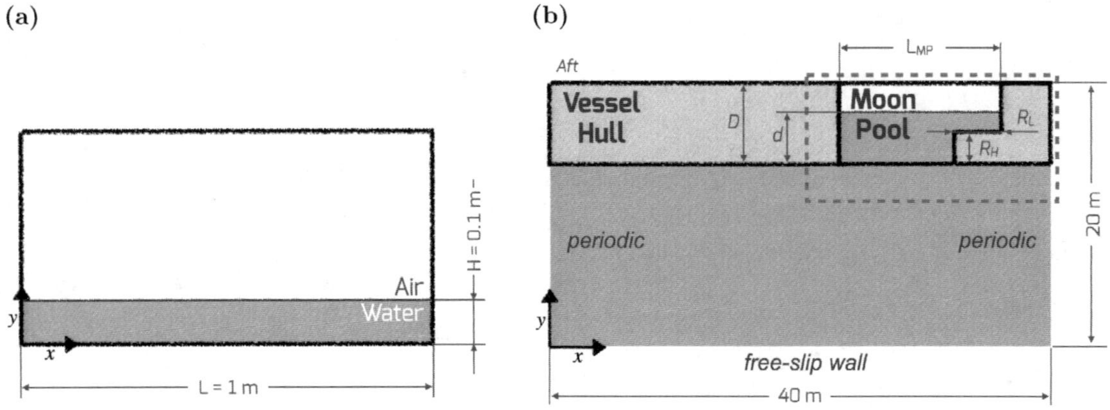

Figure 1: Models of: (a) the benchmark tank for sloshing validation; (b) design of the moon pool; the dashed red contour refers to the close-up whose time frames are shown in figure 3.

3.2. The moon pool case

The moon pool is an opening through the hull of vessels, it allows underwater activity from on-board. It is typical of drillships (e.g., Samsung Heavy Industries SHI ships), offshore platforms, semisubmersible vessels, research vessels (e.g., the RRS *Sir David Attenborough*, the RSV *Nuyina*, etc.), barges and other.

Our geometry is proportionally based on the model provided by ESTECO, but about one third of the size of a real drillship (e.g., 96 k-type deep water SHI drillship). Figure 1b shows our model: the light blue is the sea water while the air phase is white, the grey part represents the vessel hull, in the middle of which the moon pool is set. The size of the simulation domain is $20\,\mathrm{m} \times 40\,\mathrm{m}$; the hull depth is $D = 6.4\,\mathrm{m}$, the draft is $d = 4.1\,\mathrm{m}$; the moon pool total length on the deck is $L_{MP} = 13\,\mathrm{m}$, its recess height and length are respectively $R_H = 2.9\,\mathrm{m}$ and $R_L = 3.4\,\mathrm{m}$.

On the top of the pool and on the bottom of the domain is applied the free-slip condition, while on the sides the periodicity. In our code we use *ghost particles* for boundaries: the ghosts are not real particles but at each time step, whenever a real particle gets close to the boundary, it is mirrored on the other side of the boundary so the ghost is generated. Although this approach is not suitable for complex geometries, it is low-cost in computational time and easy to set up. Due to the fact that complex boundaries made of ghosts are not straightforward to design, the vessel hull is composed of immobile SPH particles (called dummies, not advected).

Beside the many difficulties found in building the geometry and setting the boundary conditions, the SPH moon pool brings some more issues. Thanks to the large scale model, we can ignore the surface tension force contribution, but the presence of gravity can increase the acoustic noise in the pressure field. Accounting for the hydrostatic pressure gradient in the initial condition decreases the spurious pressure fluctuations at the beginning of the simulation.

The moon pool presence affects the stability of the vessel, both when it is sailing and when it is not moving but subjected to sea waves. The vessels at sea have six degrees of freedom, the water inside the moon pool can have *piston motion*, vertical rise of water typical in square-shaped pool, and *sloshing motion*, horizontal oscillations of the water between the sides of the pool. In rectangular-shaped pool these two effects are mixed. Plus other problems appear in heavy sea, like *green water*, meaning large quantity of water overtopping the deck. We decided to focus more on the sloshing, hence we simplify the problem making it two-dimensional and considering a captive vessel, with no degrees of freedom.

Journal of Physics: Conference Series **2367** (2022) 012030 doi:10.1088/1742-6596/2367/1/012030

The horizontal force that acts on the fluid is the same as in equation (13) and reproduces a regular wave. For this case the displacement is $A = 33.33\,\mathrm{m}$, the frequency is $f_w = 0.4 f_0$ and the resulting natural mode is $\omega_1 = 0.779\,\mathrm{s}^{-1}$.

4. Results

4.1. Sloshing tank benchmark

The sloshing tank simulation has been performed with $N = 32768$ particles, out of which about one fifth represent the water phase, and the mean interparticle distance is $\Delta r \simeq 0.0039\,\mathrm{m}$. The results obtained from the simulation have been compared with the analytical solution [16] for the wave height $\zeta(x,t)$ and the sloshing force $\mathrm{f}_x(t)$:

$$\zeta = \sum_m^{\infty} \beta_m(t) \cos\left(\frac{m\pi}{L}x\right) \tag{15}$$

$$\mathrm{f}_x(t) = \rho H L a_0(t) + \rho \frac{L^2}{\pi^2} \sum_m^{\infty} \ddot{\beta}_m(t) \frac{1 + (-1)^{m+1}}{m^2}. \tag{16}$$

The value of $\beta_m(t)$ and its second derivative $\ddot{\beta}_m(t)$ come from solving the following ordinary differential equation using the fourth-order Runge-Kutta scheme

$$\ddot{\beta}_m(t) + \omega_m \beta_m(t) = -a_0 \left(\frac{2}{\pi m}\right) \tanh\left(\frac{\pi m}{L} H\right) [(-1)^m - 1]. \tag{17}$$

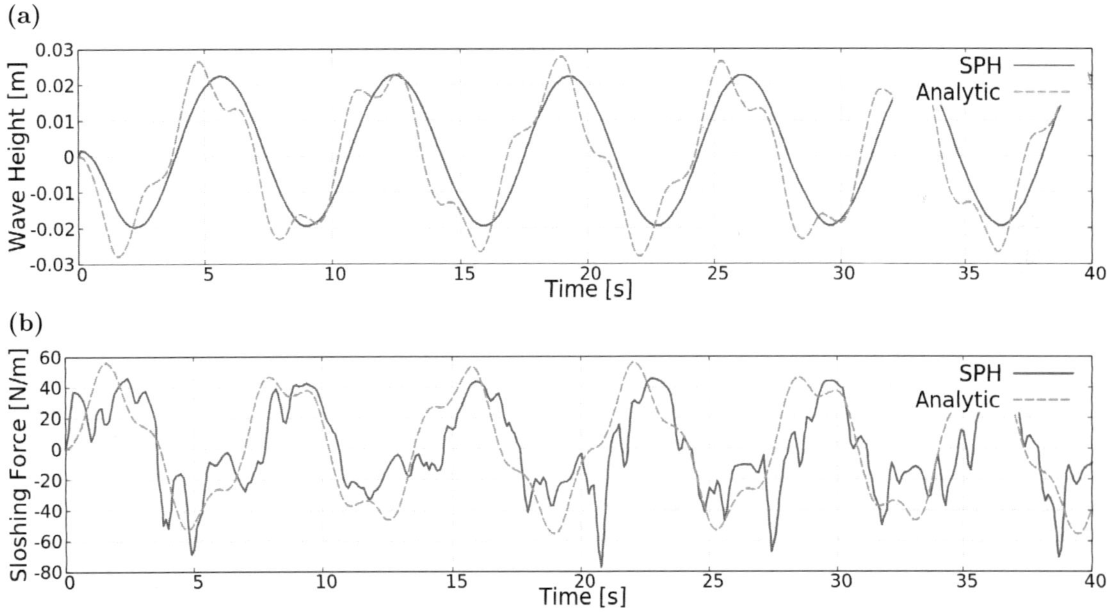

Figure 2: Sloshing tank benchmark, comparison between SPH results in blue and analytic formulation [16] in yellow: (a) wave height on the left wall of the tank, (b) sloshing force on the left wall of the tank.

In figure 2a the wave height, in figure 2b the sloshing force in a 40-second simulation are shown. The SPH behavior is close to the analytical one although the lines are not exactly overlapping; we supposed that a higher resolution could give a better matching. The spurious fluctuations visible in the sloshing force are a direct consequence of the oscillations in pressure, a well known problem of weakly compressible approaches. These inconsistencies are hardly removable, but recently some correction techniques have been proposed, for example the so-called δ−SPH [17].

4.2. The moon pool case

The moon pool simulation has been performed with $N = 131072$ particles and the mean interparticle distance is $\Delta r \simeq 0.0781\,\mathrm{m}$.

The sequence of pictures in figure 3 corresponds to the close-up indicated by the dashed red line in the the moon pool, figure 1b. These pictures are a set of time frames, from 11.0 s to 14.5 s, in quasi-periodic regime. The water phase is coloured in blue, the air phase is represented by particles coloured in different shades of red, based on their x-position at initial time. This choice was made with the aim to illustrate how SPH really works: the particles are free to move in the domain and they keep approximately uniform interparticle distances between each other during the simulation. In general, when the particle distribution is not uniform any more, the particles gather in small groups and soon after the simulation will fail. When it appears, a possible reason could be the mathematical model itself or the time settings. In our moon pool simulation, the particles distribution maintains its order; this translates in stable and long-running simulation.

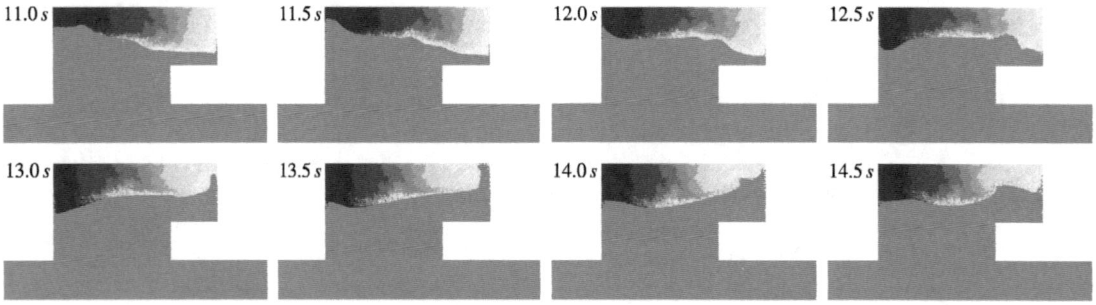

Figure 3: The close-ups referred to in figure 1b. Time frames of the sloshing phenomenon from 11.0 s to 14.5 s are included. Light blue particles for water phase; reddish for air phase where the shade of red codes the particle initial positions.

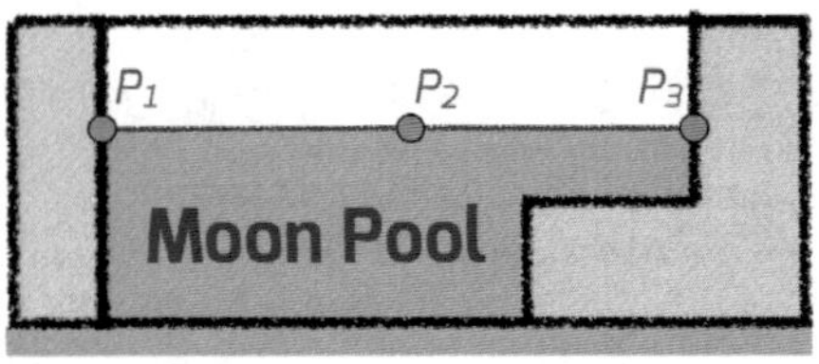

Figure 4: Probes inside the moon pool: they detect the interface (free-surface) position during the simulation.

To detect the water level inside the pool we set three probes upon the air-water interface. The probe P_1 is close to the left wall of the pool, the second probe P_2 is in the center while P_3 is on the right wall (see figure 4). The sloshing effect is evident in figure 5: after approximately about 5 s transient the three elevation records reach the steady-state regime. P_1 and P_3 have inverse crests and troughs, which is obvious because the wave travels in one direction and then moves in the opposite way. P_2 has less evident oscillatory mode. The maximum water elevation is reached in P_3 and is about 2 m, which is due to the presence of the recess. The last point leads us thinking that such a geometry could not be the best one for damping the waves and further research is warranted. We also find with a FFT that the dominant frequency inside the pool is $f_p \simeq 0.2\,\mathrm{s}^{-1}$ (see figure 6), about 75% higher than the external excitation frequency f_w.

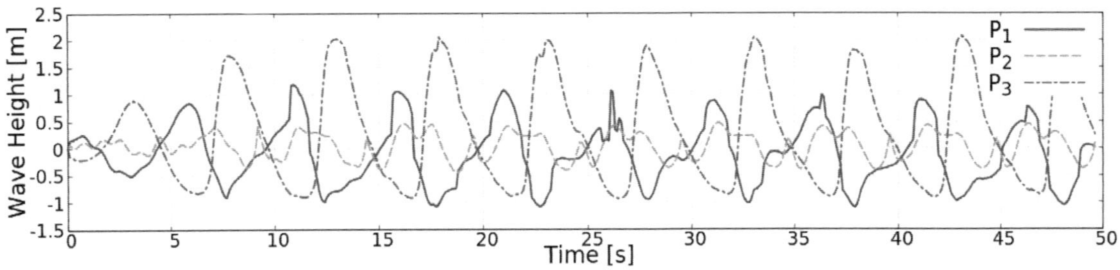

Figure 5: Wave height inside the moon pool: P_1 in purple is the left probe, P_2 is the center probe, P_3 is the right probe.

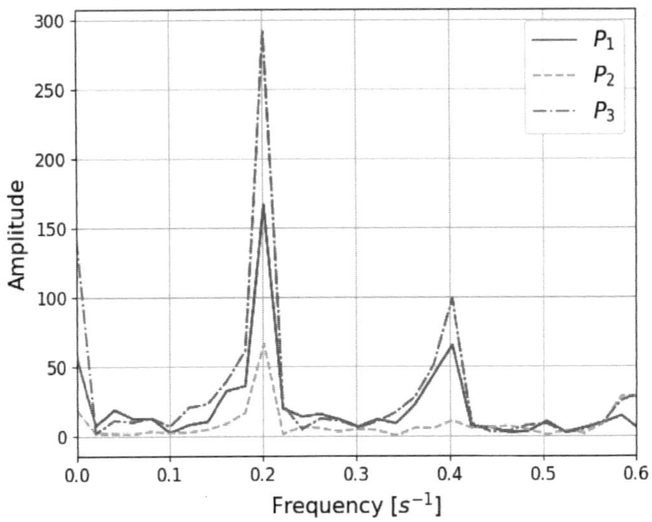

Figure 6: The Fourier transform of the wave amplitude of the three probes; the dominant frequency inside the pool $f_p \simeq 0.2\,\mathrm{s}^{-1}$.

5. Conclusions

In this work we addressed the problem of sloshing due to regular wave entering a complex geometry as the moon pool is.

We first validated our code for the sloshing effect by reproducing the rectangular tank benchmark of [15]. Although the SPH results do not match perfectly the analytical solution, the wavy trend is consistent. We think we could match better the reference data by increasing the resolution, but in the current multiphase setting it would be excessively time-costly. Indeed, we are already working on the free-surface simulation of this benchmark, which could allow us to simulate exclusively on the water phase neglecting the air phase.

Secondly, we studied the sloshing in the moon pool. In CFD, this case is often treated in three dimensions. We ran a simplified version of this complex case in our in-house code: two-dimensional captive vessel in periodic wavy sea. After a short period of transitional regime the flow reaches a steady-state, which is confirmed by the quasi-periodic oscillations detected by the three probes located inside the pool.

As the next-term work, we intend to validate the capability of the SPH approach by comparing our moon pool model to what has been studied in the literature. Of particular interest are experimental results that could confirm the work that we have done so far. Furthermore, we want to attempt the design optimisation of the moon pool using the software provided by ESTECO called modeFRONTIER [18]. This tool can be coupled with many codes and commercial software, automates the optimization of parameters and allows for a systematic analysis of the outputs.

Acknowledgments
The study has been carried out in the framework ITN-EID project COMETE No. 813948, funded by EU H2020 Maria Skłodowska-Curie Actions.

Appendix
Our in-house code was firstly developed by Dr. K. Szewc until 2016 and further extended in our group. The code has been written in C++ language and parallelized with NVIDIA® CUDA® .

All the SPH simulations presented in these paper ran on NVIDIA GPU GeForce® RTX 2080 Ti. In these conditions, an SPH simulation is about five times faster than an Ansys® Fluent® simulation with comparable resolution, computed in a Windows 10 Pro with Intel® Core™ i5-3570 CPu @ 3.40 GHz, 12 GB RAM.

References
[1] Szewc K, Tanière A, Pozorski J and Minier J-P 2012 A study on application of smoothed particle hydrodynamics to multi-phase flows *Int. J. Nonlin. Sci. Num. Simul.* **13** 383
[2] Olejnik M and Pozorski J 2020 A robust method for wetting phenomena within Smoothed Particle Hydrodynamics *Flow, Turb. Combust.* **104** 115–137
[3] Son H-J, Choi S-H, Kim M-H, Hwangbo S-M 2008 Drag Reduction of Recess Type Moonpool Under Vessel's Forward Speed *Proc. of the ASME 27th Int. Conf. on Offshore Mechanics and Arctic Engineering* (June 15-20, Estoril, Portugal) **1** 143
[4] Hammargren E and Törnblom J 2012 Effect of the Moonpool on the Total Resistance of a Drillship *MSc thesis* Chalmers University, Gothenburg, Sweden
[5] Guo X, Lu H, Yang J and Peng T 2017 Resonant water motions within a recessing type moonpool in a drilling vessel *Ocean Eng.* **129** 228
[6] Machado L 2020 Moonpool Dimensions and Position Optimization with Genetic Algorithm of a Drillship in Random Seas *PhD thesis* UFRJ/COPPE, Rio de Janeiro, Brasil
[7] Monaghan JJ and Gingold RA 1977 Smoothed particle hydrodynamics: theory and application to non-spherical stars *Monthly Notices of the Royal Astronomical Society* **181** 375
[8] Lucy LB 1977 A numerical approach to the testing of the fission hypothesis *The Astronomical Journal* **82** 1013
[9] Monaghan JJ 1992 Smoothed Particle Hydrodynamics *Annual Review of Astronomy and Astrophysics* **30** 543
[10] Price DJ 2011 Smoothed Particle Hydrodynamics: Things I wish my mother taught me *Adv. Comput. Astrophys.* Preprint arXiv:1111.1259v1
[11] Violeau D 2012 *Fluid Mechanics and the SPH Method: Theory and Applications* (Oxford, UK: Oxford University Press)
[12] Vacondio R, Altomare C, De Leffe M, Hu, XY, Le Touzé D, Lind S, Marongiu JC, Marrone S, Rogers BD and Souto-Iglesias A 2021 Grand challenges for Smoothed Particle Hydrodynamics numerical schemes *Comput. Part. Mech.* **8** 575
[13] Kajzer A 2022 Conservative weakly compressible smoothed particle hydrodynamics applied to flows with high vorticity *J. Phys.: Conf. Series* (this volume)
[14] Hu X and Adams N 2006 A multi-phase SPH method for macroscopic and mesoscopic flows *J. Comput. Phys.* **213** 844
[15] Green MD and Peiró J 2018 Long duration SPH simulations of sloshing in tanks with a low fill ratio and high stretching *Comput. Fluids* **174** 179
[16] Faltinsen O M and Timokha AN 2009 *Sloshing* (Cambridge, UK: Cambridge University Press)
[17] Molteni D and Colagrossi A 2009 A simple procedure to improve the pressure evaluation in hydrodynamic context using the SPH *Comput. Phys. Commun.* **180** 861
[18] Russo R, Clarich A, Poloni C and Nobile E 2015 Mission and Shape Optimization using modeFRONTIER: Application to Boomerang throw *ERCOFTAC Bulletin* **102** 23

Journal of Physics: Conference Series **2367** (2022) 012031 doi:10.1088/1742-6596/2367/1/012031

Analysis of the moist air transonic flow in the symmetric and asymmetric nozzle of the low expansion

P Wiśniewski[1], S Dykas[1], M Majkut[1] and K Smołka[1]

[1] Silesian University of Technology, Department of Power Engineering and Turbomachinery, Gliwice, Poland

E-mail: piotr.wisniewski@polsl.pl

Abstract. Condensation has a negative impact on turbomachinery efficiency in many energy processes. This paper compares the expansion of moist air in symmetric and asymmetric nozzles with a low expansion rate. The presented numerical study is supported by analytical and experimental research. The experimental tWesting was carried out using an in-house experimental test rig where two nozzles were examined for moist air with relative humidity of about 25%. The nozzles were tested for a supersonic outlet, as well as for the outlet condition with elevated pressure, which resulted in the occurrence of a normal shock in the test section. The asymmetric nozzle profile is congruent with the symmetric nozzle. However, the nozzle walls are shifted linearly in the flow direction. The Schlieren photography technique and static pressure measurements on the nozzle wall were used for qualitative identification of both condensation and shock waves. The presented numerical modelling was conducted using commercial computational fluid dynamics software extended with an in-house condensation model. The code was validated against in-house experiment as well as against the data available in the literature. The analysis of the flow in the considered two types of nozzles with a very low expansion rate revealed very interesting structures of pressure waves. The impact of the linear shift of the nozzle walls on the condensation process and the interaction of the condensation wave with aerodynamic oblique as well as normal shock waves were investigated.

1. Introduction

The efficiency of expansion in machines and devices is of major importance in many branches of industry, especially when the working fluid participate in the generation of mechanical work or heat. Customary, it is assumed that a 1 percentage drop in machinery efficiency results in a similar drop in the whole system efficiency. The analysed herein problem is relevant as the most commonly used working fluid in power generation is atmospheric air. In many applications there is no physical possibility to dry the working fluid at the inlet to the machinery [1], thus accounting the impact of phase transition on the flow phenomena seems necessary. Atmospheric air always contains water in the form of steam, liquid droplets or ice crystals, which in certain condition may lead to phase change phenomena. It is especially relevant in expanding flow if the supersonic velocities are reached owing the creation of oblique or normal shock waves, which interact with the condensation waves. This problem has been considered by many researchers. The progress in atmospheric air transonic flows research was possible due to the development of experimental techniques. The triger for this research was a Schlieren picture presented by Prandtl in 1935 depicting complex shock structure in a supersonic flow through converging-divergin nozzle. The two crossing oblique shock waves were assumed to appear due to the presence of initial air humiditiy. The depicted X-shaped shock structure was called a condensation shock. Since then, scientists has sterted focusing on the phase-change phenomena occurring in atmospheric air flow. The numerical study were tigerred by Schnerr in 1990s, who confirmed the X-shaped shock to be condensation wave by means of in-house CFD code [1, 2, 3]. Despite the fact this phenomena has been focus of scientific research for over 80 years, it is still topical as it concerns bot internal [3, 4] and external flows [1, 5]. The condensation phenomena in internal and external transonic humid air flows have been the focus of much research, with studies initiated over 30 years ago and continuing to present times [3]. The studies conducted by Schnerr et al. [1, 2, 6] became a base for further experimental and numerical studies. The non-stationarity of the flow was studied by Adam, who investigated the symmetrical and asymmetrical fluctuation induced by the interaction of condensation and shock waves in a parallel nozzle, i.e. nozzle which walls downstream of the throat are parallel [7]. An in-house code was developed by Dykas et al. [8, 9, 10, 11], who studied the condensation in nozzles with high expansion rate [12, 13]. This code became the basis of the model implemented to commercial software ANSYS Fluent and is presented in this study [14, 15, 16, 17].

2. Experimental study

The subject of this research are two linear convergent-divergent nozzles designed according to Witoszyński's curve [18] with a resulting expansion rate of 1000 s^{-1}. The nozzle differs from one another due to its symmetry, the nozzle W1 is symmetrical, whereas nozzle W2 has the top wall shifted linearly by 20 mm downstream the flow (see red line on Figure 1). The physical phenomena occurring in these nozzles are considered to be 2D. In Figure 1 one can see the shape of the W1 nozzle and the main dimensions, the nozzle width is 20 mm. The nozzle geometry, i.e. the top wall and bottom wall with the pressure holes, was manufactured by means of the computerized numerical control (CNC) machining from aluminium. For the lateral walls, glass windows were applied. Everything was assembled on a frame printed of PLA filament by employing the 3D printing technique.

(a)

(b)

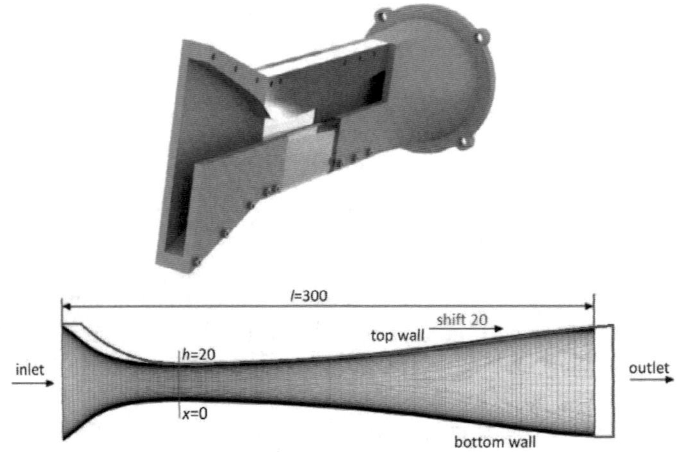

Figure 1. Experimental nozzle geometry (a) and numerical domain with mesh (b).

Figure 2 shows the scheme of the test rig available in the Department of Power Engineering and Turbomachinery (DPET) of the Silesian University of Technology (SUT) on which the experimental studies were performed. Two measurement techniques were employed here, namely the static pressure measurement on the nozzle bottom wall and schlieren photography for the flow field visualisation. These two measurement techniques have been used by us for many years and are detailed described in our previous works [19, 20].

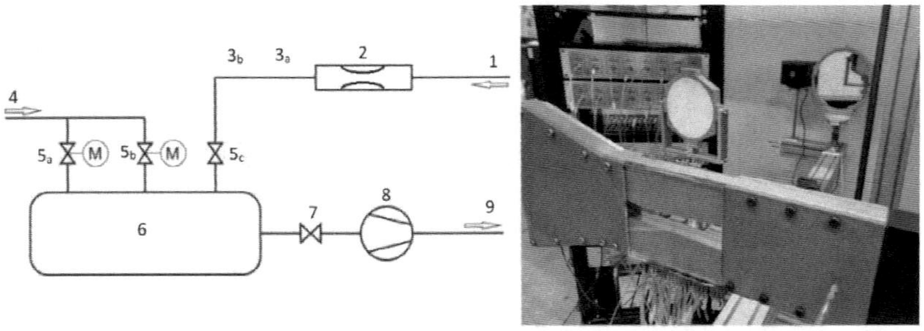

Figure 2. Scheme of vacuum installation for nozzle test: 1 – test rig inlet (T_0, p_0, φ_0 evaluation), 2 – test section, 3_a – HWA probe, 3_b – ISA orifice plate, 4 – secondary air inlet, 5_a – DN100 valve, 5_b – DN50 valve, 5_c – valve, 6 – 3 m^3 pressure vessel, 7 – cut-off valves, 8 – Roots air blower, 9 – exhaust to the environment.

The nozzle geometry, i.e. the top wall and the bottom wall with the pressure holes, was made of aluminium by means of CNC machining. For the lateral walls, glass windows were applied. Everything was assembled together on a frame made of a PLA filament using the 3D printing technique. Two geometries designed according to Witoszyński's curve were proposed, symmetrical and asymmetrical (W1 and W2), where in the asymmetrical nozzle the nozzle walls are shifted linearly in the flow direction by 20 mm. The experimental study were conducted on dedicated test rig in the Department of Power Engineering and Turbomachinery of the SUT.

The moist airflow through the W1 and W2 nozzles was analysed for two boundary conditions at the nozzle inlet and outlet (see Table 1). For all cases the same inlet total pressure and temperature values corresponding to ambient conditions can be assumed. Table 2 gathers the geometrical data and the experimental data of pressure measured at the bottom wall.

Table 1. Test conditions for W1 and W2 nozzle

	p_0, kPa	T_0, K	φ, %	p_b, kPa
		W1		
W1.0SS	99.7	296.65	0.0	-
W1.1SS			25.0	-
		W2		
W2.0SS	99.7	296.65	0	3 kPa
W2.1SS			26.1	3 kPa

Table 2. Geometrical nozzle data and experimental data of pressure measured at the bottom wall of the W1 and W2 nozzle

x, m	y, m	W1.1SS	W1.2SS
-0.050	-0.020345	-	-
-0.025	-0.011726	-	-
-0.006	-0.010080	57.884	68.3993
-0.003	-0.010020	-	-
0.000	-0.009996	52.610	62.78188
0.003	-0.010002	-	-
0.006	-0.010034	47.540	57.04555
0.009	-0.010089	-	-
0.012	-0.010161	42.507	51.62148
0.015	-0.010248	-	-

0.018	-0.010348	40.266	50.5379
0.021	-0.010457	-	-
0.024	-0.010574	37.560	46.29117
0.027	-0.010697	-	-
0.030	-0.010825	34.130	42.12767
0.033	-0.010956	-	-
0.036	-0.011090	36.046	39.20345
0.039	-0.011226	-	-
0.042	-0.011363	33.523	36.11015
0.045	-0.011502	-	-
0.048	-0.011643	32.263	33.30837
0.051	-0.011785	-	-
0.054	-0.011929	30.570	33.62938
0.057	-0.012076	-	-
0.060	-0.012225	28.351	33.5093
0.063	-0.012377	-	-
0.066	-0.012533	26.860	30.54525
0.069	-0.012694	-	-
0.072	-0.012859	27.493	28.60825
0.075	-0.013031	-	-
0.078	-0.013209	34.155	27.12425
0.090	-0.013999	-	-
0.120	-0.016680	-	-
0.150	-0.020357	-	-
0.200	-0.026678	-	-
0.235	-0.029933	-	-
0.250	-0.031885	-	-

3. Numerical study

The calculations presented herein were performed with the use of the ANSYS Fluent code extended using User-Defined Functions (UDFs) with an in-house condensation model implemented. The applied CFD method is used to study the phase-change phenomenon in the expanding flow. The model is supplemented with additional transport equations which allows to compute the liquid mass fraction and the number of droplets arising in the fluid as an effect of homogeneous condensation. The nucleation process is based on the classical theory extended with the Kantrowitz correction, whereas the droplet growth model is based on the Knudsen number (Kn) [5]. Details about the condensation model implementation can be found in previous works of the authors [15, 16, 17, 21, 22].

Figure 3 presents a comparison of the experimental visualization obtained using the Schlieren technique (a) with the numerical Schlieren (b).

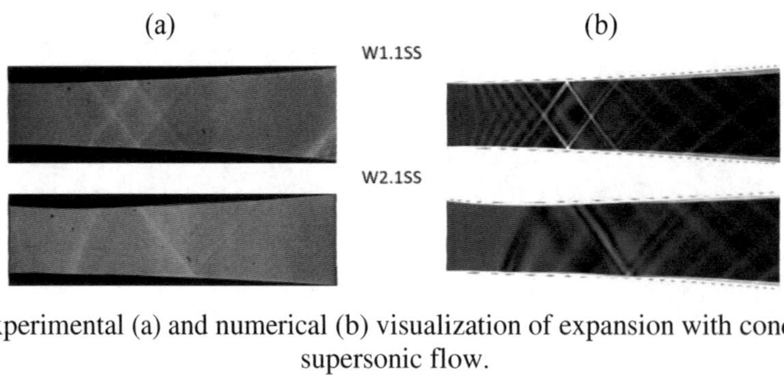

Figure 3. Experimental (a) and numerical (b) visualization of expansion with condensation in a supersonic flow.

This figure presents the structure of condensation and the aerodynamic waves arising in the expanding flow with a supersonic outlet. Although, it does not provide straight forward answer of the waves character, therefore it is important to complement the experimental study with the numerical research. The structure and position of waves obtained numerically converge with the experimental data, what lies solid basis for the study of the waves character.

Most flows in nature and engineering are asymmetric. The blade channel of a turbine, compressor or fan is curved no matter whether it is axial or radial. Due to that, the flow becomes asymmetric, which may give rise to additional physical phenomena, such as swirls or reverse flows. A numerical analysis of condensation in this type of flow reveals some interesting phenomena. This paper presents experimental and numerical testing of flows through asymmetric nozzles with a low expansion rate for different atmospheric conditions where the outlet is supersonic and the inlet is characterized by elevated pressure.

The asymmetry of the flow induces an asymmetric character of condensation and pressure waves. Figure 4 shows the numerical Schlieren for W2.0SS and the pressure distribution along the nozzle bottom and top walls. As it can be seen in the case of W2.0SS, the asymmetric pressure waves appear in the flow instead of the X-shaped waves presented in the symmetric nozzle.

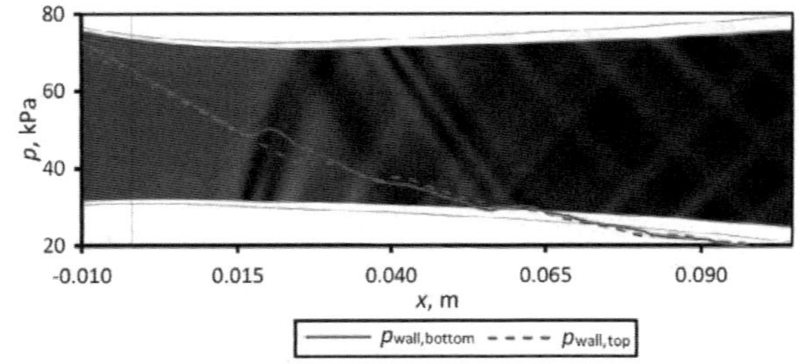

Figure 4. Schlieren photography and static pressure distribution along the top and bottom walls of the asymmetrical nozzle for the adiabatic flow.

Considering a humid flow, compression waves interact with condensation waves and can be easily confused with them, especially for low-humidity flows. The static pressure distributions on the nozzle bottom wall obtained experimentally and numerically demostrate good convergence (see Figure 5).

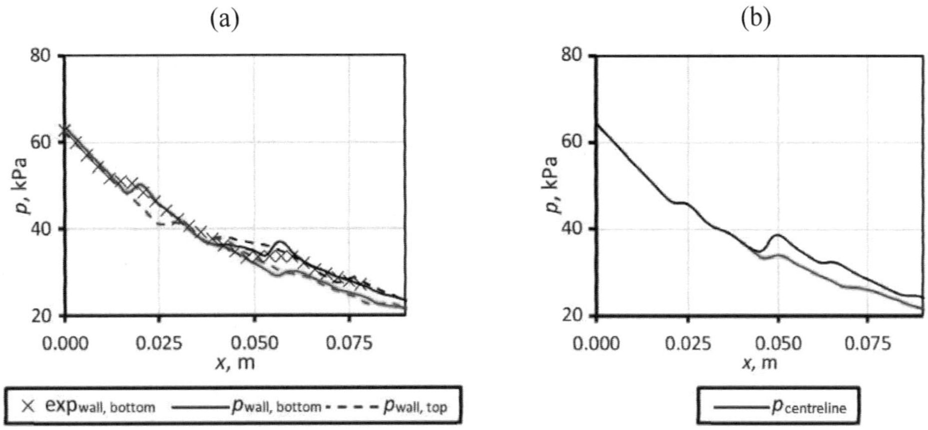

Figure 5. Static pressure distribution along the nozzle bottom wall obtained experimentally and along the top and bottom walls obtained numerically (a) and static pressure along the nozzle centre line (b) for the adiabatic (red) and diabatic conditions (black) in the asymmetric nozzle with a supersonic outlet.

The static pressure values along the nozzle wall and the centre line for the adiabatic conditions (red line) and the diabatic conditions (black line) suggest that the condensation wave position at the bottom as well as at the top wall is similar. It is due to the fact, that the condensation occurs in location where the pressure at top and bottom wall is nearly equal.

Considering Schlieren photography, numerical studies indicate a strengthening of the second compression wave (see Figure 6), which suggests that the condensation wave appears in the vicinity of the compression wave position for the adiabatic case, whereas the compression wave appears downstream the condensation wave as a reflected wave. Similar conclusions can be drawn considering experimental Schlieren photography, where the oblique wave might be considered as a condensation wave.

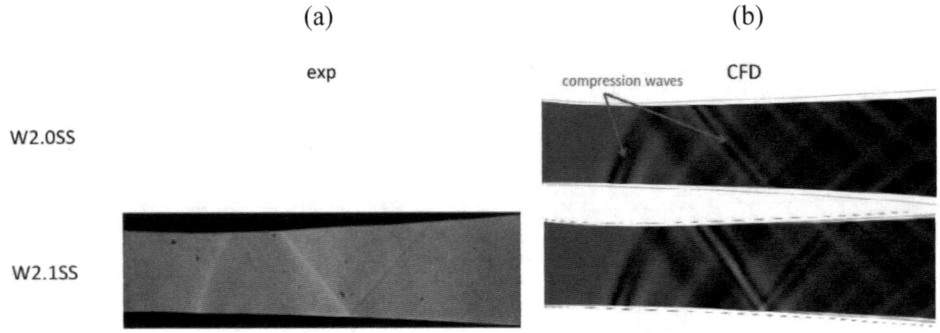

Figure 6. Experimental (a) and numerical (b) Schlieren photography for the asymmetric nozzle with a supersonic outlet.

Confronting the distribution of pressure along the nozzle walls with the Mach number and the condensate mass reveals that the waves obtained by means of visual measurements can be misleading. In Figure 7 it can be seen that condensation occurs ~0.04 m downstream the nozzle throat and is almost normal to the flow direction. This phenomenon is not visible using the Schlieren technique and the observed waves are compression waves. This indicates the need for further experimental and numerical research in a wide range of ambient conditions to achieve a sufficient understanding of condensation phenomena in the asymmetric flow.

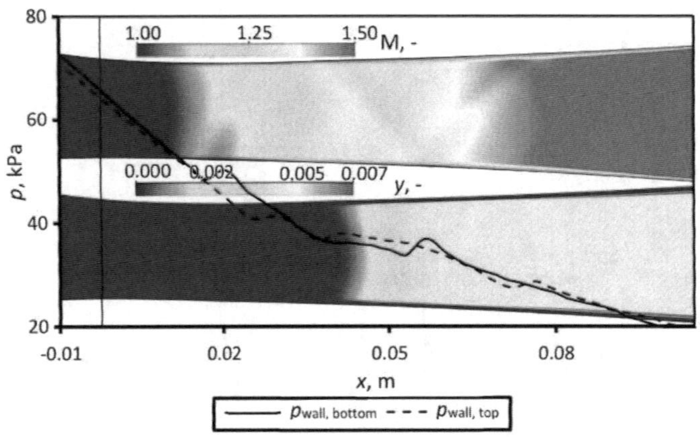

Figure 7. Confrontation of contour maps (Mach number, mass of condensate) with pressure along the nozzle top and bottom wall.

4. Summary and conclusions

The aim of the paper was to fill the gap of experimental testing of the flow of air with a low content of moisture through nozzles with a low expansion rate. The experimental tests were carried out in the DPET of the SUT using the existing in-house vacuum system to investigate condensation in symmetric and asymmetric nozzles. The nozzles were characterized by the expansion rate of $P \approx 1000\ s^{-1}$, which distinguishes them from the nozzles analysed in the literature so far. The results of the testing of the de Laval nozzle with a low expansion rate make it possible to draw the following conclusions:

- for the case of the adiabatic flow (dry air), the visible X-shaped waves have a purely aerodynamic character,
- if humidity is considered, it is set off in the vicinity of the pressure waves in the adiabatic flow,
- for low humidity values, the coexistence of a few X-shaped waves is observed. Only one of them is a condensation wave, whereas the others are compression or expansion waves. However, due to the limited comparative material, it is difficult to define their character unequivocally,
- considering the flow in the asymmetric nozzle, the condensation offset has the same position on the top and the bottom wall of the nozzle. Consequently, the condensation region intersects with the oblique aerodynamic wave,
- in the asymmetric nozzle, the condensation wave is not visible straightaway – it is visible in the form of an aerodynamic wave gaining in strength.

Taking all the above into consideration, it can be concluded that humidity has a nonnegligable impact on the expansion process in the low-expansion converging-diverging nozzle. The interaction between aerodynamic and thermodynamic waves (pressure and condensation waves) is not trivial. However, it is very difficult to distinguish the character of the occurring phenomena unequivocally because the range of the flow properties which can be measured during experiments is rather limited. A combination of experimental and numerical studies gives a deeper insight into the flow structure, but due to the limited amount of measured data, there is a need for more experimental research to obtain reliable validation data for numerical models.

Acknowledgments

The research was financed by European Union through the European Social Fund under the POWR.03.05.00-00-Z305 project, the Polish National Agency of Academic Exchange throught the NAWA 2021 BPN/BEK/2021/1/00011/U/00001 grant and by the National Science Centre under project number 2020/37/B/ST8/02369.

References

[1] Schnerr G H and Mundinger G 1993 Similarity, drag, and lift in transonic flow with given internal heat addition *European Journal of Mechanics - B/Fluids* **12(5)** pp. 597-611

[2] Schnerr G H and Dohrmann U 1994 Drag and lift in non-adiabatic transonic flow *AIAA Journal* **32** pp. 101-107

[3] Schnerr G H 1986 *Homogene Kondensation in stationaeren transsonischen Stroemungen durch Lavalduesen und um Profile* (Habilitationsschrift, Karlsruhe)

[4] Goodheart K A 2004 *3-D Transonic Flow Dynamics with Nonequilibrium Condensation* (Munich)

[5] Goodheart K A, Dykas S and Schnerr G H 2003 Numerical modelling of heterogeneous/homogeneous condensation on the ONERA M6 wing *Proceedings of the 12th International Conference on Fluid Flow Technologies* pp. 335-342

[6] Schnerr G H and Dohrmann U 1990 Transonic flow around airfoils with relaxation and energy supply by homogeneous condensation *AIAA Journal* **28(7)** pp. 1187-1193

[7] Adam S 1996 *Numerische und experimentelle Untersuchung instationarer Dusenstromungen mit Energiezufuhr durch homogene Kondensation* (Karlsruhe)

[8] Dykas S 2006 *Badania przepływów transonicznych z kondensacją pary wodnej* (Gliwice: Wydawnictwo Politechniki Śląskiej)
[9] Wen C, Ding H and Yang Y 2020 Optimisation study of a supersonic separator considering nonequilibrium condensation behaviour *Energy Conversion and Management* **222** 113210
[10] Dykas S, Majkut M, Smołka K and Strozik M 2018 Numerical analysis of the impact of pollutants on water vapour condensation in atmospheric air transonic flows *Applied Mathematics and Computation* **338** pp. 451-465
[11] Dykas S, Majkut M, Smołka K and Strozik M 2018 Study of the wet steam flow in the blade tip rotor linear blade cascade *International Journal of Heat and Mass Transfer* **120** pp. 9-17
[12] Dykas S, Majkut M, Smołka K and Strozik M 2017 Comprehensive investigations into thermal and flow phenomena occurring in the atmospheric air two-phase flow through nozzles *International Journal of Heat and Mass Transfer* **114** pp. 1072-1085.
[13] Dykas S, Majkut M and Smołka K 2019 Influence of Air Humidity on Transonic Flows with Weak Shock Waves *Journal of Thermal Science* **29** pp. 1551-1557
[14] Wiśniewski P, Dykas S and Yamamoto S 2020 Importance of Air Humidity and Contaminations in the Internal and External Transonic Flows *Energies* **13(12)** 3153
[15] Wiśniewski P, Dykas S, Yamamoto S and Pritz B 2020 Numerical approaches for moist air condensing flows modelling in the transonic regime *International Journal of Heat and Mass Transfer* **162** 120392
[16] Wiśniewski P, Majkut M, Dykas S, Smołka K, Zhang G and Pritz B 2022 Selection of a steam condensation model for atmospheric air transonic flow prediction *Applied Thermal Engineering* **203** 117922
[17] Wiśniewski P, Dykas S and Pritz B 2021 Numerical Studies of the Air Humidity and Pollution Importance in the Flow through the Gas Turbine's Rotor Compressor *Proceedings of 14th European Turbomachinery Conference, Turbomachinery, Fluid Dynamics & Thermodynamics*
[18] Witoszyński E 1924 *Über Strahlerweiterung und Strahlablenkung* (Berlin: Springer)
[19] Dykas S, Majkut M, Smołka K and Strozik M 2015 Experimental Research on Wet Steam Flow With Shock Wave *Experimental Heat Transfer* **28** pp. 417-429
[20] Dykas S, Majkut M, Strozik M and Smołka K 2015 Experimental study of condensing steam flow in nozzles and linear blade cascade *International Journal of Heat and Mass Transfer* **80** pp. 50-57.
[21] Wiśniewski P, Dykas S and Yamamoto S 2020 Importance of Air Humidity and Contaminations in the Internal and External Transonic Flows *Energies* **13(12)** 3153
[22] Wiśniewski P, Dykas S and Zhang G 2020 Numerical studies of air humidity importance in the first stage rotor of turbine compressor *Archives of Thermodynamics* **41(4)** pp. 223-234

Non-equilibrium dissipation scaling in atmospheric turbulence

M Wacławczyk[1], **J L Nowak**[1] **and S P Malinowski**[1]

[1] Institute of Geophysics, Faculty of Physics, University of Warsaw, Pasteura 5, 02-093 Warsaw, Poland

E-mail: marta.waclawczyk@fuw.edu.pl, jakub.nowak@fuw.edu.pl, szymon.malinowski@fuw.edu.pl

Abstract. This work is devoted to the detection of non-equilibrium turbulence states in atmospheric turbulence. The non-equilibrium scaling contradicts the classical Richardson-Kolmogorov cascade picture and many turbulence models do not account for it. The existence of such scaling has been discovered in various laboratory experiments. We show here that non-equilibrium states are also present in the stratocumulus-topped boundary layers, which indicates the presence rapidly changing external conditions.

1. Introduction

To describe characteristic features of turbulence, statistical theories are sought for. In this context, a number of recent research works (cf. [1] for a review) address the problem of the equilibrium Taylor's law and its failure in the presence of rapid changes of the system. The Taylor law is a cornerstone relation which connects the statistics influenced by small scales, that is, turbulence kinetic energy dissipation rate ϵ with the statistics determined mostly by large turbulence eddies: the root-mean-square of the turbulent velocity fluctuations $\mathcal{U}$ and the integral length scale $\mathcal{L}$. From the three variables the following non-dimensional coefficient can be defined

$$C_\epsilon = \frac{\epsilon \mathcal{L}}{\mathcal{U}^3}. \tag{1}$$

Above, $\epsilon = \nu\langle \partial u_i/\partial x_j \partial u_i/\partial x_j\rangle, \mathcal{U}^2 = 1/3\langle u_i u_i\rangle = 2/3k$, u_i is the velocity fluctuation component, k is the turbulence kinetic energy and ν is the kinematic viscosity.

According to the Taylor hypothesis the coefficient C_ϵ is a constant

$$C_\epsilon = const, \tag{2}$$

which does not depend on the local Reynolds number $\mathrm{Re}_\lambda = \mathcal{U}\lambda/\nu$, where $\lambda = \sqrt{15\nu/\epsilon}\,\mathcal{U}$ is the Taylor microscale. Even in the stationary turbulence value of this constant differs somewhat from case to case, as it is (weakly) affected by the scaling of largest scales. It is usually contained within the range 0.45–0.55. Decaying turbulence, on the other hand, is characterized by larger C_ϵ, which equals approximately $C_{\epsilon 1} = 1.8C_{\epsilon 0}$. As it is explained in Ref. [2] such formula follows form the fact that a finite time is needed for the energy to cascade from the largest to the smallest eddies.

Another relation, which follows directly from equations (1) and (2) reads

$$\frac{\mathcal{L}}{\lambda} = \frac{C_\epsilon}{15}\mathrm{Re}_\lambda. \tag{3}$$

The general validity of equations (2) and (3) was coming into question, as various laboratory experiments revealed that C_ϵ can vary considerably. Next, breakthrough research works [1] revealed an 'unusual', although universal dissipation scaling in laboratory experiments of decaying grid turbulence in the initial stages of the decay. This experimental dependence was later derived analytically in Ref. [3] in the form

$$\frac{C_\epsilon}{C_{\epsilon 0}} \approx \left(\frac{\mathrm{Re}_{\lambda 0}}{\mathrm{Re}_\lambda}\right)^{15/14}. \tag{4}$$

where $\mathrm{Re}_{\lambda 0}$ is the equilibrium value of the Taylor-scale based Reynolds number and $C_{\epsilon 0}$ is the equilibrium value of C_ϵ for stationary turbulence. Decaying turbulence is characterized by $C_\epsilon > C_{\epsilon 0}$. On the other hand, if turbulence production is locally larger than the dissipation, turbulence develops and $C_\epsilon < C_{\epsilon 0}$. Hence, classification of turbulence states can be made on this basis.

Moreover, instead of equation (3) the non-equilibrium turbulence states are described by a different relation between the Taylor microscale λ and the integral length scale $\mathcal{L}$, see [3]

$$\frac{\mathcal{L}}{\lambda} \approx \frac{C_{\epsilon 0}}{15}\mathrm{Re}_{\lambda 0}^{15/14}\left(\frac{1}{\mathrm{Re}_\lambda}\right)^{1/14}. \tag{5}$$

The concept of 'non-equilibrium' is often used in the context of thermodynamics, to describe time-courses of physical processes. In fluid mechanics this notion refers to a system where statistics temporarily deviate from some pre-defined universal, self-similar formula. In turbulence study this will indicate deviations of the energy spectrum from its (possibly non-stationary) self-similar form [4]

$$E(\kappa, t) = k(t)\mathcal{L}(t)\Phi(\kappa\mathcal{L}(t)), \tag{6}$$

where Φ is a universal function, $k(t)$ and $\mathcal{L}(t)$ are turbulence kinetic energy and turbulence length scale, which may, in general, depend on time. In the context of stratified multiphase flows the notion of 'non-equilibrium' was used in Refs. [5], [6] to describe states where probability density of finding one of the phases deviated from logistic distribution. In each case 'non-equilibrium' states are triggered by rapid changes of external conditions, such as e.g. changes of forcing.

The main aim of this work is to calculate C_ϵ from in-situ airborne measurements of atmospheric turbulence, where the flow conditions in the investigated region are unknown a priori and only limited information, that is 1D intersections of the flow field along the flight track, are available. We describe and discuss methods used to calculate energy dissipation rate, turbulence intensity and the longitudinal length scale from the available data. Next, by studying the dependence of C_ϵ and $\mathcal{L}/\lambda$ on Re_λ, we can identify whether turbulence is in- or out of equilibrium, see also [7]. This is possible, as the classical scaling relations (2) and (3) differ substantially from their non-equilibrium counterparts (4) and (5). Moreover, we can also recover information about temporal tendencies in the investigated region by comparing C_ϵ with its equilibrium value $C_{\epsilon 0}$.

We focus on the stratocumulus-topped boundary layer (STBL) data collected during the ACORES (Azores stratoCumulus measurements Of Radiation, turbulEnce and aeroSols) campaign in the Eastern North Atlantic [8]. The detailed study of turbulence statistics for this campaign was performed in [9], however, the temporal tendencies were not discussed therein.

Dependence of C_ϵ and $\mathcal{L}/\lambda$ on Re_λ at different altitudes in STBL will be presented in Ref. [7]. Our current study differs from Ref. [7] in that it presents explicitly changes of turbulence energy dissipation rate, turbulence intensity and the turbulence integral length scale along the flight track, provides a more detailed discussion of the procedures necessary to estimate turbulence energy dissipation rate and compares results of two different methods used for this purpose.

2. Data and methods

High-resolution wind velocity data used in this study were collected during helicopter flights in July 2017 in the eastern North Atlantic inside the 10 km by 10 km square adjacent to the Graciosa island [8]. The trajectory included vertical profiles up to 2 km and horizontal legs of the length of several kilometers. In our analysis we used data from the horizontal segments for flight no. 14 on 18 July 2017, distinctive by the stratocumulus presence. Measurements were performed with the Airborne Cloud Turbulence Observation System (ACTOS) [10] mounted 170m below the BO-105 helicopter. The sampling frequency of the sensor was $f_s = 100$ Hz and the helicopter speed $V \approx 20$ m/s. It hence follows that the small scales of turbulence were resolved down to ~ 0.5 m. This cut-off scale is still much larger than the size of smallest turbulence structures (Kolmogorov's scale) which is of the order 0.001–0.01 m in the atmospheric turbulence.

From the measurements, time series of vertical wind velocity component w recorded along the flight path were obtained. For the analysis, the signal was first decomposed into the mean and fluctuating parts

$$w(t) = \langle w(t) \rangle + w'(t)$$

using a moving average of 50 s. The same window size was used in Ref. [9]. Time series corresponding to the horizontal leg at height 992 (inside the stratocumulus cloud) and the calculated moving average are presented in the upper panel in figure 1. After detrending, another window of size 150 s was used to calculate turbulence statistics ϵ, $\mathcal{U}$ and $\mathcal{L}$ with a sufficient accuracy. The window was next moved every 5 s along the time axis. The dissipation rate of the turbulence kinetic energy can be estimated from the inertial-range part of the energy spectra or the second-order structure functions, as predicted by the Kolmogorov's theory (K41). The power spectrum calculated from the transverse velocity component reads

$$S(f) = C'_K \left(\frac{V}{2\pi}\right)^{2/3} \epsilon^{2/3} f^{-5/3}, \tag{7}$$

where $C'_K \approx 0.65$ and the transverse second-order structure function is

$$D_2(s) = \langle |w(t+s) - w(t)|^2 \rangle = C_2 \epsilon^{2/3} (sV)^{2/3}, \tag{8}$$

with $C_2 \approx 2.86$. The dissipation rate is estimated from equation (7) by fitting a line with $-5/3$ slope on a logarithmic plot within a certain range of frequencies, further called the "fitting range". Analogously, a line with 2/3 slope is fitted to the second order structure function according to equation (8). The estimates will be referred to as ϵ_{PS} and ϵ_{D2}, respectively. In theory they should be identical, however, in practice, the methods respond differently to errors due to finite size of the averaging window, finite sampling frequency and to measurement errors. Detailed analysis of this problem was presented in Ref. [11], where, additionally, methods based on the zero-crossing statistics and an iterative method (see [12]) were investigated. The conclusion which follows from those studies was that the method based on the second-order structure function (8) was less sensitive to smaller averaging window and performed better when the fitting range was moved towards smaller scales. In the present application this latter property is important, as our aim is to detect non-stationary states of turbulence. In such a case large scales are affected, while the small scales follow the K41 scaling more exactly.

Both estimates of the dissipation rate are presented in the second panel in figure 1. Although ϵ_{PS} tend to be somewhat larger than ϵ_{D2} both curves are, in this case, very similar. For further analyses we use ϵ_{D2} only, because of its preferable properties.

The turbulence intensity $\mathcal{U}$ was estimated as the r.m.s. of the vertical velocity

$$\mathcal{U} = \sqrt{\langle w'(t)^2 \rangle}, \tag{9}$$

(see the third panel in figure 1). The integral length scale $\mathcal{L}$ is defined as

$$\mathcal{L} = \int f(r) \mathrm{d}r, \tag{10}$$

where $f(r)$ is the two-point correlation coefficient of the longitudinal velocity component. As our study is based on the vertical (transverse) velocity component w, instead of f we were able to calculate the transverse correlation coefficient

$$g(r) = g(sV) = \frac{\langle w'(t+s) w'(t) \rangle}{\langle w'^2(t) \rangle}. \tag{11}$$

In the isotropic turbulence the longitudinal and transverse velocity components are related by the formula

$$g(r) = f(r) + \frac{1}{2} r \frac{\partial f(r)}{\partial r}. \tag{12}$$

In order to make use of equation (12) we assume the the longitudinal velocity correlation function is, approximately, exponential

$$f(r) = \exp\left(-\frac{r}{\mathcal{L}}\right). \tag{13}$$

Substituting (13) into (12) we obtain

$$g(r) = \exp\left(-\frac{r}{\mathcal{L}}\right)\left(1 - \frac{1}{2}\frac{r}{\mathcal{L}}\right). \tag{14}$$

To estimate $\mathcal{L}$ we calculated $g(r)$ from equation (11) and integrated it up to the distance r where $g(r)$ crosses 0. As it follows from the theoretical formula (14) this crossing occurs for $r = 2\mathcal{L}$. Integrating equation (14) over r from 0 to $2\mathcal{L}$ gives, approximately $0.57\mathcal{L}$. Hence, to estimate $\mathcal{L}$ we divided the numerically calculated integral of $g(r)$ by 0.57. Results are presented in the lowest panel in figure 1.

3. Results

Having $\mathcal{U}$, $\mathcal{L}$ and $\epsilon = \epsilon_{D2}$ in our hand we can calculate the non-dimensional coefficients C_ϵ as defined in equations (1) and $\mathcal{L}/\lambda$, as well as Re_λ. By plotting C_ϵ as a function of Re_λ we can check whether it is constant, as predicted by equation (2) or whether it is a function of Re_λ, as follows from equation (4). Analogously, plotting $\mathcal{L}/\lambda$ as a function of $\mathcal{L}/\lambda$ we can determine whether it follows the equilibrium relation (3) or its non-equilibrium counterpart (5).

We first present C_ϵ as a function of the position of the centre of the averaging window, where $x = tV$, see figure 2. In the study of STBL we found $C_\epsilon = C_{\epsilon 0} = 0.45$ describes best the stationary case. We have marked this value as a black straight line in figure 2. The upper straight line refers to self-similar turbulence decay in the absence of forcing where $C_{\epsilon 1} \approx 1.8 C_{\epsilon 0} \approx 0.81$, see Ref. [2]. It is clearly seen in figure 2 that C_ϵ is not a constant, it is large in the initial part (close to $C_{\epsilon 1}$) and next decreases towards $C_{\epsilon 0}$.

We next plotted the non-dimensional coefficients $C_{\epsilon 0}$ and $\mathcal{L}/\lambda$ as a function of Re_λ, see figures 3 and 4. Most of the points follow very well the non-equilibrium relations (4) and (5). In our

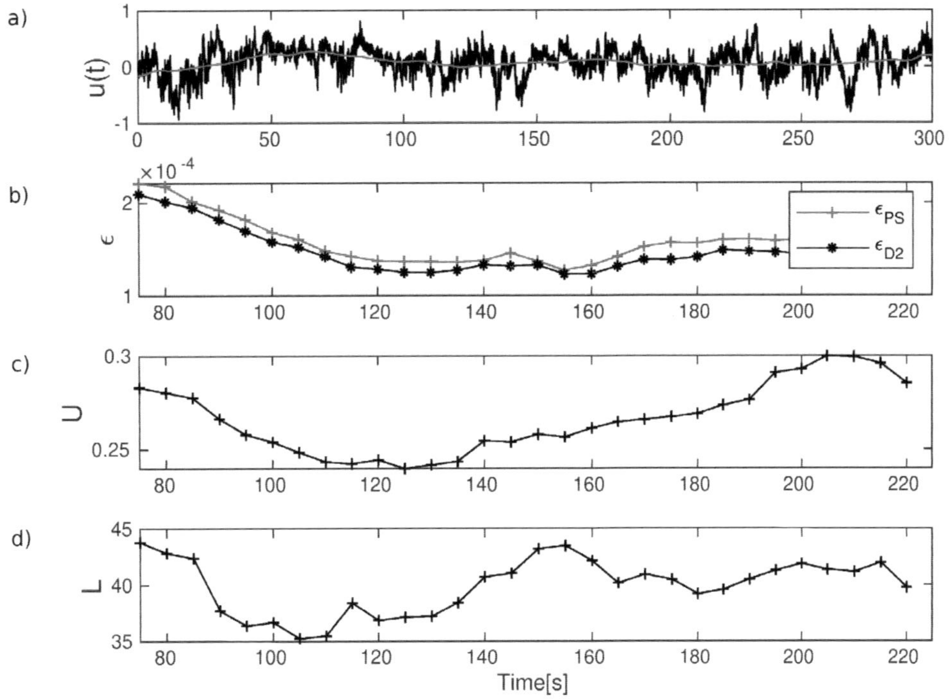

Figure 1. a) Time series of the vertical velocity recorded during flight no. 14 on 18 July 2017 by ACTOS, as a part of the ACORES campaign. Red solid line depicts the moving average calculated with the window 50s. b) turbulence kinetic energy dissipation rates ϵ_{PS} and ϵ_{D2}, calculated from equations (7) and (8), respectively, c) turbulence intensity $\mathcal{U}$ estimated according to equation (9), d) the integral length scale $\mathcal{L}$ calculated by integrating equation (14). Statistics in plots b) c) and d) were estimated with the moving average of 150s.

case the equilibrium Reynolds number $\mathrm{Re}_{\lambda 0}$ is unknown, hence it was calculated with the use of least-squares algorithm by fitting the points in figure 3 to the line $\sim \mathrm{Re}_\lambda^{-15/14}$ We obtain $\mathrm{Re}_{\lambda 0} \approx 4800$. We also use this value to calculate the constant in equation (5) and plot the red line in figure 4. It is to note that we did not perform another curve fitting for the data points in figure 4. The good match of the lines with experimental data (red stars) in figure 4 confirms validity of the formulas (4) and (5) derived analytically in Ref. [3].

Black lines in figures 3 and 4 are the equilibrium formulas (2) and (5), respectively where we substituted $C_\epsilon = C_{\epsilon 1} \approx 0.81$. For part of the experimental data, depicted as black dots, follow these formulas, in particular, $\mathcal{L}/\lambda$ increases with Re_λ.

In the investigated case $C_\epsilon > C_{\epsilon 0}$ for most of the points. This characterizes weak, decaying turbulence. Possibly, we deal here with a region of cloud dissipation, or a region of stable stratification, followed by a region of turbulence production due to buoyancy. Moreover, the non-equilibrium scaling suggest that forcing conditions changed rapidly, such that turbulence spectra did not relax towards their self-similar form (6).

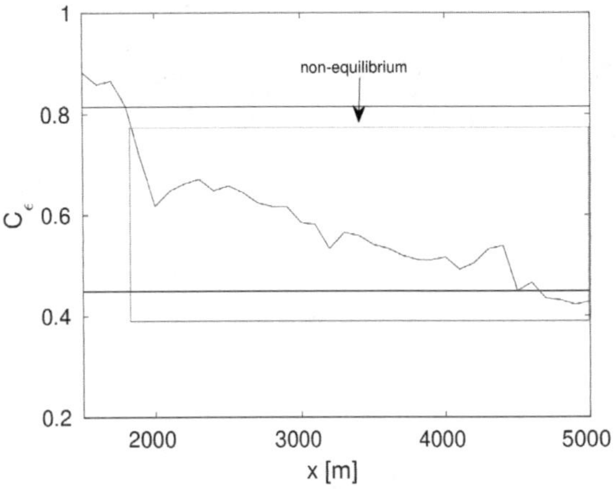

Figure 2. C_ϵ as a function of $x = tV$. Solid black lines - equilibrium scalings $C_{\epsilon 0} = 0.45$ and $C_{\epsilon 1} = 1.81 C_{\epsilon 0}$.

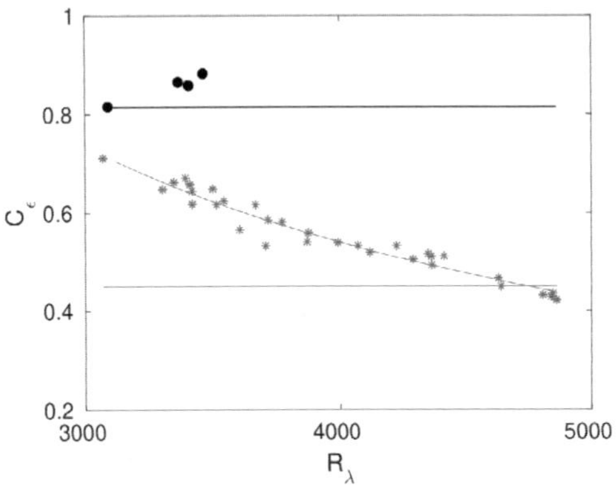

Figure 3. C_ϵ as a function of Re_λ. Solid black lines - equilibrium scalings $C_{\epsilon 0} = 0.45$ and $C_{\epsilon 1} = 1.81 C_{\epsilon 0}$, red line - non-equilibrium scaling, equation (4), calculated statistics: symbols.

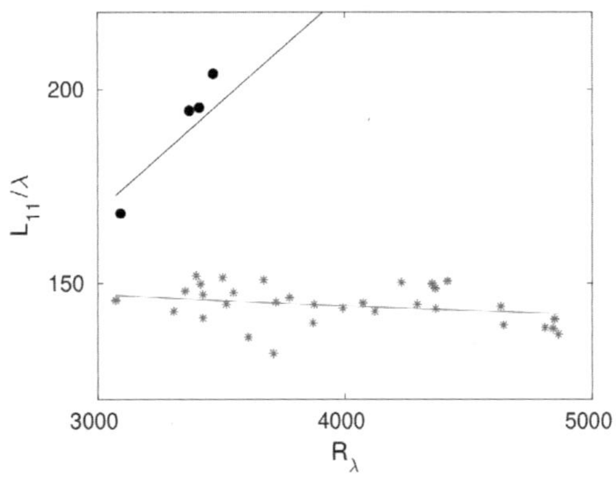

Figure 4. $\mathcal{L}/\lambda$ as a function of Re_λ. Solid black line - equilibrium scaling (3), red line - non-equlibrium scaling, equation (5), calculated statistics: symbols.

4. Conclusions

High accuracy of vertical wind velocity measurements, good resolution of the data from the ACORES campaign and relatively stable conditions along the horizontal flight legs in STBL allows the turbulence statistics to be reliably estimated. We calculated non-dimensional indicators C_ϵ and $\mathcal{L}/\lambda$ and plotted them as a function of the Taylor-scale based Reynolds number Re_λ. As it follows from our study, turbulence inside the investigated stratocumulus is described by non-equilibrium relations, which also indicates deviations from the Kolmogorov's $-5/3$ scaling of the spectra.

It should be noted that relation (2) with equilibrium $C_\epsilon = const$ is a basis of many common turbulence closures such as the k–ϵ model. Also the Smagorinsky model in Large Eddy Simulation employs the classical (statistically stationary) Richardson-Kolmogorov cascade picture. These classical relations do not describe non-equilibrium turbulence states which appear to be present inside the stratocumulus cloud. This poses a challenge for turbulence modeling, which we plan to address in future. Moreover, in a future work (in progress) we plan to repeat the current analysis for larger sets of measurement data in the atmospheric boundary layers, in order to better classify and parametrize turbulence states. In particular, more detailed analysis of C_ϵ in STBL at different altitudes and different types of stratocumulus clouds will be presented in Ref. [7].

Acknowledgments

The financial support of the National Science Centre, Poland (Project "Stably-Stratified Atmospheric Boundary Layer: Beyond the standard Monin and Obukhov Approach" No. 2020/37/B/ST10/03695) is gratefully acknowledged.

References

[1] Vassilicos J C 2015 Dissipation in turbulent flows *Annu. Rev. Fluid Mech.* **47** 95–114
[2] Bos W, Shao L and Bertoglio J-P 2007 Spectral imbalance and the normalized dissipation rate of turbulence *Phys. Fluids* **19** 045101
[3] Bos W and Rubinstein R 2017 Dissipation in unsteady turbulence *Phys. Rev. Fluids* **2** 022 601(R)
[4] Rubinstein R and Clark T T 2017 "Equilibrium" and "non-equilibrium" turbulence *Theor. Appl. Mech. Lett.* **7** 301–305
[5] Wacławczyk T 2021 Modeling of non-equilibrium effects in intermittency region between two phases *Int. J. Multiphase Flow* **134** 103459
[6] Wacławczyk T 2022 On differences between deterministic and statistical models of the interphase region *Acta Mech. Sin.* **38** 722045
[7] Wacławczyk M, Nowak J L, Siebert H, Malinowski S P 2022 Detecting non-equilibrium states in atmospheric turbulence *submitted to J. Atm. Sci.*
[8] Siebert H and Coauthors 2021 Observations of aerosol, cloud, turbulence, and radiation properties at the top of the marine boundary layer over the eastern north Atlantic Ocean: The ACORES campaign *Bull. Amer. Meteorol. Soc.* **102** E123.
[9] Nowak J L, Siebert H, Szodry K-E and Malinowski S P 2021 Coupled and decoupled stratocumulus-topped boundary layers: turbulence properties *Atm. Chem. Phys.* **21** 10965–91
[10] Siebert H, Lehmann K and Wendisch M 2006 Observations of small-scale turbulence and energy dissipation rates in the cloudy boundary layer *J. Atmos. Sci.* **63** 1451–66
[11] Wacławczyk M, Gozingan A S, Nzotungishaka J, Mohammadi M, and Malinowski S P 2020 Comparison of different techniques to calculate properties of atmospheric turbulence 762 from low-resolution data *Atmosphere* **11** 199
[12] Wacławczyk M, Ma Y-F, Kopeć J M and Malinowski S P 2017 Novel approaches to estimating the turbulent kinetic energy dissipation rate from low- and moderate-resolution velocity fluctuation time series *Atmos. Meas. Tech.* **10** 4573–85

AUTHOR INDEX

Arellano-García, H 109
Baena-Moreno, F M 109
Bergant, A 195
Błoński, Dominik 46
Bodnár, Tomáš 211
Brus, G 62
Caban, Lena 101
Cai, X 93
Doerffer, P 141
Dróżdż, A 157
Duda, D 125, 171
Duda, P. 23
Dykas, S 30, 93, 235
Elsner, W 157
Elsner, Witold 164
Filipek, R 69
Flaszyński, P 7, 141
Gajewska, Karolina 164
Gnatowska, R 157
Gnatowska, Renata 164
González-Castaño, M 109
Gumowski, K 133
Horáček, V 125
Jóźwik, Krzysztof 15
Kaczyński, P 141
Kajzer, A 203
Kajzer, Adam 54
Kaminski, P 77
Kantoch, Robert 117
Krzysiak, A 1
Kubacki, S 133, 157
Kudela, Henryk 46
Li, Songqi 179
Li, Wenpeng 179
Majkut, M 30, 93, 235
Malinowski, S P 243
Malinowski, Szymon P 149
Marchewka, Emil 15
Marek, Maciej 38, 85
Méndez, J A Medina 109
Mohammadi, Moein 149
Niegodajew, Paweł 164
Noack, Bernd R 179
Nowak, J L 243
Nowak, Jakub L 149
Obidowski, Damian 15
Piotrowicz, M 141
Płusa, T. 23
Pozorski, J 203, 227
Pozorski, Jacek 219
Prokop, T 62
Rajek, Michał 219
Rarata, Z 157
Reorowicz, Piotr 15
Sequeira, Adélia 211
Shabani, S 93
Smołka, K 30, 93, 235
Sobczak, Krzysztof 15
Sokolenko, V 157
Spricigo, E 227
Staśko, T 30
Stempka, Jakub 187
Stosiak, M 195
Strzelecka, Katarzyna 46
Suchanek, G 69
Suresh, T 7
Szmyd, J 62
Szulc, O 7
Szwaba, R 141
Szymanek, Ewa 85
Tyliszczak, A 77
Tyliszczak, Artur 85, 101, 117, 187
Urbanowicz, K 195
Uruba, V 125, 171
Wacławczyk, M 243
Wawrzak, Agnieszka 117
Wiśniewski, P 235
Yanovych, V 125, 171